SPECIALIZED CURING METHODS FOR COATINGS AND PLASTICS

Specialized Curing Methods for Coatings and Plastics

Recent Advances

M. William Ranney

NOYES DATA CORPORATION
Park Ridge, New Jersey, U.S.A.
1977

Library of Congress Catalog Card Number: 77-71928
ISBN: 0-8155-0660-0
Printed in the United States

Published in the United States of America by
Noyes Data Corporation
Noyes Building, Park Ridge, New Jersey 07656

FOREWORD

The detailed, descriptive information in this book is based on U.S. patents issued since early 1975 that deal with specialized curing methods for coatings and plastics.

This book serves a double purpose in that it supplies detailed technical information and can be used as a guide to the U.S. patent literature in this field. By indicating all the information that is significant, and eliminating legal jargon and juristic phraseology, this book presents an advanced, technically oriented review of specialized curing methods as depicted in U.S. patents. To round out the complete technological picture, we have included seven reissues and four applications published under the trial voluntary protest program initiated by the Commissioner of Patents and Trademarks in January 1975.

The U.S. patent literature is the largest and most comprehensive collection of technical information in the world. There is more practical, commercial, timely process information assembled here than is available from any other source. The technical information obtained from a patent is extremely reliable and comprehensive; sufficient information must be included to avoid rejection for "insufficient disclosure." These patents include practically all of those issued on the subject in the United States during the period under review; there has been no bias in the selection of patents for inclusion.

The patent literature covers a substantial amount of information not available in the journal literature. The patent literature is a prime source of basic commercially useful information. This information is overlooked by those who rely primarily on the periodical journal literature. It is realized that there is a lag between a patent application on a new process development and the granting of a patent, but it is felt that this may roughly parallel or even anticipate the lag in putting that development into commercial practice.

Many of these patents are being utilized commercially. Whether used or not, they offer opportunities for technological transfer. Also, a major purpose of this book is to describe the number of technical possibilities available, which may open up profitable areas of research and development. The information contained in this book will allow you to establish a sound background before launching into research in this field.

Advanced composition and production methods developed by Noyes Data are employed to bring our new durably bound books to you in a minimum of time. Special techniques are used to close the gap between "manuscript" and "completed book." Industrial technology is progressing so rapidly that time-honored, conventional typesetting, binding and shipping methods are no longer suitable. We have bypassed the delays in the conventional book publishing cycle and provide the user with an effective and convenient means of reviewing up-to-date information in depth.

The Table of Contents is organized in such a way as to serve as a subject index. Other indexes by company, inventor and patent number help in providing easy access to the information contained in this book.

15 Reasons Why the U.S. Patent Office Literature Is Important to You –

1. The U.S. patent literature is the largest and most comprehensive collection of technical information in the world. There is more practical commercial process information assembled here than is available from any other source.

2. The technical information obtained from the patent literature is extremely comprehensive; sufficient information must be included to avoid rejection for "insufficient disclosure."

3. The patent literature is a prime source of basic commercially utilizable information. This information is overlooked by those who rely primarily on the periodical journal literature.

4. An important feature of the patent literature is that it can serve to avoid duplication of research and development.

5. Patents, unlike periodical literature, are bound by definition to contain new information, data and ideas.

6. It can serve as a source of new ideas in a different but related field, and may be outside the patent protection offered the original invention.

7. Since claims are narrowly defined, much valuable information is included that may be outside the legal protection afforded by the claims.

8. Patents discuss the difficulties associated with previous research, development or production techniques, and offer a specific method of overcoming problems. This gives clues to current process information that has not been published in periodicals or books.

9. Can aid in process design by providing a selection of alternate techniques. A powerful research and engineering tool.

10. Obtain licenses – many U.S. chemical patents have not been developed commercially.

11. Patents provide an excellent starting point for the next investigator.

12. Frequently, innovations derived from research are first disclosed in the patent literature, prior to coverage in the periodical literature.

13. Patents offer a most valuable method of keeping abreast of latest technologies, serving an individual's own "current awareness" program.

14. Copies of U.S. patents are easily obtained from the U.S. Patent Office at 50¢ a copy.

15. It is a creative source of ideas for those with imagination.

CONTENTS AND SUBJECT INDEX

PART II. ELECTRON BEAM CURING

INTRODUCTION

Interest in ultraviolet and electron beam curing techniques is evidenced by the number of patents issued in this field. Over two hundred such patents, issued since April 1975, are excerpted in this book.

Among the advantages of these energy sensitive curing systems are speed of curing, freedom from pollution and improved products.

Although ultraviolet systems are initially more expensive than conventional equipment, savings are realized in costs of pollution control systems and in costs of fuel for solvent evaporation and heat curing necessary in conventional systems. In some instances, particularly the labeling of cans and cartons, ultraviolet is said to be the more economical system.

Electron beam curing techniques offer the same advantages as ultraviolet but the initial cost is much higher and more extensive and expensive safety shielding for workers is required. Electron beam curing is, however, in use in the manufacture of heat- and abrasion-resistant plastic wrap, plastic insulation of wire and cable and in the curing of paint.

In the ultraviolet curing section of this book there are chapters on coatings, imaging systems, plastics intended for uses other than coatings and photoinitiators. The electron beam section has a chapter on coatings and one on plastics which includes processes for specific applications.

In many of the patents examples of both ultraviolet and electron beam curing are given. These patents were usually assigned to the ultraviolet or electron beam section according to the technique used in the larger number of examples.

Part I. Ultraviolet Curing

COATINGS

COATING COMPOSITIONS

Glycidyl Ester of α,β-Unsaturated Carboxylic Acid and Anhydride

A photocurable composition described by *T. Nishikubo, M. Kishida and M. Imaura; U.S. Patent 3,878,076; April 15, 1975; assigned to Nippon Oil Seal Industry Co., Ltd., Japan* includes (a) a polyester having a photopolymerizable group in its side chains, obtained by a ring-opening alternating copolymerization of a glycidyl ester of a photopolymerizable α,β-unsaturated carboxylic acid and a carboxylic acid anhydride, and (b) a photopolymerization initiator. The composition may also include a photopolymerizable monomer.

As photopolymerization initiators there may be employed 2-alkylanthraquinone, benzoin, benzoin ether, benzoin ester, benzoin thioether, decyl chloride, benzoin ether-Michler's ketone or N,N-dialkylaniline-alkyl halide, used independently or in combination. The amount of these photopolymerization initiators to be employed is preferably from about 0.01 to 5% by weight of the photocurable ingredients, and when coloring agents such as pigments are added to the photocurable composition, an amount of the initiator of from about 7 to 15% by weight of the curable ingredients is preferable.

Example: 74 g of phthalic acid anhydride (0.5 mol) and 64 g of glycidyl acrylate (0.5 mol) were added to 60 g of trimethylolpropane triacrylate, and 0.5 g of benzyltriethylammonium chloride as a catalyst and 0.1 g of hydroquinone monomethyl ether as a thermal polymerization inhibitor were added to the mixture, after which the components were reacted at from 100° to 120°C for two hours. The obtained varnish had an acid value of 23.6 and a viscosity of 3,200 poises at 24°C.

The photocurable composition was completed in the following manner: 1 part by weight of trimethylolpropane triacrylate (in Experiment No. 1 this monomer was not added, and in Experiment No. 7, 1 part by weight of ethylene glycol diacrylate was added in place of trimethylolpropane triacrylate), 1 part by weight of benzoin ethyl ether (in Experiment No. 1, an amount used of benzoin ethyl ether was 2% by weight of the varnish, and in Experiment No. 7, 1 part by weight of benzoin isobutyl ether was added in place of benzoin ethyl ether) and 1 part by weight of the coloring agent were added to 7 parts by weight of the varnish.

After the obtained composition was coated on a tin plate to give a film thickness of 30 μ, it was exposed to a 500 watt high pressure mercury lamp at a distance of 30 cm, and the curing time was measured. The obtained results are shown in the following table.

Experiment No.	Coloring Agent	Curing Time, sec
1	-	0.3
2	Carmine lake	0.2-0.5
3	Hansa Yellow	0.2-0.5
4	Victoria Blue	0.2-0.5
5	Phthalocyanine Blue	0.7-1.0
6	Carbon Black	0.7-1.0
7	Carmine lake	0.2-0.5

Further, the photocurable composition comprising 9.8 parts by weight of the varnish and 0.2 part by weight of benzoin ethyl ether was prepared, and this composition was coated on the end of one sheet of glass of 24 mm width to give a film thickness of 30 μ and another sheet of the same size was placed over a length of 24 mm on the end of the sheet.

Then these were exposed to a 500 watt high pressure mercury lamp for 5 seconds. As a result, the two sheets were firmly adhered to each other to the extent that the two sheets did not easily peel off each other by pulling them from both sides longitudinally by hand.

Oligoacrylates

T. Ogasawara, Y. Senzaki, H. Kato and H. Tatemichi; U.S. Patent 3,888,830; June 10, 1975; assigned to Toagosei Chemical Industrial Co., Ltd., Japan provide curable compositions consisting essentially of a radical polymerization initiator and oligoacrylates containing a few acryloyl groups and at least one cyclohexene nucleus in the molecule and being curable by other means than the action of ionizing radiation even in the presence of a molecular oxygen-containing gas.

Reference Example 1: A reactor provided with an agitator, thermometer and water separator, was charged with 76 g (0.5 mol) of tetrahydrophthalic anhydride, 134 g (1 mol) of trimethylolpropane, 144 g (2 mols) of acrylic acid, 1,000 cc of toluene, 98% sulfuric acid in the amount of 2.5% by weight of the total of the ingredients excluding the 98% sulfuric acid, and phenothiazine in the amount of 0.08% by weight of the acrylic acid.

The resulting liquid mixture was agitated at 110°C while removing water produced by the esterifying reaction, as an azeotropic mixture with the toluene from the reaction system. After removal of the theoretical amount of water produced in eight hours, the reaction was terminated and the reaction mixture cooled. The reaction mixture was washed with 600 cc of an aqueous solution containing

3% by weight of ammonia and then washed with an aqueous solution containing 20% by weight of ammonium sulfate, after which the toluene layer was incorporated with 0.05 g of hydroquinone and then distilled at 50°C under a reduced pressure of 6 mm Hg to cut the toluene, leaving the oligoacrylate as the bottoms. The oligoacrylate obtained as the bottoms was a yellow-colored viscous liquid [viscosity: 54,000 cp (21°C)] and had an acryloyl group equivalent of 155.

Reference Example 2: The procedure of Reference Example 1 was followed except that the trimethylolpropane was substituted by 92 g (1 mol) of glycerine. The oligoacrylate obtained was a light-brown colored, viscous liquid having an acryloyl group equivalent of 135.

Reference Example 3: The procedure of Reference Example 1 was followed except that the trimethylolpropane was substituted by 134 g (1 mol) of 1,2,6-hexanetriol, to obtain an oligoacrylate. The oligoacrylate so obtained was a yellow-colored, viscous liquid having an acryloyl group equivalent of 154.

Reference Example 4: The procedure of Reference Example 1 was followed except that the trimethylolpropane and acrylic acid were substituted by 136 g (1 mol) of pentaerythritol and 216 g (3 mols) of acrylic acid, respectively. The oligoacrylate obtained was a yellow-colored, viscous liquid [viscosity: 26,250 cp (25°C)] with an acryloyl group equivalent of 122.

Examples 1 through 4: Each of the oligo(meth)acrylates obtained in Reference Examples 1 through 4 was combined with benzoin ethyl ether in the amount of 2% by weight of the acrylic compound used, to form a composition which was thoroughly mixed and then coated to a thickness of 10 to 15 μ on a 0.5 mm thick iron plate.

The composition-coated plates were each placed on a conveyor which was moved 13 cm below high pressure mercury lamps (H 2000 TQ, 2 kw, 2 lamps) at such a speed that the coating of the composition was irradiated by the mercury lamps in the presence of the air for the predetermined period of time as indicated in the table below. The results are shown in the extreme right-hand column of the table.

	Oligo(meth)acrylate Used	Time of Irradiation by UV Rays (sec)	Appearance of Coating
Example 1	Reference Example 1	15	Excellent*
Example 2	Reference Example 2	15	Excellent*
Example 3	Reference Example 3	15	Excellent*
Example 4	Reference Example 4	15	Excellent*

*Appearance of coating was hard, colorless, transparent and smooth.

Isocyanate-Containing Prepolymers

H. Hisamatsu, K. Takahashi and M. Takase; U.S. Patent 3,891,523; June 24, 1975; assigned to Dainippon Ink & Chemicals Inc., Japan provide a photopolymerizable composition comprising (a) 100 parts by weight of a photopolymerizable prepolymer having isocyanate groups as well as one or more unsaturated addition-polymerizable double bonds, the prepolymer being one obtained by reacting a polyisocyanate compound, a polyhydroxy compound and an unsaturated addition-polymerizable compound having a hydroxyl group, and (b) 0.001 to 15 parts by weight of a photosensitizer. The photopolymerizable composition finds use as a paint, a printing ink, and a plate making material, as well as an adhesive.

For maximum adhesion, it is necessary when carrying out the reaction of forming the photopolymerizable prepolymer to blend the polyisocyanate compound, polyhydroxy compound and unsaturated addition-polymerizable compound having a hydroxyl group so that the number of isocyanate groups exceeds the number of hydroxyl groups.

Example: 104.4 g (0.6 mol) of tolylene diisocyanate (2,4-tolylene diisocyanate/2,6-tolylene diisocyanate at a 95/5 weight ratio) were heated to 70°C, and while ensuring that this temperature was not exceeded, a homogeneously mixed solution heated to 70°C consisting of 26.8 g (0.2 mol) of trimethylolpropane, 79.0 g of methyl methacrylate and 0.02 g of p-benzoquinone was added dropwise during a two-hour period, following which the reaction was contained for a further six hours.

This was followed by adding dropwise 57.2 g (0.44 mol) of 2-hydroxyethyl methacrylate during a period of 0.5 hour, after which the reaction was continued for a further 10 hours, followed by the addition of 43.0 g of methyl methacrylate to obtain a methyl methacrylate solution of a light yellow, transparent photopolymerizable prepolymer A containing 1.8% by weight of isocyanate groups.

3.1 g of benzoin methyl ether were added to the so obtained prepolymer solution followed by homogeneously mixing to prepare the photopolymerizable composition of the process.

The so prepared photopolymerizable composition was applied to a wooden test piece having a smooth surface at the rate of 100 g per square meter. The coated surface was covered with a 30-mm-thick untreated polyethylene film exercising care to ensure that no air bubbles were permitted to enter. This was then exposed to actinic rays with a 2 kw high pressure mercury-vapor lamp for 30 seconds at a radiation distance of 30 cm.

The polyethylene film was then stripped and a smooth cured coating was obtained. The Sward hardness of this coating was 28 and the fastness of adherence of the coating was excellent, its adhesion strength being 10.5 kg/cm^2. Its resistance to abrasion and chemicals as well as other properties were also satisfactory.

Separately, a methyl methacrylate solution of a light yellow prepolymer containing 0.1% by weight of isocyanate groups was obtained by following the procedure described above, except that the amount of 2-hydroxyethyl methacrylate used was 78 g (0.6 mol). A control photopolymerizable composition was prepared by adding 3.3 g of benzoin methyl ether to this prepolymer solution and homogeneously mixing.

Using the so prepared photopolymerizable composition, a smooth cured coating was formed on a wooden test piece following the same procedure described above.

While the coating properties such as hardness (Sward hardness of 26), and resistance to abrasion and chemicals of this cured coating were either comparable or only somewhat inferior to the cured coating previously described, the fastness of adherence of this coating was inferior to that of the hereinbefore described cured coating, the adhesion strength in this case being only 2.2 kg/cm^2. It was thus found that the photopolymerizable composition of the process was greatly superior in this respect.

Air-Dryable Polyesters

The process of *H. Rudolph, H.-J. Traenckner, K. Fuhr, W. Deninger and M. Patheiger; U.S. Patent 3,898,144; August 5, 1975; assigned to Bayer AG, Germany* relates to unsaturated polyesters and to their use in air-drying, high-curing, molding and coating compositions.

The photopolymerizable composition comprises:

(1) 70 to 10% by weight of an unsaturated polyester which has an acid number of from 5 to 55 and is the condensation product at a temperature of from 170° to 200°C of
  - (a) 30 to 50% by weight of α,β-unsaturated dicarboxylic acid,
  - (b) 10 to 30% by weight of a diallyl ether of a polyhydric alcohol, the alcohol being selected from the group consisting of trimethylolpropane, trimethylolethane, trimethylolbutane, pentaerythritol and glycerin,
  - (c) 10 to 50% by weight of a monohydric or polyhydric alcohol containing saturated ether groups which is selected from the group consisting of diethylene glycol, triethylene glycol, tetraethylene glycol, dipropylene glycol, the monoalkyl ethers of the glycols having from 1 to 10 carbon atoms in the monoalkyl moiety, methyl, ethyl, propyl, butyl, isopropyl, sec-butyl, t-butyl, n-pentyl, neopentyl, n-hexyl, isohexyl,
  - (d) 0 to 10% by weight of an acid selected from the group consisting of phthalic acid, adipic acid, terephthalic acid, malonic acid, tetrahydrophthalic acid, hexahydrophthalic acid, succinic acid, glutaric acid and norbornenedicarboxylic acid and
  - (e) 0 to 30% by weight of a polyalcohol selected from the group consisting of ethylene glycol, propanediol-1,2, butanediol-1,3, butanediol-1,4, glycerine, trimethylolpropane, pentaerythritol, neopentyl glycol, cyclohexanediol-1,2,2,2-bis(p-hydroxycyclohexyl)-propane and 1,4-bismethylolcyclohexane;

(2) 0.1 to 4.0% by weight of a photoinitiator of the formula:

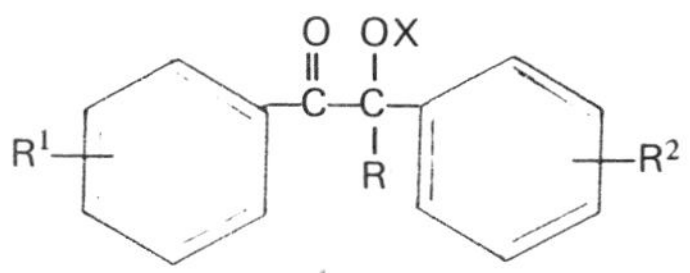

wherein R is hydrogen, alkyl having 1 to 4 carbon atoms, phenyl, benzyl or hydroxymethyl, X is hydrogen, alkyl having 1 to 4 carbon atoms, phenyl or trimethylsilyl with the proviso that R and X are not simultaneously hydrogen and R^1 and R^2 are each selected from the group consisting of hydrogen, alkyl having 1 to 4 carbon atoms, methoxy and halogen or 0.1 to 4.0% by weight of a photoinitiator of the formula:

$$R^1-C_6H_4-C(=O)-C_6H_4-R^2$$

wherein R^1 is CH_2-X; $CH-X_2$; CH_3; R^2 is H; CH_3; CH_2-X; $CH-X_2$; CX_3 and X is chlorine, bromine, iodine and

(3) 10 to 70% by weight of a copolymerizable vinyl monomer selected from the group consisting of styrene, vinyltoluene, divinylbenzene, vinyl acetate, acrylic acid and its esters with alkanols having from 1 to 6 carbon atoms, acrylonitrile, methacrylic acid and its esters with alkanols having from 1 to 6 carbon atoms, allyl acetate, allyl acrylate, phthalic acid diallyl ester, triallyl phosphate, triallyl cyanurate, N-vinylpyrrolidone, N-vinylpyridine.

Example 1: 2,552 parts by weight of fumaric acid and 451 parts by weight of 1,2-propylene glycol are heated to 50°C in a four-neck flask with stirrer, thermometer, nitrogen inlet and descending condenser, while passing nitrogen over the mixture. The internal temperature is raised to 150°C at the rate of 10°C per hour and 1,441 parts by weight of diethylene glycol, 757 parts by weight of glycerin diallyl ether, 428 parts by weight of diethyleneglycol monobutyl ether and 0.32 part by weight of hydroquinone are added.

The mixture is condensed, while the internal temperature rises to 170°C, until a viscosity of 19.5 seconds in the DIN-4 cup and an acid number of 32 are reached. A 69% strength solution in styrene is prepared and 2.5% of benzoin-isopropyl-ether are added. The solution thus obtained is spread as a 500 μ thick layer on glass by means of a film spreader and irradiated, at a distance of 8 cm, for 90 seconds with a super-actinic fluorescent tube and subsequently for 30 seconds with a mercury high pressure lamp.

After one hour, the pendulum hardness is measured according to Albert-Koenig (a measure of the curing of the polyester). A film having a pendulum hardness, according to Albert-Koenig, of 84 seconds, is obtained.

Example 2: The following polyester is produced as described above, employing the following components and quantities, the parts being parts by weight: 2,552 parts of fumaric acid, 451 parts of propylene glycol-1,2, 1,558.2 parts of diethyleneglycol, 941.6 parts of trimethylolpropane diallyl ether and 0.76 part by weight of hydroquinone. The acid number is 35.

68 parts of these polyesters are dissolved in 29.5 parts of styrene and 2.5 parts of benzoin isopropyl ether are added to the solution. The viscosity is about 1,030 cp. After irradiating a film as obtained according to Example 1 for 90

seconds under a super-actinic fluorescent tube and for 30 seconds under a mercury high pressure lamp, each time at a distance of 8 cm, this pendulum hardness according to Albert-Koenig after one hour storage in the dark is 122 seconds.

Example 3: Another polyester is produced as described above employing the following components and quantities, the parts being parts by weight: 2,552 parts fumaric acid; 451 parts of propylene glycol-1,2; 2,204.4 parts of triethylene glycol, 941 parts of trimethylolpropane diallyl ether, 0.86 part of hydroquinone. Acid number is 28.5; pendulum hardness is 102 seconds.

Acrylyl-Terminated Prepolymer plus Alkyl Acrylate Melamine Ether

R.P. Hall; U.S. Patent 3,899,611; August 12, 1975; assigned to SCM Corporation provides an actinic radiation curing process whereby a liquid coating composition is applied to a substrate and polymerized by actinic, especially ultraviolet, irradiation at room temperature without substantial solvent release to provide a smooth, tough, adherent, chemical and stain-resistant, heat-resistant, nonyellowing, noncurling protective and decorative coating.

The coating composition comprises in parts by weight the following components:

(1) about 10 to 70 parts of an acrylyl terminated prepolymer having a molecular weight from about 170 to 30,000;

(2) about 5 to 75 parts of an alkyl acrylate melamine ether containing an average of more than 3 acrylate or lower alkacrylate groups per triazine ring and miscible with the acrylyl terminated prepolymer;

(3) about 0 to 75 parts of a monoethylenically unsaturated viscosity-reducing diluent miscible with, and copolymerizable under actinic irradiation with, the prepolymer and the melamine ether; and

(4) an actinic radiation-sensitive polymerization initiator in a proportion effective for initiating polymerization of the components (1), (2) and (3) upon exposure of the composition to actinic radiation, the proportion being from about 0.01 to 5 parts;

wherein the parts of the components (1), (2), (3) and (4) total 100.

Example: (A) Preparation of an Acrylyl Terminated Prepolymer – Into a stirred, heated reaction vessel 30.19 lb of isophorone diisocyanate (3-isocyanatomethyl-3,5,5-trimethyl cyclohexyl isocyanate having a molecular weight of 222.3) are charged at a temperature of 140°F under an air sparge with moderate agitation. 4.0 g of dibutyl tin dilaurate (catalyst) are then added. 8.01 lb of 1,6-hexanediol at 140°F are then slowly added over a 10-hour period to the reaction vessel with agitation. This reaction product is an isocyanate terminated urethane oligomer.

7.47 lb phenoxyethyl acrylate (ethylenically unsaturated diluent) and 9.23 g of methyl ether of hydroquinone (inhibitor) are added to the reaction mixture to reduce viscosity.

The temperature of the reaction mixture is then brought to 175° to 180°F with agitation and 29.27 lb of 2-hydroxyethyl acrylate are added over a period of about four hours. At the end of this period, the reaction product is a water white, viscous liquid solution of urethane oligomer diacrylate in phenoxyethyl acrylate.

(B) Preparation of the Unsaturated Melamine Acrylate Resin — Into a suitable reaction vessel equipped with a thermometer and stirrer are charged 13 parts of hexamethoxymethyl melamine (0.033 mol), 24 parts of 2-hydroxyethyl acrylate (0.2 mol), 0.13 part of toluene sulfonic acid, and 0.018 part of the methyl ether of hydroquinone.

An air sparge is begun through the reaction mixture and 25 inches of vacuum is applied to the reaction vessel to assist in the removal of volatiles. The reaction mixture is stirred while the temperature is raised to between about 145° and 165°F.

This temperature is maintained for about one to four hours after which the reaction product (a water white, viscous liquid) is cooled to room temperature. The weight fraction of monomeric melamine compound is reduced from about 85% in the starting material to about 30 to 50% in the product, as shown by gel permeation chromatography.

(C) Preparation and Application of the Actinic Light Curable Coating Composition — Thirty parts of the diacrylyl terminated prepolymer of part (A), 15 parts of the unsaturated melamine acrylate resin of part (B), 10 parts of 2-hydroxyethyl acrylate, 40 parts of phenoxy ethyl acrylate, 5.4 parts of silica flatting agent (Syloid 74, a precipitated silica having an average particle size of about 10 to 11 microns), 3.2 parts of pigment grade Al_2O_3, and two parts of the butyl ether of benzoin are blended together at room temperature to form a coating composition.

A conventional vinyl-asbestos composition for forming vinyl floor tiles is extruded from a heated extrusion chamber in the form of a sheet about one-eighth inch in thickness. The sheet is immediately passed through a conventional roll coater at the rate of 200 feet per minute where the above described coating composition is applied at a thickness of 2 to 3 mils. The coated sheet is then immediately passed under a source of ultraviolet radiation.

The source of ultraviolet radiation is a plasma arc radiation torch optically directed by a reflector system to irradiate the freshly coated sheet workpiece passing below a rectangular irradiation window on an enclosed, horizontal conveyor moving at 200 feet per minute (providing about 0.1 second of irradiation).

The atmosphere around the workpiece during its irradiation is kept essentially inert by purging it with nitrogen. Radiation energy supplied by such apparatus at the workpiece surface is about 35 kw/ft^2, with about 10% of this radiation being in the ultraviolet spectrum.

At the end of this radiation curing treatment, the vinyl asbestos sheet has a clear, hard, tough, nonyellowing, chemically durable coating thereon of about 2 mils thickness. The coated vinyl asbestos sheet is cut into one foot squares and is ready for use in floor tile applications.

Coated floor tiles made by the procedure of this example were subjected to occasional water splashes at room temperature and humidity for a period of several days. After this period, the amount of curl was determined by measuring the perpendicular distance from a line joining two opposite corners of the square. The average distance for the samples was well below the 30-mil distance allowed by U.S. Government specification SS-T-312.

Several of the floor tiles were also subjected to repeated cleansing with conventional household cleansing fluids. Even after several of such cleansings, the floor tiles had not yellowed or stained. To test the hardness of the coating on the floor tiles, attempts were made to scratch the coating by forcefully rubbing the edge of a 1964 nickel against it. The edge of the coin could not be made to penetrate the coating.

Polyene-Polythiol Overcoat

J.L. Guthrie and F.J. Rendulic; U.S. Patents 3,877,971; April 15, 1975; 3,900,594; August 19, 1975 and 3,908,039; September 23, 1975; all assigned to W.R. Grace and Co. provide a method for overcoating which includes applying to a substrate a photocurable lacquer composition having:

(a) a polythiol component containing at least two —SH groups per molecule,
(b) a polyene component selected from particular polyenes (subsequently set forth herein) which contain two or more terminally positioned unsaturated carbon-to-carbon bonds per molecule, contain at least one of the following: N, O, and six-membered carbon-containing ring, and are characterized in that they will react with the polythiol to form tough and durable coatings, and
(c) a resin component.

Optionally, a sensitizer component may be included if desired to decrease curing time and improve efficiency. Curing of the applied composition is effected by exposure to a free radical generator such as actinic radiation to form, typically, clear, tough, durable polythioether coatings.

Polyenes which are useful herein include triallyl isocyanurate; bisphenol A diallyl ether; the diadduct of diallyl amine and bisphenol A 4,4'-diglycidyl ether; the urethane diadduct of a member selected from the ene-ol group consisting of allyl alcohol, diallyl malate, and trimethylolpropane diallyl ether and a member selected from the cyclic C_6-containing diisocyanate group consisting of toluene diisocyanate, 3,3'-dimethyl-4,4'-biphenylene diisocyanate, 3-isocyanatomethyl-3,5,5-trimethylcyclohexyl isocyanate, 4,4'-methylenebis(cyclohexyl isocyanate), and Solithane 291 (a polyester isocyanate having a molecular weight of about 2,800 and prepared by reacting ethylene glycol, propylene glycol, adipic acid and toluene diisocyanate).

Preferred polyenes are triallyl isocyanurate, bisphenol A diallyl ether, and the diadduct of toluene diisocyanate and diallyl malate, and 2,4,6-tris(allyloxy)-s-triazine.

Resins useful herein include normally solid, solvent soluble linear or branched polyesters, preferably linear, having molecular weight from about 500 to 100,000 which may be prepared by reacting a polybasic acid or a polybasic acid anhydride having the general formulas shown below, with a polyhydric alcohol having the general formula $[R_5](OH)_y$.

$$[R_4]\!-\!(\overset{\overset{\displaystyle O}{\|}}{C}\!-\!OH)_x \qquad \text{and} \qquad \text{O}\!\left\langle\begin{matrix}\overset{\overset{\displaystyle O}{\|}}{C}\\ \underset{\underset{\displaystyle O}{\|}}{C}\end{matrix}\right\rangle\!Z\!-\!(\overset{\overset{\displaystyle O}{\|}}{C}\!-\!OH)_z$$

In the formulas, x and y are numerals of 2 or more and preferably 2 to 4, and R_4 and R_5 are saturated or unsaturated hydrocarbyl or oxyhydrocarbyl groups having valence of x and y respectively and which may have from 2 to about 21 and from 2 to about 30 carbon atoms respectively, and preferably from 2 to about 10 and from 2 to about 8 carbon atoms respectively; z is 0 or 1; Z is a saturated or unsaturated hydrocarbyl group having a valence of 2 when z is 0, a valence of 3 when z is 1, and from about 2 to 12 carbon atoms and preferably from 2 to about 6 carbon atoms. Typically, the reaction is carried out using about one equivalent weight of ol functionality per one equivalent weight of acid functionality.

Other soluble resins which are useful herein are polyester-polyenes which may be prepared by capping the polyester resins described hereinabove with a member having the general formula

$$[HO\!-\!(\overset{\overset{\displaystyle O}{\|}}{C})_p]_q\!-\!(R_6)_r\!-\!(R_7)_s]$$

wherein R_6 is a hydrocarbyl member free of remotely internal nonaromatic ene and yne functionality and having a valence of q plus s and from about 2 to 10 carbon atoms; p is 0 or 1; q is a numeral from 1 to 10; r is 0 or 1; s is a numeral from 1 to 10; and R_7 is selected from the group consisting of

$$-(\overset{R}{\underset{R}{C}})_g\!-\!\overset{R}{C}\!=\!\overset{R}{C}\!-\!R, \qquad -O\!-\!(\overset{R}{\underset{R}{C}})_h\!-\!\overset{R}{C}\!=\!\overset{R}{C}\!-\!R, \qquad -S\!-\!\overset{R}{C}\!=\!\overset{R}{C}\!-\!R,$$

$$-\overset{R}{N}\!-\!\overset{R}{C}\!=\!\overset{R}{C}\!-\!R, \qquad -(\overset{R}{\underset{R}{C}})_g\!-\!C\!\equiv\!C\!-\!R \qquad \text{and} \qquad -\overset{\overset{\displaystyle O}{\|}}{C}\!-\!O\!-\!(\overset{R}{\underset{R}{C}})_g\!-\!\overset{R}{C}\!=\!\overset{R}{C}\!-\!R$$

where g is 0 or an integer from 1 to 9; h is 0 or 1; and the various R radicals are selected from the group consisting of hydrogen, fluorine, chlorine, furyl, thienyl, pyridyl, phenyl and substituted phenyl, benzyl and substituted benzyl, alkyl and substituted alkyl, alkoxy and substituted alkoxy, and cycloalkyl and substituted cycloalkyl, where the substituents on the substituted members are selected from the group consisting of nitro, chloro, fluoro, acetoxy, acetamide, phenyl, benzyl, alkyl, alkoxy and cycloalkyl, the alkyl and alkoxy groups having from 1 to 9 carbon atoms and the cycloalkyl groups having from 3 to 8 carbon atoms. The resin may be included in an amount from about 1 to 50 parts by weight per 100 parts by weight of polyene-polythiol combination.

Example 1: 275 g (1.86 mols) of phthalic anhydride, 60.8 g (0.62 mol) of maleic anhydride, 90.5 g (0.62 mol) of adipic acid, and 328 g (3.09 mols) of diethylene glycol were heated in a vessel for 3 hours at 200°C. The resulting polyester resin product was cooled and removed from the vessel. A photocurable lacquer composition was prepared by mixing at 50°C 10 g of the resin (a solid at 25°C) with 10 g (0.04 mol) of triallyl isocyanurate, 14 g (0.03 mol) of pentaerythritol tetrakis(3-mercaptopropionate), 0.5 g of benzophenone, 0.025 g of phosphorous acid, and 0.05 g of Irganox 1076 (a hindered phenol antioxidant). The mixture was heated to 70°C to dissolve the benzophenone producing a clear homogeneous mixture having a viscosity in the range of 12,000 to 18,000 cp and a pH in the range of 4.5 to 5.5.

The photocurable lacquer was applied at 25°C to steel can body stock having a red printing ink printed thereon by conventional techniques. The lacquer was applied as a liquid film having a thickness of 0.2 to 0.4 mil using a conventional roller coater. The area of application included the printed area plus adjacent nonprinted portions of the steel.

The applied liquid film was exposed to ultraviolet light using a 5000-watt Hanovia lamp positioned 4 inches from the film. Exposure was for 3 to 5 seconds, during which time the liquid film cured to a clear solid coating which was found to be characterized with high gloss, strong bonding to both the printed ink and the steel, and excellent abrasion resistance.

Example 2: A polyester resin was prepared at 200°C using the resin preparation procedure of Example 1 except that prior to cooling, 4.3 g (0.02 mol) of trimethylolpropane diallyl ether was added and the temperature of the reactants was increased to 240°C, which temperature was maintained for 1 hour. The resin product was cooled and removed from the vessel. Analysis of the product showed it to be a polyester-tetraene which was a solid at 25°C.

A curable lacquer composition was prepared by mixing 10 g of this polyester-tetraene at 60°C with 10 g (0.04 mol) of triallyl isocyanurate, 14 g (0.03 mol) of pentaerythritol tetrakis(3-mercaptopropionate), 0.5 g of benzophenone, 0.025 g of phosphorous acid and 0.05 g of Irganox 1076. This lacquer composition was applied as a film on printed steel can body stock and thereafter cured using the application and curing procedures of Example 1. The cured lacquer coating was found to be characterized with good clarity, high gloss, strong bonding to the printed ink and the steel, and excellent abrasion resistance.

Example 3: 12 g of commercially available polyethylene glycol having a molecular weight of 400 was mixed with 10 g of triallyl isocyanurate, 14 g of pentaerythritol tetrakis(3-mercaptopropionate), 0.5 g of benzophenone, 0.025 g of phosphorous acid, and 0.05 g of Irganox 1076. The resulting photocurable lacquer composition was roller-coated at 20° to 30°C onto paperboard having ink printed thereon. The area of lacquer application included printed and nonprinted areas.

The coated paperboard was exposed to ultraviolet light using a 5000 watt Hanovia lamp positioned 3 inches from the paperboard. Exposure was for 3 to 5 seconds, during which time the lacquer cured to a clear solid coating which was found to be characterized with high gloss, strong bonding to both the printed ink and the paperboard, and excellent abrasion resistance.

Opacifying Gas Bubbles

A. de Souza; U.S. Patent 3,907,656; September 23, 1975; assigned to SCM Corporation provides an improvement in the process for ultraviolet curing of a paint film containing titanium dioxide pigment and an ultraviolet sensitizer, and having an acrylic binder polymerizable by free-radical-induced addition polymerization when irradiated. The improvement comprises providing between 25 to 90% of the opacification of the paint film by the inclusion of granules to entrap preformed opacifying gas bubbles within the paint film, the paint film having a pigment-binder ratio of 0.8/1.

The art describes several ways of so entrapping opacifying gas bubbles within the paint film. Entrapped opacifying gas bubbles are preferably preformed in the paint film by the inclusions of particles having hollow interiors such as glass microballoons or vesiculated irregular edges. The vesiculated particles can be irregular microspheres or chunks of polymer or tiny pieces of porous minerals which generally are adapted to entrap gas bubbles within the paint film.

Differences in the index of refraction between the entrapped gas, which is generally air, and the cured vehicle provides the opacification (as does such difference between an opacifying pigment and such vehicle). Optical opacity of films is achieved either by absorption of the incident light or by scattering of the incident light, or a combination thereof.

Example 1: A polymerizable binder consisting of equal parts of 2-hydroxyethylacrylate, diglycidyl ether of bisphenol A diacrylate, and trimethylolpropane triacrylate was ground with sufficient TiO_2 to make a pigment-binder ratio of 0.8/1. To this pigmented binder composition was added 1% thioxanthone and 3% methyldiethanolamine sensitizer combination. Several 2-mil paint films of the sensitizer composition were applied to a steel test panel and cured as follows:

(a) Exposure under nitrogen blanket of 0.2 second to a plasma-arc torch (PARS) ultraviolet source gave a thoroughly and completely cured opacified paint film.

(b) Exposure to 8,000-watt Ashdee 2 mercury lamp ultraviolet source for ten seconds produced a similar fully cured paint film.

(c) Exposure to PARS for 0.1 second in the manner set forth in (a) produced an undercured paint film having a surface cure only and no resistance to methyl ethyl ketone (MEK).

(d) Exposure to the Ashdee lamp for five seconds in the manner set forth in (b) produced an undercured paint film having a surface cure only and no resistance to MEK.

Example 2: A polymerizable binder similar to Example 1 was compounded to provide a ratio of 0.6 TiO_2 + 0.2 voids/1 binder. The preformed voids were incorporated into the paint film by the inclusion of glass microspheres or plastic vesiculated particles (polystyrene). The paint films had equivalent opacity to the paint films in Example 1. The ultraviolet sensitizer was the combination in Example 1. Exposure of a 2-mil paint film to ultraviolet sources gave the following results.

(a) A paint film containing glass microspheres was exposed to PARS for 0.1 second and resulted in a thoroughly cured film having resistance to methyl ethyl ketone.

(b) A paint film similar to (a) but containing the plastic particles was exposed to PARS for 0.1 second and resulted in a completely cured methyl ethyl ketone resistant film.

(c) A paint film containing glass microspheres and exposed for five seconds to Ashdee ultraviolet source resulted in fully cured methyl ethyl ketone resistant films.

(d) A paint film containing plastic particles was exposed for five seconds to Ashdee ultraviolet source resulting in a fully cured methyl ethyl ketone resistant film.

Silicone-Modified Polyester

The process of *M.P. Urkevich; U.S. Patent 3,919,438; November 11, 1975; assigned to General Electric Company* is concerned with silicon-modified polyesters, their method of preparation and use. Many of the coatings employing the materials of the process are curable by mere exposure to radiation including ultraviolet light, and/or are curable at room temperature with a nonradiation free-radical-type curing agent.

The composition aspect of the process is concerned with a polymerizable copolymer comprising the reaction product of:

(a) a hydroxyl terminated unsaturated polyester of:

  (1) a polycarboxylic acid reactant wherein at least about 65 mol % of the polycarboxylic acid reactant is an α,β-ethylenically unsaturated polycarboxylic acid reactant and up to about 35 mol % of the polycarboxylic acid reactant is a polycarboxylic acid reactant free from nonbenzenoid unsaturation; and

  (2) an alcohol reactant containing two terminal hydroxyl groups and allyl ether groups and/or methallyl ether groups in an amount sufficient to provide at least 0.1 mol of allyl ether groups and/or methallyl ether groups per mol of the ethylenically unsaturated polycarboxylic acid reactant; and

  (3) wherein the polyester has an acid number from about 10 to 35; and

(b) an organopolysiloxane having the average unit formula:

$$(OH)_q(OR')_rR_sSiO_{\frac{4-q-r-s}{2}}$$

wherein R is lower alkyl radical having 1 to 8 carbon atoms; and/or cycloalkyl radical having 5 to 7 carbon atoms in the ring; and/or lower alkenyl radical having 2 to 8 carbon atoms; and/or mononuclear aryl radical; and/or mononuclear aryl lower alkyl radical having 1 to 6 carbon atoms in the alkyl group; and/or halogenated derivatives of the above radicals; R' is alkyl containing from 1 to 8 carbon atoms per radical; and/or acyl of 1 to 8 carbon atoms; s has a value of 1 to 2; q has a value of 0 to 1.0; r has a value of 0 to 1.0; and the sum of q + r has a value of 0.01 to 1; and containing at least 0.25% by weight of silicon-bonded OH and/or OR' groups.

Example 1: To a reaction vessel equipped with a stirrer, a fractionation column packed with glass helices, a Dean-Stark trap, and a condenser are added 213.30 parts of maleic anhydride, 202.74 parts of trimethylolpropane monoalkyl ether, and 183.96 parts of diethylene glycol. The reaction mass is heated under a nitrogen atmosphere to a maximum temperature of 200°C. After about 4½ hours, a total of 33 parts of water of esterification are collected. The reaction mass is cooled to room temperature and about 567 parts of an unsaturated polyester having an acid number of 32 are obtained.

169 parts of this polyester and 60 parts of xylene are added to a reaction vessel equipped with a stirrer, a fractionation column packed with glass helices, a Dean-Stark trap, and a condenser. The reaction mixture is heated to 110°C in about 10 minutes at which time 76 parts of a methoxy chain-stopped linear polysiloxane containing phenyl and methyl groups; 7% by weight silicon-bonded methoxy groups, with a viscosity of 1,500 to 3,000 cs at 25°C; 25 parts of xylene and 0.38 part of tetraisopropyl titanate are added to the reaction vessel.

The reaction mass is heated under a nitrogen atmosphere to a maximum temperature of 152°C. After about 2½ hours, 6 parts of methanol are collected and external heating of the reaction is stopped. The reaction mass is distilled under vacuum at a pressure of 20 mm Hg to a maximum temperature of 145°C to remove the xylene. About 239 parts of a clear copolymer containing 70 weight percent of the unsaturated polyester and 30 weight percent of the organopolysiloxane are obtained.

Example 2: Copolymer of Example 1 is admixed with styrene to provide a composition containing 30% by weight styrene and 70% by weight copolymer. To this composition is added 1% by weight of methyl benzoin ether. The composition is coated on a glass panel at a thickness of about 1 mil.

The coated panel is subjected to radiation from an ultraviolet light lamp for about 1 minute. The coating obtained has excellent mar resistance, good adhesion, is extremely hard and is water white. The above example is repeated except that the composition is coated in thicknesses of 2 mils and 10 mils. Similar results are obtained.

Example 3: The same composition as employed in Example 2 is coated on a glass panel at a thickness of about ½ inch. The coating is then placed under a household sunlamp and cures within 20 minutes. The cured material is water white, hard, has excellent mar resistance, and demonstrates good adhesion.

Pyromellitic Acid Dianhydride plus Polyol

The composition of *T. Nishikubo, M. Kishida and M. Imaura; U.S. Patent 3,923,523; December 2, 1975; assigned to Nippon Oil Seal Industry Co., Ltd., Japan* includes the polyester obtained from a polyol having a photopolymerizable α,β-unsaturated ester group and a polycarboxylic acid anhydride such as pyromellitic acid dianhydride, along with a photopolymerization initiator and, optionally, a photopolymerizable monomer.

The polyols which are employed in the process are polyols having at least two hydroxy groups and a photopolymerizable α,β-unsaturated carboxylic acid ester group such as acrylate, methacrylate, crotonate, sorbate, cinnamate group, etc.

Example 1: 43.4 g (0.1 mol) of the reaction product of neopentyl glycol diglycidyl ether and acrylic acid (molar ratio 1:2) was dissolved in 70 g of N-vinylpyrrolidone, and 21.8 g (0.1 mol) of pyromellitic acid dianhydride was added to the solution with agitation. Then the mixture was reacted at room temperature for 24 hours. A polyester solution having a viscosity of 13,200 poises at 25°C was obtained.

The photocurable composition was prepared by adding 37.8 parts (by weight of the composition) of trimethylolpropane triacrylate, 2 parts of benzoin methyl ether, and 0.2 part of hydroquinone monomethyl ether to 60 parts of the obtained polyester solution. The composition was coated on a tin plate to give a film of a thickness of 0.1 mm, and was exposed to a 500-w high-pressure mercury lamp from a distance of 30 cm. The required curing time was 0.3 second.

Example 2: 46.2 g (0.1 mol) of the reaction product of neopentylglycol diglycidyl ether and methacrylic acid (molar ratio 1:2) was dissolved in 100 g of N-vinylpyrrolidone, and 0.2 g of hydroquinone monoethyl ether was added to the solution, and 19.6 g (0.09 mol) of pyromellitic acid dianhydride was added to the solution with agitation. Then the mixture was reacted at room temperature for 24 hours. A polyester solution having a viscosity of 5,300 poises at 25°C was obtained.

The photocurable composition was prepared by adding 28 parts of ethylene glycol diacrylate and 2 parts of benzoin ethyl ether to 70 parts of the obtained polyester solution, with curing carried out in the same manner as described in Example 1. The required curing time was 0.4 second.

Epoxide plus Carboxylic Acid Anhydride

The process of *G. Karoly and J.L. Gardon; U.S. Patent 3,930,971; January 6, 1976; assigned to M & T Chemicals Inc.* pertains to liquid, pigmented coating compositions comprising one or more epoxides, carboxylic acid anhydrides, a catalyst, and up to 50% by volume of one or more pigments, the coating composition being rapidly polymerizable in the presence of ultraviolet light to yield solid products. The coating compositions are characterized by:

(a) the presence of one or more epoxides selected from the group consisting of:

(1) liquid polyglycidyl ethers of the general formula

$$Ar(O{-}CH_2{-}\underset{\diagdown O \diagup}{CH{-}CH_2})_n$$

and low molecular weight liquid polymers thereof wherein Ar represents an aryl or alkaryl hydrocarbon radical and n is the integer 2 or 3 and

(2) liquid compounds containing two or more divalent radicals of the formula

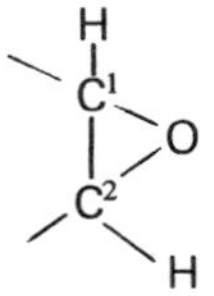

$$\begin{array}{c} \text{H} \\ | \\ -\text{C}^1 \\ | \quad \rangle \text{O} \\ -\text{C}^2 \\ \quad \backslash \text{H} \end{array}$$

wherein C^1 and C^2 are part of a five- or six-membered carboxylic ring structure, which may in turn be part of a larger molecule;

(b) the presence of one or more carboxylic acid anhydrides selected from the group consisting of liquid anhydrides of dicarboxylic acids, polycarboxylic acids and liquid mixtures containing two or more anhydrides of dicarboxylic or polycarboxylic acids, with the proviso that the composition contains 1.8 or more equivalent weights of epoxide per equivalent weight of acid anhydride;

(c) the presence of a catalyst selected from the group consisting of aryl diazonium compounds corresponding to the general formula:

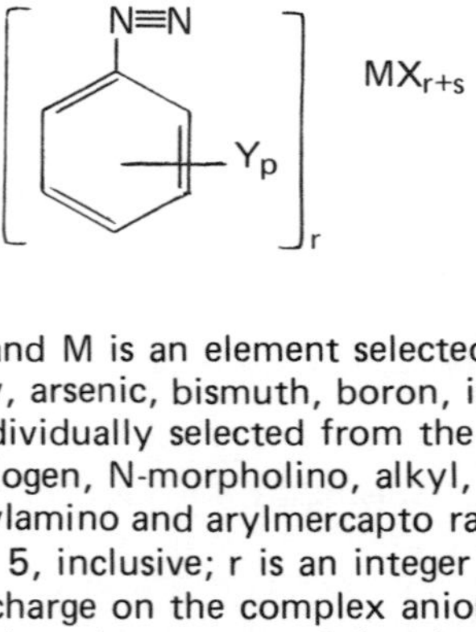

$$\left[(N{\equiv}N)\text{-}C_6H_{5-p}\text{-}Y_p \right]_r \quad MX_{r+s}$$

wherein X is halogen and M is an element selected from the group consisting of antimony, arsenic, bismuth, boron, iron, phosphorus and tin. Each Y is individually selected from the group consisting of nitro, hydroxyl, halogen, N-morpholino, alkyl, alkoxy, aryl, amino, arylamino, alkylamino and arylmercapto radicals; p is an integer between 1 and 5, inclusive; r is an integer equal to the absolute value of the charge on the complex anion MX_{r+s} and s is an integer equal to the valence state of the element M.

A light-curable, substantially solventless coating composition containing sufficient pigment to obtain acceptable levels of hiding power is achieved by the presence of the carboxylic acid anhydride.

Example 1: A homogeneous dispersion was prepared using 120 g of titanium dioxide and 80 g of 3,4-epoxy cyclohexylmethyl-(3,4-epoxy) cyclohexane carboxylate (ERL-4221). An 8.25 g portion of this material was blended with a mixture containing 1.25 g of methyltetrahydrophthalic anhydride. To this composition was added a solution of 0.1 g of 2,5-diethoxy-4-(p-tolyl mercapto) benzenediazonium hexafluorophosphate in 0.2 g of acetonitrile. A portion of this composition was coated on a steel plate using a number 2 wire-wound K-bar, and the coated side exposed for ten seconds at a distance of 3 inches (7.5 cm) from a 430-watt mercury vapor lamp. During the exposure the liquid coating was converted to a white, opaque and glossy solid.

Acceptable quality coatings were prepared using the foregoing procedure and ingredients, the only change being the replacement of the diazonium catalyst with an equal weight of either o- or p-hydroxybenzene diazonium hexafluorophosphate.

Example 2: A liquid coating composition was prepared as in Example 1, with the exception that 2.5 g of methyltetrahydrophthalic anhydride and 1.09 g of dodecenylsuccinic anhydride was used as the acid anhydride component. The exposure time required to obtain a solid coating was ten seconds.

Hydrophilic Nitrite Copolymers

J. Vacik and J. Kopecek; U.S. Patent 3,931,123; January 6, 1976; assigned to Ceskoslovenska akadamie ved, Czechoslovakia describe a hydrophilic copolymer with improved mechanical properties which contains 2 to 50 mol percent of methacrylonitrile or acrylonitrile and 50 to 98 mol percent of glycol monoesters of acrylic or methacrylic acid, where the glycol is selected from the group comprising ethylene glycol, diethylene glycol, triethylene glycol and their mixtures.

0.1 to 30 mol percent of the glycol monoesters may be substituted by a crosslinking agent selected from the group comprising 1,4-cyclohexanedimethyl dimethacrylate, ethylene glycol dimethacrylate, diethylene glycol dimethacrylate, 1,6-hexamethylene-bis-acrylamide, triethylene glycol dimethacrylate, tetraethylene glycol dimethacrylate, methylene-bis-acrylamide, ethylene-bis-acrylamide.

Example 1: n-Butanol (20 weight percent) was mixed with 80 weight percent of a mixture consisting of 87 weight percent of 2-hydroxyethyl methacrylate, 10 weight percent of methacrylonitrile and 3 weight percent of ethylene dimethacrylate and polymerized with 0.05 weight percent of methyl azobisisobutyrate (on the total weight of monomers) for 16 hours by UV irradiation at the ambient temperature.

Example 2: Methyl Cellosolve (80 weight percent) was mixed with 20 weight percent of a mixture consisting of 15 weight percent of methacrylonitrile and 85 weight percent of 2-hydroxyethyl methacrylate and the resulting mixture was polymerized at 80°C with 0.6 weight percent of azobisisobutyronitrile (on the total weight of monomers) for 10 hours in an inert atmosphere.

The soluble polymer was obtained, which may be used in preparation of coatings for poly(vinyl chloride), using evaporation of its solution followed by crosslinking with $(NH_4)_2Cr_2O_7$ under irradiation with a UV lamp for 30 minutes.

The polymer can be also applied in a preparation of lacquers for glass using the additional crosslinking by diisocyanates (0.5 weight percent on the polymer) or melamine resins (at concentrations up to 2.5 weight percent on 2-hydroxyethyl methacrylate in the polymer).

Cycloacetal

The process of *T. Hanyuda and E. Takiyama; U.S. Patent 3,931,353; January 6, 1976; assigned to Showa High Polymer Co., Ltd., Japan* relates to a polymerizable cycloacetal resinous composition, characterized by reacting 1 equivalent of diallylidenepentaerythritol with 0.8 to 2.0 equivalents of polyhydric alcohol-unsaturated monocarboxylic acid ester mono-ol having both a hydroxyl group and a polymerizable or copolymerizable unsaturated bond in the same molecule, with or without a solvent, in the presence of a polymerization inhibitor and an addition reaction catalyst.

Example 1: 260 weight parts of hydroxylethylmethacrylate, 212 weight parts of diallylidenepentaerythritol, 1.5 weight parts of para-toluenesulfonic acid and 0.25 weight part of hydroquinone were placed in a one-liter three-necked flask equipped with a condenser, stirrer, thermometer and inlet tube for nitrogen.

The mixture in the flask was stirred in the atmosphere of nitrogen, and reacted at a temperature of 80° to 85°C for 10 hours. It was proven by infrared ray absorption spectrum determination that more than 80% of the reactants were reacted to produce a light yellowish resin having a low viscosity of 1.6 to 1.7 poises at 27°C. The reaction product had the following chemical formula:

$$CH_2{=}\underset{\displaystyle CH_3}{\underset{|}{C}}{-}COOCH_2CH_2OCH_2CH_2CH\left\langle\begin{matrix}O{-}CH_2 & CH_2{-}O\\ & C & \\ O{-}CH_2 & CH_2{-}O\end{matrix}\right\rangle CHCH_2CH_2OCH_2CH_2OOC{-}\overset{\displaystyle CH_3}{\overset{|}{C}}{=}CH_2$$

Example 2: 1 weight part of benzoin methyl ether was dissolved in 100 weight parts of the polymerizable cycloacetal compound of Example 1, and to the resultant mixture was added 10 weight parts of 30% styrene solution of polyvinyl acetate. The resultant composition was coated by a knife coater on a steel plate so as to make a thickness of a coated film 0.2 mm.

This coated film was placed under a mercury lamp positioned 10 cm from the coated film. The film was cured by 1 to 2 seconds' exposure to ultraviolet rays, and the pencil hardness of the cured surface reached 4 H after 5 seconds' exposure. Thus, a resinous composition of this process proved to be useful for a paint to be cured by ultraviolet rays.

Flexible Vinyl Ester Resin

Thermosettable vinyl ester resins which provide improved flexibility and abrasion resistance are prepared in the process of *G. Zachariades and R.Q. Davis; U.S. Patent 3,933,935; January 20, 1976; assigned to The Dow Chemical Company.* In the process a polyepoxide is reacted with a dicarboxylic acid half ester of a monomer having the formula

$$CH_2{=}\underset{\displaystyle R}{\underset{|}{C}}{-}\overset{\displaystyle O}{\overset{\|}{C}}{-}OR_1(OCH_2\underset{\displaystyle R_2}{\underset{|}{C}H})_nOH$$

where R is H or methyl, R_1 is a two to four carbon alkylene group, R_2 is H, methyl or ethyl and n has an average value of 3 to 6.

Example 1: The preparation of flexible vinyl ester resins according to this process is illustrated as follows. A three-necked 500-ml flask equipped with a stirrer was charged with 148 g (1 mol) of phthalic anhydride and 300 g (1 mol) of an acrylate monomer which was a reaction product of hydroxyethyl acrylate (HEA) and 3 mols of propylene oxide. To the mixture was added 0.1 g of hydroquinone inhibitor. The mixture was heated slowly to 118°C and reacted for about 6 hours.

The mixture was then cooled to 100°C and 200 g of a polyglycidyl ether of bisphenol A having an epoxide equivalent weight (EEW) of 190 (D.E.R. 331) was added along with 0.4 g of DMP-30 catalyst. The mixture was slowly heated to 115°C and reacted for four hours until the residual acid and epoxy value was less than about 0.6%. The resin may then be diluted with a copolymerizable monomer and cooled.

Example 2: By the procedure of Example 1 a vinyl ester resin was prepared using an acrylate monomer which was a reaction product of HEA and 6 mols of propylene oxide (PO) having an average molecular weight of 510.

Example 3: An acrylate monomer was prepared by reacting 3 mols of ethylene oxide (EO) with HEA. This monomer having an average molecular weight of 272 was used to prepare a vinyl ester resin according to the procedure of Example 1.

The curability of the vinyl ester resins of this process by exposure to ultraviolet light is shown in the following tests. The resins of Example 1 (3 PO/HEA) and Example 3 (3 EO/HEA) were each diluted with 45% acetoxypropyl acrylate, coated on Bonderite 37 steel panels and cured by exposure to UV light.

In each case 2% of n-butyl benzoin ether (Trigonal 14) was added as a photoinitiator. The coated steel panel was passed at a speed of 100 ft/min under a 200 watt lamp and the number of passes required to cure to a tack-free state was determined. The results are shown below.

Resin	No. of Passes to Cure	Pencil Hardness	MEK Resistance	Taped Adhesion
Ex. 1	8	HB	Pass 20	Excellent
Ex. 3	5	H	Pass 20	Excellent

In a similar manner the resin of Example 2 (6 PO/HEA) and a comparable resin prepared with ethylene oxide instead of propylene oxide (6 EO/HEA) were cured by exposure to UV light. In each case no copolymerizable monomer was employed.

Resin	No. of Passes to Cure	MEK Resistance
6 PO/HEA	8	Pass 30
6 EO/HEA	4	Pass 45

The above results show that the resins, per se, are curable and do not require the presence of a copolymerizable monomer.

Epoxide Blend

W.R. Watt; U.S. Patent 3,936,557; February 3, 1976; assigned to American Can Company provides an improved blend of epoxide materials, fluid at room temperature. It consists essentially of at least one epoxidic prepolymer material having an epoxy equivalent weight below 200, constituting between about 10 and 85% of the weight of the blend, and selected from the group consisting of an epoxy resin prepolymer consisting predominantly of the monomeric diglycidyl ether of bisphenol A, a polyepoxidized phenol or cresol novolak, a polyglycidyl ether of a polyhydric alcohol, and a diepoxide of a cycloalkyl or alkylcycloalkyl hydrocarbon or ether.

The blend consists additionally of at least about 15% by weight of an epoxidic ester having two epoxycycloalkyl groups, and from 0 to 15% by weight of a monoepoxide having a viscosity at 23°C of less than 20 cp. Polymerizable compositions advantageously consist essentially of the above-specified ingredients and a radiation-sensitive catalyst precursor which decomposes upon application of energy to provide a Lewis acid catalyst effective to initiate polymerization of the above-mentioned epoxidic materials.

Such compositions are especially useful in providing rapidly curable coatings, which may contain no more than a few percent by weight of unpolymerizable materials. Thus, in accordance with the process, an epoxidized polymer is produced by forming a mixture of the epoxidic materials mentioned above and the catalyst precursor, applying the mixture so formed to a substrate, and subsequently applying energy to the mixture on the substrate to release the Lewis acid catalyst in sufficient amounts to effect substantial polymerization of the epoxidic materials.

Catalyst precursors ordinarily will be present in the polymerizable compositions in amounts ranging from about 0.5 to 2% of the total weight of the compositions less than 0.1 or more than 5% seldom being called for. Preferred photosensitive Lewis acid catalyst precursors are aromatic diazonium salts of complex halogenides, which decompose upon application of energy to release a halide Lewis acid.

Example 1: An epoxide blend was made up of the following epoxides in parts by weight:

Vinylcyclohexene dioxide	2 parts (28.6%)
(3,4-Epoxycyclohexyl)methyl 3,4-epoxycyclohexanecarboxylate	5 parts (71.4%)

To 10 g of this blend there were added 60 mg (0.60% of the total weight) of 2,5-diethoxy-4-(p-tolylthio)benzenediazonium hexafluorophosphate, and stirring was continued until dissolved. The resulting light-sensitive formulation was used to coat a plate of chromated steel, using a No. 3 drawbar. The steel plate was exposed to a 360-watt mercury arc at a distance of 2 inches for 2 seconds, then placed in an oven at 110°C for 10 minutes.

The coating cured to a hard, glossy protective finish. To test adhesion the coated plate was immersed for 30 minutes in a hot water bath maintained at 70°C, removed, blotted dry, scratched with a sharp blade and impressed with a piece of highly adhesive plastic tape. Pulling the tape off sharply failed to remove the coating from the metal plate.

Example 2: The bis(epoxycyclohexyl) ester may provide a very high proportion of the epoxide materials, as illustrated by the following blend:

Prepolymer consisting substantially of diglycidyl ether of bisphenol A	40 g (10%)
(3,4-Epoxycyclohexyl)methyl 3,4-epoxycyclohexanecarboxylate	360 g (90%)

The prepolymer product used was specified to have a viscosity at 25°C in the range of 6,500 to 9,500 cp and an epoxy equivalent weight of 180 to 188. To this weight of the blend was added a solution of 4 g (0.97% of the total weight) of p-chlorobenzenediazonium hexafluorophosphate in 8 ml (9.65 g or 2.33%) of propylene carbonate.

The resulting mixture was placed in the fountain of a Heidelberg printing press of the dry offset type and applied to paper as a thin coating. The paper was then placed on a conveyor passing beneath two 1200-watt high-pressure mercury arcs. Shortly after exposure to the mercury arcs, the coating had cured to a hard, glossy finish.

Low Viscosity Vinyl Acetate Polymers

Low viscosity radiation curable compositions are produced by *G.W. Borden and C.H. Carder; U.S. Patent 3,943,103; March 9, 1976; assigned to Union Carbide Corporation* by the addition of certain low molecular weight vinyl resins to one or more functional reactive solvents.

It was a completely unexpected and unobvious finding that such compositions cured to yield coatings and inks having the excellent properties that they possess since previous experience with vinyl resins in coating and ink compositions had indicated that their properties were improved by increasing the molecular weight of the vinyl resin added thereto.

The low molecular weight vinyl resins that are used in the curable compositions are those vinyl acetate polymers containing from 5 to 100 mol percent vinyl acetate therein, preferably at least about 35 mol percent vinyl acetate, having an inherent viscosity below about 0.25, measured at 30°C using a solution of 0.2 g of resin per 100 ml of cyclohexanone and having a glass transition temperature above about room temperature preferably from about 25° to 55°C.

Included are poly(vinyl acetate), copolymers of vinyl acetate with vinyl chloride, and polymers of vinyl acetate and vinyl chloride with one or more other polymerizable comonomers present in minor molar amounts, wherein the vinyl acetate content is as set forth above. Any of the known polymerizable ethylenically unsaturated monomers containing the polymerizable $>C=C<$ group can be used as the other polymerizable comonomer that is optionally present in the vinyl acetate polymer.

The functional reactive solvents can be monofunctional or polyfunctional. A single polyfunctional reactive solvent can be used or a mixture of two or more or a combination of one or more monofunctional reactive solvents and one or more polyfunctional reactive solvents can be used. The latter mixtures are generally more desirable.

The monofunctional reactive solvent can be any polymerizable ethylenically unsaturated monomer containing a single >C=C< group. The polyfunctional reactive solvents serve primarily to confer crosslink density to the cured composition so that it is rendered less soluble and less thermoplastic than the vinyl resin alone. These polyfunctional compounds contain at least two polymerizable >C=C< groups in the molecule.

The radiation curable compositions of this process contain from 5 to 70 weight percent of the low molecular weight vinyl resin; from 0 to 60 weight percent of the monofunctional reactive solvent; and from 10 to 70 weight percent of the polyfunctional reactive solvent.

Example: A polymerization reactor was charged with 232 grams of vinyl chloride, 216 grams of vinyl acetate, 32 grams of 2-hydroxypropyl acrylate, 1,108 grams of acetone and a solution of 0.9 gram of isopropylperoxydicarbonate in 12 grams of acetone as the catalyst. The polymerization was carried out over a period of about 25 hours at a temperature of about 73°C and a pressure of from 41 to 69 psig.

During this period there was added, during the first 4 hours, a mixture of 467 grams of vinyl chloride, 170 grams of vinyl acetate and 213 grams of 2-hydroxypropyl acrylate. During the next 21 hours, a mixture of 4,035 grams of vinyl chloride, 2,265 grams of vinyl acetate, 1,345 grams of 2-hydroxypropyl acrylate and 6,500 grams of acetone was added to the polymerization reactor while simultaneously removing reaction mixture at a rate to keep the volume in the reactor substantially constant.

Further, an additional quantity of 33 grams of the catalyst in 410 grams of acetone was added during the entire polymerization. At the end of the polymerization reaction the contents of the reactor were combined with the product recovered during the reaction; the total solids in the varnish was 33%. A small portion of the reaction product was precipitated by adding isopropanol and water to the varnish. The terpolymer was filtered, washed and dried. This dried vinyl terpolymer had an inherent viscosity of 0.222.

The coating composition was prepared by adding 100 parts of the vinyl terpolymer solution prepared above to 67 parts of neopentyl glycol diacrylate and then removing unreacted vinyl chloride and vinyl acetate, and acetone under vacuum. To this composition there was added 2 weight percent of benzophenone and 3 weight percent of methyldiethanolamine as the photoinitiator. This composition was identified as Composition A.

A similar composition, Composition B, was prepared using 2 weight percent of n-butyl ether benzoin as the photoinitiator. The compositions were coated on birch plywood and their properties evaluated. The coatings were applied by sanding the wood, applying a 1-mil coating of the composition and curing as shown below, sanding the coating, and applying a top coat of the same material using a No. 20 wire-wound rod. Composition A was cured in air by exposure to a 4.4-kw mercury arc radiation source for 20 seconds. Composition B was cured by initially purging with nitrogen for 40 seconds and then curing under nitrogen by exposure to the same radiation source for 12 seconds. The cured coatings had the following properties.

	A	B
Sward hardness (glass = 100)	14-18	14-16
60° Gardner gloss	77-80	88-90
Taber wear factor*	8.2	5.4
Crosshatch (intercoat) adhesion (%)	85	40
Resistance to:		
Nail polish remover	9	9
Schaeffer's No. 32 ink	10	10
3% aqueous ammonia	10	10
Mercurochrome	10	10
Magic bleach	10	10

*CS-17 wheels, 1,000-gram weights, 200 cycles, reported as milligrams weight loss per hundred cycles.

Hydantoin Glycol Derived Polyene-Polythiol

C.R. Morgan; U.S. Patents 3,945,982; March 23, 1976 and 3,984,606; October 5, 1976; both assigned to W.R. Grace & Company provides a curable polymer composition prepared from compatible polyene and polythiol components derived from a hydantoin glycol. This polyene and polythiol mixture is a highly reactive composition which is capable of being photocured when exposed to actinic radiation in the presence of a UV sensitizer to insoluble polythioether-containing materials which exhibit excellent physical and chemical properties.

The cured product has remarkable flexibility and high elongation at failure. The desirable characteristics of the cured materials make the hydantoin glycol derived polyene-polythiol curable composition particularly useful as coatings on wire and formable metals.

The curable composition is comprised of a polyene component containing at least 2 reactive carbon to carbon unsaturated bonds per molecule which is a reaction product of N,N'-bis(2-hydroxyethyl)dimethylhydantoin and at least one unsaturated organic compound such as an ene acid, or ene isocyanate; and a polythiol component containing at least two thiol groups, which is the reaction product of N,N'-bis(2-hydroxyethyl)dimethylhydantoin and a mercaptocarboxylic acid.

In forming the composition comprised of the polythiol and the polyene, it is desirable that the photocurable composition contain a photocuring rate accelerator from about 0.005 to 50 parts by weight based on 100 parts by weight of the polyene and polythiol. When UV radiation is used for the curing reaction, a dose of 0.0004 to 6.0 watts/cm^2 is usually employed.

Example 1: To a 3,000-ml resin kettle equipped with stirrer, thermometer, nitrogen inlet and outlet and vented addition funnel was charged under a nitrogen blanket 959.4 grams of commercially available trimethylolpropane diallyl ether and 0.98 gram of stannous octoate catalyst. 1,000 grams of commercially available isophorone diisocyanate was charged to the addition funnel and added dropwise to the kettle with stirring over a 4½-hour period while maintaining the temperature below 70°C. After isophorone diisocyanate was completely added, the temperature was allowed to drop to room temperature (24°C), the nitrogen blanket discontinued and the reaction was stirred for 48 hours.

To a separate 1,000-ml resin kettle equipped with stirrer, thermometer, nitrogen inlet and outlet and vented addition funnel was charged 535 grams of the reaction product from above along with 16 drops of stannous octoate. 301.3 grams of commercially available pentaethoxylated N,N'-bis(2-hydroxyethyl)dimethylhydantoin was charged to the addition funnel and thereafter added dropwise to the resin kettle with stirring while maintaining the temperature below 70°C. The resultant polyene product will be referred to as Polyene A.

Example 2: To a 2,000-ml 3-necked flask equipped for distillation with stirrer and nitrogen inlet was charged 432 grams (2 mols) of commercially available N,N'-bis(2-hydroxyethyl) dimethylhydantoin, 445 grams of mercaptopropionic acid and 17.5 grams p-toluene sulfonic acid. One hundred milliliters of ethylene dichloride was added to the flask. The mixture was heated with stirring for 8 hr and the evolved water was continuously removed by azeotropic distillation at 71° to 75°C returning the ethylene dichloride to the reaction flask.

The solution in the flask was then washed once with about 1,000 ml water, twice with about 1,000 ml of 5% sodium bicarbonate and finally with about 1,000 ml of water. The solution was dried over anhydrous magnesium sulfate, mixed with 10 grams of decolorizing carbon and filtered. The solvent was removed by vacuum distillation, affording 672 grams of product or an 86% yield of a polythiol which will be referred to as Polythiol Z.

Example 3: The following formulation was prepared in a 200-ml brown sample bottle. Accurately weighed amounts of the polyene, stabilizers, photosensitizer and polythiol were added to the bottle and admixed until homogeneous before use. Polyene A, a solid at room temperature, was heated to 80°C in order to facilitate weighing and handling.

Component	Weight, g
Polyene A	49.2
Polythiol Z	30.0
Benzophenone (photosensitizer)	1.58
H_3PO_4 (stabilizer)	0.039
Irganox 1076 (stabilizer)	0.158
Ionol (stabilizer)	0.079

The formulation was poured on a glass plate and drawn down to a 20-mil-thick film. The film on the plate was exposed to UV radiation for 2 minutes under a UV Ferro lamp at a surface intensity of 7,000 $\mu w/cm^2$. The cured sample had an elongation at failure of 133%.

Dyes as Photosensitizers

J.D. LaBash and V.D. McGinniss; U.S. Patent 3,966,574; June 29, 1976; assigned to SCM Corp. provide a formulation particularly suitable for food-associated uses and capable of photopolymerization by a source of ultraviolet radiation comprising a photopolymerizable ethylenically unsaturated compound, about 0.5 to 5% of a dye formulation consisting essentially of FD&C red No. 3 (phthalein) in combination with FD&C red No. 2, FD&C red No. 40, or both FD&C red No. 2 and red No. 40, and an amount of an organic amine activator. Particularly improved results are obtained by employing in combination FD&C red No. 3 and FD&C red No. 40, with or without FD&C red No. 2. Preferably the dyes are used in equal proportions. Preferably, about 2 to 10% of the amine activator is employed.

Example: The polymerizable composition of each test run consisted of ⅓ 2-ethylhexyl acrylate, ⅓ trimethylol-propane triacrylate and ⅓ diacrylate of the diglycidyl ether of bisphenol A. Methyldiethanolamine was employed in varying amounts between about 2 and 10% based on the weight of the total formulation. The dyes used were 1% each of FD&C red No. 40 and FD&C red No. 3 based on the weight of the total formulation.

Each sensitized coating composition was poured onto a steel panel and drawn down with a No. 8 wound wire rod to a film thickness or coating of approximately 0.5 mil. The coated but wet panels were then placed on a conveyor belt and exposed to sources of ultraviolet radiation. Two separate sources of ultraviolet radiation were used, one being a plasma arc radiation source (PARS) and the other a conventional ultraviolet light source having two 4,000-watt mercury lamps. Tack-free coatings were obtained after a 5-second exposure to the conventional UV source and after a 0.10-second exposure to the PARS source.

Unsaturated Polyester with Epoxy Diacrylate

M. Wismer; U.S. Patent 3,968,016; July 6, 1976; assigned to PPG Industries, Inc. describes wax-free polyester resin compositions that may be admixed with an epoxy-based diacrylate to form a composition which cures rapidly when subjected to actinic light in the presence of air.

The composition comprises: (a) an ethylenically unsaturated polyester of an alpha, beta ethylenically unsaturated polycarboxylic acid and a polyhydric alcohol; (b) from about 10 to 80% by weight, based on the ethylenically unsaturated polyester, of the reaction product of a member of the group consisting of acrylic acid and methacrylic acid with a polyglycidyl ether of a polyphenol or polyhydric alcohol; and (c) from about 0.01 to 10% by weight, based on the weight of the composition, of a photosensitizer.

Example: A container was charged with 9.25 grams of an unsaturated polyester resin formed by mixing 30 parts of styrene with 70 parts of a polyester comprising 5 mols of maleic anhydride, 5 mols of phthalic anhydride and 10.6 mols of propylene glycol with 0.01% by weight of methyl hydroquinone inhibitor, 7.5 grams of vinyl toluene, 2.2 grams of Triganol 14 (a mixture of the benzoin ethers of butyl and amyl alcohols) and 50 grams of the reaction product of acrylic acid and the diglycidyl ether of bisphenol A (Epon 828).

The above composition was mixed and subjected to ultraviolet light from a high intensity mercury vapor lamp (330 watts per square inch) at a line speed of 15 feet per minute for two passes in the presence of air. The above film was compared to a film formed by mixing 98.5 grams of the polyester resin and 1.5 grams of Triganol 14 alone and subjecting to ultraviolet light from the same source for three passes at 15 feet per minute in the presence of air.

The above films were tested for solvent resistance by rubbing with a cloth soaked with acetone and mar resistance by scratching with a fingernail. The film formed using the epoxy diacrylate was not affected by 100 rubs in acetone (each rub consists of a backward and forward motion while exercising firm pressure on a solvent soaked swab) while the film formed without the epoxy diacrylate lifted after only 50 rubs. The film formed using the epoxy diacrylate was not affected by the fingernail test while the film formed without the epoxy diacrylate marred.

Modified Acrylate Resin

An ultraviolet curable acrylate resin composition with improved adhesion to metal surfaces is described by *C.W. Uzelmeier and P.D. Jones; U.S. Patent 3,971,834; July 27, 1976; assigned to Shell Oil Company*. It consists of a reaction product of (1) a secondary monoamine selected from the group consisting of dimethylamine, diethylamine, methylethylamine, diisopropylamine, piperidine, toluidine, diallylamine, methylcyclohexylamine, N-methylhydroxylamine, methylethanolamine, butylethanolamine, N-acetylethanolamine and phenylethanolamine, (2) acrylic acid and (3) a polyepoxide in a ratio of between about 0.2 and 0.9 chemical equivalent of the monoamine per chemical equivalent of the polyepoxide and between about 0.8 and 0.1 chemical equivalent of the acrylic acid per chemical equivalent of the polyepoxide where the polyepoxide is a glycidyl ether of 2,2-bis(4-hydroxyphenyl)propane. The reaction product is prepared by contacting components (1), (2) and (3) at reaction temperature of about 125° to 200°F.

Illustrative Embodiment 1: Preparation of 25% Modified Acrylate Resin — 4 equivalents of Epon 828 resin (752 grams) were contacted with 1 equivalent of diisopropanolamine (133 grams) at a temperature of about 130°F under an air sparge. The reaction mixture exothermed to a temperature of about 200°F and was then held at that temperature for about ½ hour.

To this mixture was added 3 equivalents of glacial acrylic acid (216 grams) and 200 ppm hydroquinone (0.22 gram). The reaction mixture exothermed to about 250°F and was held at that temperature for about 1½ hours until the acidity had fallen below about 0.01 eq/100 g. The final resin composition had an acidity of 0.0076 eq/100 g and a Gardner color of 7.

Comparative Example 1: Preparation of Unmodified Acrylate Resin — One equivalent each of Epon 828 resin (188 grams) and glacial acrylic acid (72 grams) were heated to 200°F under an air sparge in the presence of 220 ppm hydroquinone (0.052 gram). Then 0.7% of 50% aqueous tetramethylammonium chloride catalyst was added (1.82 grams). The reaction was allowed to exotherm to 240° to 250°F, whereupon cooling was initiated to hold the temperature in that range. After 2 hr, 45 min, when the acidity had fallen below 0.01 eq/100 g, the mixture was cooled to 170°F and removed.

Illustrative Embodiment 2: The resin composition prepared in Illustrative Embodiment 1 and Comparative Example 1 were photocured by adding 30% 2-hydroxypropyl acrylate diluent and 5 phr Trigonal 14 photoinitiator to the resin, coating Q steel panels with the resin mixture at 0.2 to 0.3 mil film thickness, and subjecting the coated panels to an ultraviolet lamp of 200 watts per linear inch at a line speed of 200 ft/min.

The degree of cure and adhesion to metal surfaces was measured by two standard tests employed by the paint and coating industry, MEK rubs and Scotch tape adhesion, respectively. Some of the UV-cured panels were pasteurized for 45 min in water at 160°F prior to testing. Adhesion was retained after pasteurization. The results are shown below.

Resin	UV Cured: MEK Rubs	UV Cured: Scotch Tape Adhesion	UV Cured-Pasteurized: MEK Rubs	UV Cured-Pasteurized: Scotch Tape Adhesion
100% acrylate (C.E. 1)	54	Poor	100	Poor
25% DIPA (I.E. 1)	9	Excellent	14	Excellent

Polybutadiene in Two-Step Reaction

In the process of *H.E. De La Mare; U.S. Patent 3,974,129; August 10, 1976; assigned to Shell Oil Company* a curable resin composition is prepared by (a) reacting a low molecular weight polybutadiene resin having an average molecular weight of between about 500 and 10,000, a viscosity of between about 50 to 3,000 centipoises, a cis content of between about 50 and 95% and containing less than about 2% vinyl bonds with a polycarboxylic acid anhydride of the general formula

$$\begin{array}{c} H\text{—}C{=}C\text{—}R_1 \\ \text{(cyclic anhydride: } O{=}C\text{—}O\text{—}C{=}O) \end{array}$$

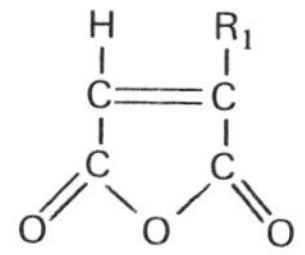

where R_1 is selected from the group consisting of H, CH_3, and Cl in the presence of an inhibitor and in a weight ratio of anhydride to polybutadiene resin of between about 5:95 and about 50:50 thereby forming a modified polybutadiene resin moiety with attached anhydride groups; and (b) reacting the modified polybutadiene resin moiety with a hydroxyalkyl acrylate of the general formula

$$CH_2{=}C(R_2)\text{—}C({=}O)\text{—}O\text{—}R_3\text{—}OH$$

where R_2 is a hydrogen or methyl group and R_3 is an alkyl group of 1 to 10 carbon atoms, in a molar ratio of hydroxyalkyl acrylate to attached anhydride group of between about 25:100 and 100:100.

The resin composition is cured by any conventional curing technique including ultraviolet radiation, thermal baking, metal-catalyzed air drying and the like. Because of the ease with which it cures thermally the resin could also be applied as a water-borne self-curing coating. Where it is desired to employ an ultraviolet curing system it is necessary to added a photosensitizer. Because of their excellent adhesion to metal surfaces, the resin compositions are particularly useful as metal coatings.

Opaque Coatings

Opaque coatings or films having a high covering capacity produced from coating compositions containing synthetic resins, fillers, solvents, common catalysts, sensitizers, additives and auxiliary agents are provided in the process of *H. Brose, K.D. Depping, D. Hentschel, B. Kostevc and K. Schmidt; U.S. Patent 3,984,584; October 5, 1976; assigned to BASF Farben & Fasern AG, Germany.*

Radiation permeable mixtures of coating compositions are used containing the following components: (a) 14 to 20 parts by weight, preferably 15 to 18 parts by weight, photosensitized light-curing condensation resins having double bond values of 0.180 to 0.450; (b) 55 to 75 parts by weight, preferably 65 to 70 parts by weight, of fillers having a low covering capacity; (c) 0.5 to 10 parts by weight, preferably 1 to 6 parts by weight, of film-forming cellulose derivatives or mixed polymerizates of vinyl chloride with other monomers which are compatible with A and D; and (d) 8 to 30 parts by weight, preferably 10 to 20 parts by weight, of nonreactive solvents in which optionally up to 5 parts by weight of the component D may be substituted by monomeric copolymerizable solvents. The coating compositions are applied to a substrate and cured by irradiation with UV light, sunlight or powerful ionizing rays.

Example: (A) An unsaturated polyester resin is produced in a conventional manner as a condensation product from the following substances: 20 parts diglycol, 9 parts dipropylglycol, 3 parts trimethylolpropane, 19 parts maleic anhydride, and 15 parts tetrahydrophthalic anhydride. 0.003 part hydroquinone is added to this mixture and a resin having an acid number of 35 and a double bond value of 0.314 is obtained which is then further diluted with ethylglycol at a temperature of between 90° and 95°C to a solid material content of 60%.

(B) A coating composition is produced from the following constituents: 26 parts of the solution obtained according to (A), 9 parts ethylglycol, 3 parts of a 25% solution of cellulose acetobutyrate in ethanol/toluene in a ratio of 2:1, 1.3 parts benzoin isopropyl ether, 41 parts permanent white, 8 parts talcum, grain size 5 to 10 microns, 0.7 part zinc stearate, and 11 parts precipitated calcium carbonate.

Approximately 130 g/m^2 of this coating composition are applied to a chipboard by means of a machine for applying a coating. The chipboard bearing the layer of coating composition is then passed through a drying channel in which the mercury vapor high-pressure lamps are arranged in series at intervals of 25 cm. The distance of the layer to be hardened from the mercury vapor high-pressure lamps is about 10 to 20 cm. The curing process takes 60 seconds. A pure white layer which is opaque and can be polished, is obtained.

Modified Prepolymer plus Acrylate Monomer Ester

In the process of *K. Brack; U.S. Patent 3,989,609; November 2, 1976; assigned to Dennison Manufacturing Company* a prepolymer containing unsaturated hydrocarbon groups is prepared and mixed on a roller mill with one or more acrylic ester monomers and various additives to make a coating formulation of a desired viscosity.

In general, low viscosity formulations are used for overprint varnishes, on paper or foil, or, with pigments, for certain types of printing inks. Higher viscosity formulations are used to apply thick films on panels, tiles or other bodies. Thin films are cured to hardness by brief exposure to ultraviolet light. Thicker films require more energetic radiation such as plasma arc and electron beam radiation. The prepolymers particularly useful for making such radiation curable coatings are the reaction products of polyether polyols and bis- or polyisocyanates and hydroxy alkenes or acrylic (or methacrylic) hydroxy esters, and, likewise, reactive polyamides modified with dicarboxy alkenes, their anhydrides or esters. A small amount of wax incorporated in the coating formulations results in coatings with release characteristics similar to those of PTFE coatings.

Example 1: 522 parts 2,4-toluene diisocyanate, 500 parts dried toluene and 0.5 part stannous octoate are placed in a vessel equipped with an agitator and blanketed with nitrogen. With stirring, a solution of 735 parts poly(propylene oxide) triol (Dow Voranol, CP-700, Hydroxyl No. 229; 0.02% H_2O) in 750 parts dried toluene is added at such a rate that the temperature of the reaction mixture does not exceed 50°C. The reaction mixture is stirred at 50°C for 1 hr. Then the nitrogen blanket is replaced by a dry air blanket and 383 parts hydroxyethyl acrylate (10% excess), mixed with 100 parts of dried toluene, are added at such a rate that the temperature of the mixture does not rise above 75°C.

After the addition is completed, the reaction mixture is kept stirring at 90°C for 1 hr. One part p-methoxy phenol (polymerization inhibitor) is added and the solvent is removed under reduced pressure at or below 30°C. 1,650 parts of a clear, colorless, resinous oil, containing a trace of toluene solvent are obtained. Analysis of this material shows 0.06% isocyanate content.

Example 2: UV-Curable Overprint Varnish — 100 parts of the product of Example 1, 235 parts trimethylolpropane triacrylate, 30 parts hydroxyethyl acrylate, 3 parts stearyl acrylate and 25 parts benzoin isobutyl ether are mixed well on a roller mill. Films of 0.4 mil thickness are applied with a wire-wound coating rod onto paper, aluminum foil, vinyl-coated aluminum foil, polyester-coated Mylar, and steel. The coated substrates are exposed for 1/10 second to the UV radiation given off by a medium-pressure mercury vapor lamp at a distance of 5". After this exposure all samples are cured to hard, glossy coatings with a pencil hardness of at least 2H and a rub resistance of at least 40 rubs, using methyl ethyl ketone as the solvent.

Opaque Coating from Normally Transparent Polyester Paste Filler

H. Brose, K.D. Depping, D. Hentschel, B. Kostevc and K. Schmidt; U.S. Patent 3,993,798; November 23, 1976; assigned to BASF Farben & Fasern AG, Germany provide cured, opaque coatings with high hiding power from a normally transparent and radiation-transmitting unsaturated polyester paste filler composition containing customary additives, curing catalysts and optionally fillers of low hiding power and nonreactive solvents and/or plasticizers.

The filler has a weight ratio of unsaturated polyester to the sum of copolymerizable monomers and optionally nonreactive solvents and plasticizers of from about 10:12 to 10:40. It may further have up to 95 wt % of the copolymerizable monomers replaced by the nonreactive solvents and/or plasticizers. The composition may be cured by UV rays, sunlight or other ionizing radiation.

Example: An unsaturated polyester resin is made from 26 parts of 1,2-propylene glycol, 16 parts of maleic anhydride and 20 parts of phthalic anhydride. 0.003 part of hydroquinone is added to the bath as a stabilizer and the resin is diluted with styrene, at temperatures of between 90° and 95°C to a solids content of 65%. The resulting polyester resin has an acid number of 35.

A paste filler composition is made from 21.5 parts of the polyester solution obtained above, 1.0 part of benzoin isopropyl ether, 3.0 parts of cellulose acetobutyrate solution (25% solids content in a 1:2 mixture of alcohol and styrene), 41.0 parts of blanc fixe, 7.9 parts of microtalc, average particle size 5 to 10 μ, 0.7 part of zinc stearate, 11.3 parts of a precipitated $CaCO_3$, 4.8 parts of vinyltoluene, and 4.8 parts of ethyl glycol.

130 g/m^2 of this composition are applied to a chipboard sheet and passed through a drying tunnel in which mercury vapor high-pressure lamps are 25 cm apart. The distance of the layer from the lamps is 25 to 30 cm and curing time is 40 seconds. A white coating results which is opaque, susceptible to grinding, and can serve as a paste filler coating for subsequent lacquering build-up.

INKS

Solvent-Free Printing Ink

G.R. Buckwalter; U.S. Patent 3,881,942; May 6, 1975; assigned to Borden, Inc. provides both a fast drying, solvent-free printing ink vehicle comprising; (a) an ester of an aliphatic alcohol and a C_{12} to C_{20} unsaturated fatty acid, (b) a film forming resin, and (c) a metal salt of peroxydiphosphoric acid and a fast drying, solvent-free printing ink composed of a major proportion of the foregoing printing ink vehicle and 5 to 45 weight percent of a pigment. The composition of the vehicle will comprise 5 to 80 weight percent of the ester of an aliphatic alcohol and a C_{12} to C_{20} unsaturated fatty acid; 5 to 40 weight percent of a film forming resin; and 1 to 10 weight percent of a metal peroxydiphosphate.

When the printing ink vehicle contains a polyunsaturated material and a photosensitizer in addition to the ester, the film forming resin, and the metal peroxydiphosphate, the vehicle may be dried either by heat alone or heat in conjunction with actinic radiation. By employing two means for drying the vehicle, as well as inks based on the vehicle, even faster drying times may be achieved.

Example 1: A printing ink having the following composition is prepared:

	Parts by Weight		Parts by Weight
Heat treated tung oil	30.0	Barium peroxydiphosphate	4.0
Pentaerythritol tetraacrylate	10.0	Methyl ester of rosin acids	10.0
Aluminum octoate gel of		Benzophenone	3.0
rosin ester	16.5	Morpholine	3.0
Methyl eleostearates	13.5	Phthalocyanine blue pigment	10.0

When a substrate printed with this ink is heated within the range of about 100° to 175°C while simultaneously being irradiated with ultraviolet light, the ink dries in about 0.5 to 0.8 second.

Example 2: A printing ink having the same composition as that of Example 1, except the 10.0 parts of pentaerythritol tetraacrylate are replaced by 10.0 parts of trimethylol propane triacrylate, is prepared.

When a substrate printed with this ink is heated at a temperature within the range of about 100° to 175°C while simultaneously being irradiated with ultraviolet light, the ink dries in about 0.5 to 0.8 second.

Tertiary Beta Amine as Oxygen Scavenger

The process of *J.E. Gaske; U.S. Patents 3,914,165; October 21, 1975; and 3,925,349; December 9, 1975; both assigned to DeSoto, Inc.* relates to radiation curable coating compositions, and certain tertiary amine compounds upon which the same are based, it being intended to avoid the oxygen inhibition which normally characterizes such compositions. It is particularly intended to enable rapid cure of 100% solids systems with ultraviolet light using a minimum of energy while minimizing or eliminating effluent fumes or vapors. As a feature of the process, inks are provided which cure at high speed using ultraviolet radiation and which are easily degraded in dilute aqueous caustic solution to permit recycling of the printed paper which is produced.

In accordance with this process, an amine containing at least one amino hydrogen atom is adducted with a stoichiometric excess of ethylenic material comprising a polyacrylate using a Michael addition to form an adduct containing unreacted acrylate groups and at least one tertiary beta amino group resulting from the addition.

It has been found that these Michael addition products are rapidly polymerized or copolymerized through vinyl polymerization under the influence of ionizing radiation. In the presence of appropriate photosensitizers, which are well known, they cure rapidly to form solid films under ultraviolet radiation. The point of importance is that the radiation polymerization of ethylenic materials is extended and speeded when the unsaturated tertiary beta amines of this process are present, and when ultraviolet radiation is used, the tertiary beta amine content of the adduct appears to scavenge oxygen preventing air inhibition of coating compositions, which is unusual and of obvious importance. The product is a viscous clear liquid stable at room temperature in the dark or in filtered fluorescent light.

From the standpoint of proportions in the Michael addition reaction, at least about 0.5% of the materials reacted together should be constituted by the amine, but larger amounts are preferably used (at least 1%) to enhance resistance to air inhibition during an ultraviolet cure. The maximum proportion of amine in the Michael addition reaction is determined by the need for a stoichiometric excess of ethylenic material comprising polyacrylate. This provides unreacted acrylate groups in the product which insures that the amine component is nonvolatile and chemically incorporated by polymerization into the radiation cured product.

Example 1: 1 mol of pentaerythritol triacrylate is incrementally added to 1 mol of diethylamine under reflux conditions with constant stirring. The heat of this Michael addition reaction is limited to a maximum of 70°C by moderating the rate of amine addition and by the use of an external water bath. When the reaction is complete the temperature begins to decrease.

Example 2: The liquid of Example 1 was used with 3% (weight) benzophenone to disperse colored pigments. Dispersion was made using a glass mortar and pestle followed by mulling with a glass muller on a glass plate.

Yellow Ink	**Grams**
Lead chromate Primrose Yellow	50.0
Photosensitized Example 1 fluid	40.0
Black Ink	
Hydrite 10 clay	3.0
Lampblack	1.0
Photosensitized Example 1 fluid	10.0

Blue Ink	Grams
Microfine marble dust (calcium carbonate)	28.6
Phthalocyanine blue	3.3
Photosensitized Example 1 fluid	68.1
Gold Bronze Metallic Ink	
MD-650-B pale gold powder (Alcan Metal Limited)	1.0
Photosensitized Example 1 fluid	8.0

The metallic ink was prepared in each instance by simple stirring. All the inks were applied to paper by the use of a No. 6 wire wound rod draw down or a hand brayer. All the inks converted with ease in the presence of air upon 1/12 second exposure to the 5 kilowatt arc except the metallic ink which required an exposure of equal duration on the back side of the paper as well as on the front side. The yellow ink converted at ½ power input to the arc. Satisfactory conversion of the inks is viewed as the ability of the converted inks to resist transference of color by rubbing with a piece of uncoated white paper. None of these converted inks would offset onto uncoated white paper immediately after being photoconverted with extreme thumb pressure applied.

Also, all the inks could be immediately recoated with each other, or with the clear photosensitized Example 1 liquid, and converted in 1/12 to 1/24 of a second ultraviolet exposure. There was no visible bleed-through of color and in all cases intercoat adhesion was excellent. Further, all of the converted ink surfaces had a glossy tack free surface.

Five square inches of paper coated with these compositions were cut into ¼ inch squares and placed in 1 inch diameter by 2¾ inch tall screw cap vials. 10 cc of 3% aqueous sodium hydroxide was added. The vials were not shaken but allowed to stand undisturbed at room temperature. Within 30 minutes the coatings had disintegrated and the paper was observed to be white with no coating on it. The fluid around the paper was colored (where color is involved). No remnant of coating could be observed with 3 power optical magnification. The fluid is slightly gelatinous in nature. This was not so with the gold metallic ink only. The film had parted from the paper but was intact as a curled metallic colored sheet. With slight shaking this broke down into very fine suspended metallic particles.

The criterion for selection of pigment of dyestuff has been found to be very fine particle size with extreme dispersion needed, light must be able to be transmitted or an inert light transparent fine particle (such as clay or calcium carbonate) must be used in conjunction with the opaque pigment such as carbon black, the colored component must not be reactive with the fluid Michael addition adduct or monomers used, and the colored component must not be itself photoactive, consuming ultraviolet light energy and thereby retarding the desired photoconversion.

Built-In Sensitizer

In the process of *G. Rosen; U.S. Patent 3,926,640; December 16, 1975; assigned to Sun Chemical Corporation* carboxy-substituted benzophenones are reacted with hydroxyl-containing polyethylenically unsaturated esters, resulting in compounds that have built-in sensitizers and are useful for printing inks, coating compositions, adhesives, and the like with or without a secondary sensitizer. The carboxy-substituted benzophenones have the following formula:

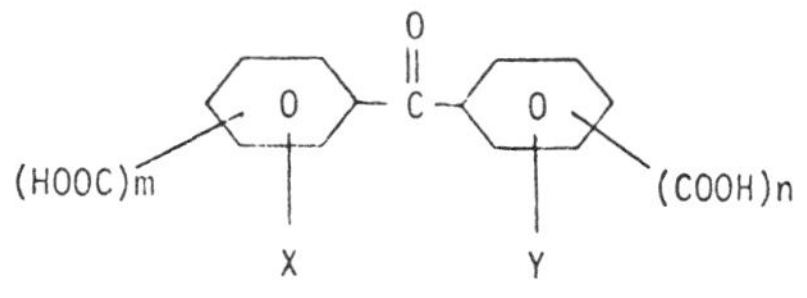

where m and n each is an integer from 0 to 3 and the sum of m and n is in the range of 1 to 6; and X and Y may each be 1 to 4 halogen atoms, e.g., chlorine, bromine, or iodine; dialkylamino groups having 1 to 4 carbon atoms; or other groups which confer desirable properties to the product, such as for example, mercaptan, disulfide, alkene, peroxy, alkoxy, carbonyl, amide, amine, nitro, hydroxy, ether, aryl, or the like; X and Y may be the same or different and either or both may be omitted.

Example 1: A mixture of 747 parts of pentaerythritol-3,5-acrylate (1 equivalent OH) and 120 parts of benzophenone tetracarboxylic dianhydride (BTDA) was heated at 80° to 90°C in the presence of phosphoric acid as catalyst. The product was a half-ester adduct of the pentaerythritol-3,5-acrylate and BTDA.

Example 2: The products of this process were formulated into inks and tested as follows. A mixture of 85 parts of the product of Example 1 and 15 parts of phthalocyanine blue was printed onto coated paper by letterpress and dried by passing it under three 200-watt/inch mercury vapor lamps at the rate of 1,200 feet per minute. The product dried rapidly.

Example 3: The procedure of Example 2 was repeated with each of the following colorants instead of phthalocyanine blue: lithol rubine red, carbon black, milori blue, and phthalocyanine green. The results were comparable.

Synergistic Mixture of Sensitizers

In the printing field, it is required that a relatively thick liquid coating film of, for example, ultraviolet light-curable type printing ink or coating varnish be solidified (dried or cured) upon irradiation with ultraviolet light in quite a short period of time, and that the coating composition must be stable without undergoing a dark reaction on a printing or in a coating machine or in a container.

M. Sumita, R. Konno and T. Takeuchi; U.S. Patent 3,945,833; March 23, 1976; assigned to Toyo Ink Manufacturing Company, Ltd., Japan found that a synergistic effect which would not be expected from the performance of components used separately can be obtained when a specific combination of sensitizers is incorporated into an ink or a coating varnish. The synergistic effect can be obtained even though each of the sensitizers is inferior in sensitization and/or poorer in storage stability due to the dark reaction.

The process provides a photosensitive coating composition comprising a prepolymer containing radically crosslinkable ethylenically unsaturated double bonds, and a sensitizer which comprises a mixture of: (a) benzophenone, a halogenated benzophenone or combination thereof, and (b) 4,4'-bis(diethylamino)benzophenone in a weight ratio of (a) to (b) of about 1:2 to 10:1. A suitable molecular weight for the prepolymers ranges from 800 to 15,000. Combinations of p-chlorobenzophenone or o-chlorobenzophenone with 4,4'-bis(diethylamino)benzophenone show especially remarkable effects. In the examples which follow all parts, percentages, ratios and the like are by weight, unless otherwise indicated.

Example 1: 717 parts of Epikote 828 (a condensation product of bisphenol A and epichlorohydrin; mean molecular weight, 370; epoxy equivalent, 184 to 194), 283 parts of acrylic acid, 0.1 part of hydroquinone and 0.3 part of triethylenediamine were charged in a four-necked flask equipped with a refluxing condenser and a stirrer, and the resulting mixture was reacted at 90° to 120°C for 15 to 20 hours while bubbling air into the reaction mixture until the acid value became <1 and then the reaction mixture was removed to obtain an epoxyacrylate. To 254 parts of the epoxyacrylate thus obtained were added 156 parts of trimethylolpropane triacrylate and 290 parts of an acrylated isocyanate obtained from an equimolar reaction between β-hydroxyethylacrylate and toluene diisocyanate.

The resulting mixture was charged into a four-necked flask equipped with a refluxing condenser and a stirrer, and allowed to react at 60°C for several hours while bubbling air into the reaction mixture. When the reaction ratio exceeded 80%, β-hydroxyethyl acrylate was added to the reaction mixture in an amount equimolar to the remaining unreacted isocyanate. The reaction was continued until the unreacted isocyanate groups decreased to less than 0.2% by weight and the reaction mixture was then removed.

Example 2: A magenta ink for offset printing was prepared using the prepolymer obtained in Example 1 and each of sensitizers indicated in the table below according to the following standard formulation: 15.0 parts Carmin 6BH (a red pigment), 45.8 parts prepolymer, 20.0 parts sensitizer, 4.0 parts white petrolatum, 0.2 part hydroquinone; and 15.0 parts trimethylolpropane triacrylate.

Each of the resulting printing inks was printed onto an art paper (coated paper) in an amount of 30 mg/100 cm^2 using an RI-tester. Immediately after the printing, the printed article was placed on a belt conveyer and passed at a distance of 13 cm below a 2 kw-high pressure mercury lamp having an input energy of 60 w/cm. The sensitivity was determined in terms of the conveyor speed (m/min) required for drying each of the printing articles. The results are shown below.

Sensitizer	Ratio (%)	Sensitivity (m/min)	Storage Stability (day)
Benzoin Ethyl Ether	100	24	1
Michler's Ketone	100	8 (not dried)	More than 30
Tetraethyl-p,p'-diaminobenzophenone	100	"	"
p-Chlorobenzophenone	100	"	"
o-Chlorobenzophenone	100	"	"
p-Bromobenzophenone	100	"	"
Benzophenone	100	"	"
Tetraethyl-p,p'-diaminobenzophenone	50	20	"

(continued)

Sensitizer	Ratio (%)	Sensi-tivity (m/min)	Storage Stability (day)
Benzophenone	50		
Tetraethyl-p,p'-	50		
diaminobenzophenone		30	"
p-Chlorobenzophenone	50		
4,4'-bis(Diethylamino)-	10		
benzophenone		35	"
p-Chlorobenzophenone	90		
4,4'-bis(Diethylamino)-	30		
benzophenone		32	"
p-Chlorobenzophenone	70		
4,4'-bis(Diethylamino)-	60		
benzophenone		25	"
p-Chlorobenzophenone	40		
4,4'-bis(Diethylamino)-	10		
benzophenone			
p-Bromobenzophenone	70	28	"
Benzyl	20		

It was found from the above results, sensitizers which showed poor results when used individually exhibited extremely excellent effects when used in combination.

Conductive Ink for Circuit Board

D.A. Bolon, G.M. Lucas and R.L. Bartholomew; U.S. Patent 3,968,056; July 6, 1976; assigned to General Electric Company provide a radiation curable ink convertible to a conductive coating exhibiting a specific resistivity of less than 10 ohm-cm when cured on the surface of a substrate using actinic radiation at a temperature of up to about 60°C within 2 minutes or less. The radiation curable ink comprises by volume (a) from about 10 to 60% of an organic resin binder having a viscosity of from 50 to 10,000 cp at 25°C, and (b) from about 90 to 40% of a particulated electrically conductive metal containing material substantially free of metal containing material having an aspect ratio of diameter to thickness of a value of greater than 20.

In instances where it is desired to make UV curable inks, UV sensitizers can be employed when the organic resin binder is in the form of a polyester or polyacrylate or other polymerizable UV curable material. There can be employed from about 0.5 to 5% by weight of the UV sensitizer based on the weight of resin.

Example: A polyester prepolymer was prepared by effecting reaction between about 35.3 parts of fumaric acid, 11.9 parts of dicyclopentadiene and 25.3 parts of propylene glycol. The resulting prepolymer was blended with about 24.4 parts of styrene containing 100 ppm of tert-butylhydroquinone and 1.8 parts of benzoin sec-butyl ether along with 0.7 part of 135°F paraffin wax and warmed until a solution was obtained.

A photocurable ink was prepared by blending the above organic resin binder with 67 parts of silver coated glass spheres having an average diameter of about 10 to 50 microns. On a volume basis, there was employed about 2 volumes of conductive filler per volume of resin. The above photocurable ink was printed in pattern form onto a 2 by 6 inch polystyrene substrate. The treated polystyrene substrate was then placed at a distance of about 8 inches from the arc tube of a General Electric H3T7 lamp which had been ballasted to permit operation at about 960 watts input. There were employed two quartz filters below the lamp having dimensions of about 5 x 10 inches. The filters were supported on steel supports which formed a channel through which air was blown. The upper filter support was in contact with a 6 foot copper coil having an average diameter of about ⅜ inch through which water was passed at about 25°C. The full intensity of the lamps was measured at about 20,000 μw/cm^2 and the temperature of the substrate did not exceed about 50°C.

After a 2 minute cure, the ink on the panel was tested for continuity. Cure of the ink on the panel was determined by a bake cycle of 60 minutes at 70°C after irradiation. If after 2 minutes exposure, the ink strip is tack free and it shows no more than a 2% weight loss based on the weight of tack-free ink, the ink is considered cured. It was found that the resulting conductive cured ink had a specific resistivity of 0.015 ohm-cm.

The cured ink in the circuit board was then evaluated for adhesion by flexing it at least 5 times sufficient to produce a distance of 1 inch between the center of the arc to an imaginary straight line drawn between the two ends of the board. No significant change in conductivity of the cured ink strip was found.

An abrasion test was also run on the connecting tabs by attaching the edge of the circuit board to a steel clamp at least 16 times, where the clamp spring had a compressive force of at least five times the weight of the board freely suspended. Although the clamp contacted the cured strip connecting tabs, no adhesive separation of the strip was noted and the specific resistivity of the cured strip remained substantially the same.

A further evaluation of the circuit board was made by exposing it to 96% relative humidity at a temperature of 120°F for 14 days without allowing condensation of water on the surface of the board. It was found that the specific resistivity of the circuit board remained substantially unchanged.

A printing ink was made consisting of 33 parts of the above organic resin binder and 67 parts of silver flake having an aspect ratio greater than 20. A screen printed pattern from this ink did not photocure following the above described conditions. An irradiated strip had a specific resistivity of greater than 1000 ohm-cm. In addition, the strip was unsuitable as a circuit board material because it failed all of the above shown tests.

In related work *D.A. Bolon, G.M. Lucas and M.S. Jaffe; U.S. Patent 3,989,644; November 2, 1976; assigned to General Electric Company* found that if 0.5 to 10% by weight of certain nonionic surfactants based on the weight of the organic resin binder are employed in the above described radiation curable ink then the tendency of the particulated electrically conductive metal containing filler to settle is substantially reduced. It has also been found that the conductivity of the resulting cured printing ink is also substantially enhanced.

Included by the nonionic surfactants which can be used are, for example, polyoxyethylenes, ethoxylated alkylphenols, ethoxylated aliphatic alcohols, carboxylic esters, carboxylic amides, polyoxyethylene fatty acid amides, polyalkyleneoxide block copolymers, etc.

Example: An organic resin binder was prepared by dissolving in the ingredients of the organic resin binder of the example of U.S. Patent 3,968,056, above, 5% by weight of a nonionic surfactant in the form of a polyalkylene oxide block copolymer. The photocurable ink of that example was designated A. Another photocurable ink B was prepared following the same procedure using the organic resin binder containing the nonionic surfactant.

The above photocurable inks were printed and irradiated as described above. After 2 minutes cure, the panels were tested for continuity. The cured strips were about 3 mils by 40 mils by 5 inches. It was found that the strip made from the conductive cured ink derived from the radiation curable ink A had a specific resistivity of 0.015 ohm-cm. However, the radiation cured strip derived from the nonionic surfactant-containing ink B had a specific resistivity of 0.006 ohm-cm.

Radiation curable inks A and B were stirred vigorously and then allowed to rest for 48 hours. Ink B containing the surfactant remained substantially uniform after 48 hours. However, the ink A free of surfactant appeared to be in two phases after 24 hours on the shelf.

Both cured polystyrene panels were then subjected to an atmosphere of 96% humidity at a temperature of 120°F for 2 weeks. The panels were then tested again and it was found that the resistance of the cured ink did not substantially change.

SPECIAL ADDITIVES AND SOLVENT

Acrylic Acid as Flatting Agent

C.H. Carder; U.S. Patent 3,966,572; June 29, 1976; assigned to Union Carbide Corporation found that the addition of a small amount of acrylic acid to a photocurable coating composition containing silica increases the ability of the silica to produce a lower gloss film. This improvement is generally obtained without any appreciable increase in viscosity or thixotropy. In some instances, however, it has been observed that viscosity is lowered to some degree by the addition of the acrylic acid. The 100% coatings compositions that are useful in this process are the acrylate-based compositions. The amount of acrylic acid added to the 100% solids coating composition is from about 0.1 to 10 weight percent, preferably 1 to 5 weight percent.

In the example, adhesion was measured by noting the percentage of film remaining after a tape pull (Scotch No. 610) on a grid of razor blade cuts, 10 in each direction, ⅛ inch apart. Resistance to stains was measured by placing drops of the staining liquid on the coated panel, then covering the liquid with a small inverted dish or bottle cap. After the designated period, the liquid was wiped away with a wet towel and the imparted stain was rated on a scale of 1 to 10 with 10 representing no stain and 1 designating very severe staining.

Acetone resistance, a measure of through-cure, was determined by placing a small swatch of cotton or absorbent paper towel saturated with acetone on the cured film, then measuring the time required in seconds for the film to be lifted off the substrate. During the test the cotton or towel was kept saturated with acetone by the addition of small amounts via a dropper.

Example: A series of formulations were prepared as follows:

	Parts by Weight		
	Control	A	B
Acrylated urethane adduct	26.7	26.7	26.7
2-phenoxyethyl acrylate	15.3	15.3	15.3
Neopentyl glycol diacrylate	18.0	18.0	18.0
Acrylic acid	–	1.2	3.0
n-Butyl ether of benzoin	1.2	1.2	1.2

The acrylated urethane adduct used was produced by adding 98 grams of phenoxyethyl acrylate, 410 grams trimethylhexamethylene diisocyanate, and 1 gram of dibutyltin dilaurate to a flask. This mixture was heated to 40°C, and over a period of about 2 hours 473 grams of 2-hydroxyethyl acrylate was added at a temperature of from 40° to 45°C. The mixture was then heated for an additional hour, cooled and transferred to a container until used. The control and formulations A and B had added to each of them 10 weight percent of 2 different silica flatting agents. In the first series the flatting agent used was Syloid 74. In the second series the flatting agent used was Syloid 308. The flatting agent was dispersed in the formulation by stirring for 60 seconds with a high speed stirrer.

Films were applied with a No. 20 wire wound rod to sanded birch panels and cured by exposure to radiation from 2.2 kw mercury lamps under nitrogen by exposure for 1.8 seconds. The panels were sanded and a second top coat of the same material was applied and cured in the same fashion. The properties of the cured films are set forth below and compared with those of the control for each series produced.

First Series	Control	A	B
60° Gardner gloss*	77-81	56-60	46-65
Sward hardness	28	34	28
Crosshatch adhesion, %	30	60	75
Taber wear factor	8.6	5.7	8.6
Resistance to			
Nail polish remover	10	10	10
Black ink	10	10	10
3% ammonia	10	10	10
Mercurochrome	10	10	10
Bleach	10	10	10

*One measurement was taken in the direction of the drawdown (usually higher), and the second taken in the opposite direction.

Second Series	Control	A	B
60° Gardner gloss	40-50	34-35	40-42
Sward hardness	38	34	32
Crosshatch adhesion, %	0	0	75
Taber wear factor	11.6	17.2	7.5
Resistance to			
Nail polish remover	10	10	10
Black ink	10	10	10
3% ammonia	10	10	10
Mercurochrome	10	10	9
Bleach	10	10	10

The lower gloss values of the compositions of this process are clearly shown.

Sulfones as Solvent for Diazonium Catalyst

Stable diazonium catalyst solutions are provided by *J.H. Feinberg; U.S. Patent 3,960,684; June 1, 1976; assigned to American Can Company* in which the catalyst is dissolved in an organic sulfone. The catalyst solutions have greatly extended shelf-life while at the same time retain the ability to rapidly cure epoxy resins upon exposure to an energy source. Polymerizable compositions of the two-package type, comprising polymerizable epoxy materials and such catalyst solutions are also provided.

Example: Several large batches were prepared by mixing together the following epoxides in the indicated proportions:

Epoxide	Epoxy Equivalent Weight	25°C Viscosity, cp	Parts by Weight
(1) Diglycidyl ether of bisphenol A	172-178	4,000-6,000	55
(2) (3,4-epoxycyclohexyl)methyl 3,4-epoxycyclohexane-carboxylate	131-143	350-450	30
(3) Alkyl glycidyl ether in which alkyl groups are predominantly dodecyl and tetradecyl	286	8.5	15

A number of (aliquot) samples were withdrawn from these and employed in the examples which follow in which parts are by weight and temperature in degrees Centigrade unless indicated otherwise.

Samples were coated on paperboard, using a drawbar to provide a coating of the order of about 0.0003 inch thick, when dry, after which they were exposed to a mercury vapor lamp at a distance of 2 inches unless otherwise indicated. The relative rates of cure were determined by noting the exposure time which was necessary to produce a finish which was hard to the touch. These cure rates were compared to the rate of cure of a freshly catalyzed resin mixture utilizing a freshly prepared solution of catalyst and having a cure rate of 1.0. A value less than 1.0 denotes a slower rate of cure than that of the freshly catalyzed formulation on a scale of 1.0 to 0.0.

(A) A catalyst solution was prepared by adding 1 part p-methoxybenzene diazonium hexafluorophosphate to 2 parts of a solvent containing 98% sulfolane and 2% propylene carbonate. The solution was stored for 12½ months after which a 0.3 gram aliquot was added to 10 grams of the epoxide blend set forth above. The catalyzed epoxy formulation was light yellow in color, had an initial viscosity of 376 cp at 25°C and was found to cure at the same rate expected of a freshly prepared catalyst solution. After standing for 24 hours, the formulation was found to have a viscosity of 432 cp at 25°C.

(B) A catalyst solution was prepared by adding 3.53 grams p-methoxybenzene diazonium hexafluorophosphate to 7.06 grams dimethyl sulfolane and 2.0 grams sulfolane. The solution was stored for 10½ months after which a 0.3 gram aliquot was added to 10 grams of the epoxide blend of (A). The initial viscosity of the catalyzed formulation was found to be 371 cp at 25°C. The formulation was light yellow in color. After standing for 24 hours, the viscosity was found to be 431 cp at 25°C. The cure rate relative to that of a freshly prepared sample was found to be 0.9 when measured in the same manner under the same conditions.

(C) A yellow catalyst solution was prepared by adding 5.0 grams p-chlorobenzenediazonium hexafluorophosphate to 10.0 grams sulfolane containing 3% water. After storage for one week, the catalyst solution was still yellow in color, a result that is quite unusual since solutions of the catalyst in other solvents, for example propylene carbonate, were found to turn dark in less than 24 hours.

1.5 grams of the aged solution was employed to catalyze 50.0 grams of an epoxide blend described in (A) but containing 20 parts epoxide (1), 10 parts of epoxide (2) and 3 parts epoxide (3). The catalyzed formulation was light yellow in color and was found to cure to a hard finish after about one second exposure to a 360 watt medium pressure mercury arc.

The above catalyst solution after aging for three weeks was found to be reddish brown in color but was still clear with no sediment present. When 1.5 grams of this aged solution was added to 50.0 grams of the epoxide blend, the catalyzed formulation was yellow-orange in color and was found to cure to a hard finish after a 1 to 1½ second exposure to a 360 watt medium pressure mercury arc at a distance of 2 inches.

(D) 10.0 gram samples of polymerizable resin mixtures were prepared as follows: (a) 10.0 grams of the epoxide blend of (A); (b) 9.0 grams of the epoxide blend of (A) and 1.0 gram gamma butyrolactone; (c) 9.5 grams of the epoxide blend of (A) and 0.5 gram dodecyl vinyl ether.

To each of the above compositions was added 0.3 gram of a 10½ month aged catalyst solution prepared by mixing a 6.0 gram aliquot containing 3.53 grams p-methoxybenzene diazonium hexafluorophosphate, 7.06 grams dimethyl sulfolane and 2.0 grams sulfolane with 0.04 gram of a solution containing 95 parts sulfolane and 5 parts of a substituted poly(vinylpyrrolidone) available commercially as Ganex V-816.

Each of the three catalyzed compositions (a), (b) and (c) were found to be yellow in color and were found to cure at the same rate as the same compositions catalyzed with freshly prepared catalyst solutions. After standing overnight, these three compositions were found to still be fluid and yellow in color.

It will be apparent from the results of examples (A) through (D) catalyst solutions of the process have greatly extended storability and yet retain the ability to effect rapid cure of the catalyzed formulation upon exposure to an energy source since the aged solutions cure at substantially the same rate as the freshly prepared catalyst solutions. Moreover, the solvent is compatible with the polymerizable material, e.g., epoxide, as shown by the examples in which the aged epoxy-catalyst solutions were still fluid after 24 hours.

Antioxidants for High Vinyl Butadienes

An ultraviolet curable resin provided by *M.R. Roodvoets; U.S. Patent 3,899,406; August 12, 1975; assigned to The Firestone Tire and Rubber Company* contains antioxidant compounds selected from the class consisting of 2,2-methylene bis-(4-R_1,6-R_2-phenol) where R_1 is an aliphatic group having from 1 to 4 carbons and R_2 is an aliphatic group having 3 to 4 carbons, and at least one material selected from the class consisting of the dibutyl para-cresols and from the N,N'-di(1-R_3,1-R_4-methyl)-p-phenylenediamine where R_3 is an aliphatic group having from 1 to 4 carbons and R_4 is an aliphatic group having from 4 to 6 carbons. The antioxidants increase the shelf life of ultraviolet curable high vinyl polybutadiene and have little effect on or actually increase the cure rate of the polybutadiene.

Example: Various high vinyl polybutadiene ultraviolet curable compositions containing various antioxidants and combinations thereof were prepared in accordance with the formulations set forth in Table 1.

TABLE 1

Compound	Parts by Weight							
	A	C	G	A/C/G	A/G	A/C	C/G/H	Control
90+% vinyl polybutadiene homopolymer	70	70	70	70	70	70	70	70
Divinyl benzene	30	30	30	-	-	-	-	30
Vinyl toluene	4.0	4.0	4.0	4.0	4.0	4.0	4.0	4.0
Benzoin ethyl ether	-	-	-	30	30	30	30	-
2,2-methylenebis(4-ethyl, 6-tert-butylphenol)	0.15	-	-	0.10	0.10	0.10	-	-
Dibutyl-p-cresol	-	0.15	-	0.05	-	0.05	0.05	-
N,N'-di(1-ethyl-3-methylpentyl)-p-phenylenediamine	-	-	0.025	0.025	0.025	-	0.025	-
2,2-methylenebis(4-methyl,6-tert-butylphenol	-	-	-	-	-	-	0.10	-

The samples were mixed in bulk using an air stirrer. The polybutadiene was diluted with the coupling agent monomer to reduce the viscosity so that the wound wire rod technique could be used to coat aluminum foil.

A Hanovia 125 watt low pressure mercury vapor lamp was used as the ultraviolet source to cure the approximately 1 millimeter coating of the aluminum foil samples at intervals while they were aging. The standard cure interval was 6 minutes at 4 inches. Benzoin ethyl ether at a level of 4 parts was used as the ultraviolet initiator. Cure rate was determined from the degree of cure after 6 minutes. The degree of cure was determined by rubbing the coating with a toluene soaked rag and observing the number of strokes at which the coating began to dissolve. A higher degree of cure or faster cure rate corresponds to a higher test value. The results are set forth in Table 2. In the table the numbers give degree of cure after 6 minutes at 4 inches (40 = completely cured, 0-5 = completely uncured).

TABLE 2

COMPOUND	UN-AGED	3 DAYS	6 OR 7 DAYS
A	4	3	33
C	10	7	32
G	8	10	19
A/C/G	28	34	35
A/G	27	29	40+
A/C	20	23	36
C/G/H	18	21	40
Control	20	Gelled	Gelled

As is apparent from Table 2, generally the combinations of this process gave higher cure rates than any of the antioxidants individually. The combination containing all three classes of antioxidant according to the process gave the best overall properties (increased time to gellation and cure rate). Additionally, many other antioxidants and combinations were investigated and not found to approach the results of this process.

Additionally, combinations of antioxidant compounds A, C, and G were prepared and tested in a manner set forth above with the exception that the amount of antioxidant A was doubled, the amount of antioxidant C was increased to 0.15 and also to 0.25. The results obtained after a cure of 6 minutes at 4 inches are set forth in Table 3.

TABLE 3

	A/C/G AMT. OF A DOUBLED	A/C/G AMT. OF C INCREASED TO .15	A/C/G AMT. OF C INCREASED TO .25	TIME PERIOD
Coloration	clear	clear	clear	0 days
Cure	cure: 14	cure: 17	cure: 18	
Coloration	clear	clear	clear	2 days
Cure				standard darkened
Coloration	clear	clear	clear	5 days
Cure				standard gelled
Coloration	clear	clear	clear	8 days
Cure	cure: 15	cure: 21	cure: 23	
Coloration	clear	clear	clear	1 month accel-erated aging
Cure	cure: 15	cure: 19	cure: 9	

As apparent from Table 3, when the amount of antioxidant A (first class of antioxidant) was doubled, although the coloration remained clear, the amount of extent of curing was medium. Considering the cases where antioxidant C (dibutyl para-cresol) was increased, the coloration also was generally clear and the cure was only moderate. Thus, these results indicate that compositions containing amounts of antioxidants at their upper range tended to slightly lessen their synergistic effect.

PROCESSES

Two-Step Process

In the process of *O.W. Smith, C.H. Carder and D.J. Trecker; U.S. Patent 3,935,330; January 27, 1976; assigned to Union Carbide Corporation* coating compositions are cured by initially treating them for a brief period with ionizing or nonionizing radiation whereby they are partially cured and then completing the cure by a thermal treatment. This two-step process decreases the amount of volatiles escaping into the atmosphere and it also permits the use of compounds not useful in the past and allows significant improvement in film properties such as adhesion to metal. The initial radiation cure can be with electron beam, ultraviolet mercury lamp, plasma arc or other radiation means; the thermal cure is carried out by any heat means.

The coating compositions that are useful in this process are those that contain: (1) a reactive thermoset crosslinker that is thermally cured, and (2) a reactive solvent that responds to both radiation curing and thermal curing. The compositions can also contain (3) reactive components which are radiation sensitive or radiation curable reactive solvents. Part or all of the required reactive sites can be combined in a single molecule or resin type.

The coating composition contains from about 0.1 to 50 weight percent, preferably from 10 to 30 weight percent of (1) the reactive thermoset crosslinker; from 0.1 to 50 weight percent preferably from 15 to 25 weight percent of (2) the reactive solvent sensitive to both thermal and radiation crosslinking, with the balance of the composition comprising (3) the radiation sensitive reactive component. The percentages are weight percent based on the final coating composition. The reactive thermoset crosslinkers can be polyfunctional epoxides having at least two oxirane units or their polyoxirane-containing low molecular weight polymers, or urea/formaldehyde resins, or melamine/formaldehyde resins.

The reactive solvents sensitive to both thermal and radiation crosslinking or curing are those which contain both (a) a vinyl polymerizable double bond which is sensitive to radiation polymerization and (b) a functional group which is sensitive to and reacts during the thermal crosslinking step. These compounds are the acrylates and methacrylates of monofunctional and polyfunctional alcohols and the acrylamides and methacrylamides of monofunctional and polyfunctional amines.

The radiation sensitive reactive component which is primarily responsive to radiation curing can be represented by the formulas:

$$CH_2{=}\overset{X}{\overset{|}{C}}{-}aryl \quad \text{or} \quad CH_2{=}\overset{X}{\overset{|}{C}}CN \quad \text{or} \quad \left(CH_2{=}\overset{X}{\overset{|}{C}}COO\right)_z{-}Y$$

where X is hydrogen or methyl; aryl is phenyl or a substituted phenyl; Y is the monovalent radical, cycloalkyl of 5 to 12 carbons, cycloalkenyl of 5 to 12 carbons, $-C_mH_{2m}H$, $-C_mH_{2m}CN$, $-C_mH_{2m}Cl$, $-C_mH_{2m}NR''_2$, $-C_mH_{2m}COOR''_2$, $-C_mH_{2m}COOCH_2OCH_2CH_3$, $-(C_pH_{2p}O)_rC_pH_{2p}H$, or the polyvalent radical, $-C_qH_{2q}-$, $-(C_pH_{2p}O)_xC_pH_{2p}-$, $-(C_yH_{2y}COO)_xC_yH_{2y}-$, $-C_qH_{2q-1}-$, or $-C_qH_{2q-2}-$ where R'' is hydrogen or alkyl of 1 to 5 carbons; m has a value of 1 to 10; r has a value of 0 to 4; p has a value of 2 to 4; q has a value of 2 to 8; x has a value of 1 to 5; y has a value of 2 to 5; z is 1 to 4 and when z is 1, the group Y is monovalent and when z is 2 to 4 the group Y is polyvalent.

Example: A coating composition was prepared by admixing 30 parts of acrylic acid, 50 parts of 2-ethylhexyl acrylate, 20 parts of bis-glycidyl ether of 2,2-propylidene-4,4'-bisphenol, 0.5 part of triethanolamine and 1 part of benzoin butyl ether. This coating was applied on steel panels using a No. 20 wire-wound rod. The coated panels were cured by this process and also by conventional processes for comparative purposes as shown below.

Panel A – This coated panel was weighed, then placed in an enclosed box and purged with nitrogen for 15 seconds. The panel still in the N_2 atmosphere was then exposed to the nonionizing high intensity predominantly continuum light radiation emanating from a 50-kilowatt argon swirl-flow plasma arc for 5 seconds at a distance of 2 feet from the arc for the initial radiation cure. Subsequently, the panel was placed in an oven at 210°C for 30 minutes for the second stage thermal cure. At the end of this period the panel was reweighed and it was found that only 4 percent of the initially applied coating had volatilized during the two-stage cure. The cured coating had a Sward hardness of 30 and resisted breakthrough by acetone-soaked cotton for greater than 60 minutes.

Panel B – This panel was coated and then given the initial radiation cure by exposure to the nonionizing high intensity predominantly continuum light radiation in the same manner as was Panel A. However, it was not subsequently oven cured or subjected to the second stage thermal cure. The resulting coating was undercured, still tacky and resisted breakthrough by acetone-soaked cotton for only 60 minutes. Though it was acetone-resistant it was not a commercially acceptable cured coating.

Panel C – This panel was coated and then thermally cured in the oven under the same conditions described in Panel A for the second stage thermal cure only. It was not exposed to the initial nonionizing high intensity predominantly continuum light radiation cure. It was found that 80 percent of the initially applied coating composition had volatilized and that the remaining finish had an acetone resistance of less than 5 minutes.

The data show that curing the coating by initially radiating and then thermally curing by the two stage process produces a superior coating. A comparison of Panel A with Panel B shows that under the same radiation conditions Panel B was not fully cured; a comparison of Panel A with Panel C shows that under the same thermal curing conditions an intolerable loss of material is experienced when curing is accomplished solely by thermal means.

Two Phase Flash Photolysis Process

A process and apparatus for polymerizing oxygen-inhibited ultraviolet photopolymerizable resin-forming material such as a film is described by *M. De Sorga and V.D. McGinniss; U.S. Patent 3,943,046; March 9, 1976; assigned to SCM Corp.* The apparatus comprises a pair of UV light sources, one being a flash photolysis source, the other a sustained photolysis source, both disposed for irradiating the mass as it abides in an atmosphere such as air which tends to inhibit such polymerization. The process has two essential phases, a superficial phase and a profound phase, performed simultaneously or one in advance of and as preparation for the other.

The profound phase is performed with sustained irradiation effective for completely polymerizing the material except for inhibition of polymerization at the surface thereof due to the atmosphere. The superficial phase is performed with flash irradiation effective for forming a tack-free skin over the material. The skin helps to protect the less fully polymerized material below from oxygen inhibition when such superficial phase is performed first or simultaneously with the other phase. When such superficial phase is performed after the profound phase, the superficial phase acts to complete polymerization of the material throughout its thickness.

Example: Paint compounded as follows is applied at 1 mil wet thickness to an aluminum panel 1 inch by 1 inch. This paint is air-inhibited and when irradiated by the sustained UV energy in accordance with the second step of the example, it will yield a film having a tacky surface.

Component	Parts by Weight
The reaction product of 1 mol of isophorone diisocyanate and 2 mols of hydroxyethyl acrylate	40
Hydroxyethyl acrylate	25
2-phenoxyethyl acrylate	15
Melamine acrylate	15
Sensitizing mixture:	
Benzophenone	2
Methyldiethanolamine	1

The freshly coated side of the panel is subjected to irradiation from a Xenon Corporation flash photolysis lamp operating to emit a substantially continuous spectrum of UV energy in the 2000 to 4000 A range of wavelength. The gap of atmospheric air at room temperature between the lamp and the panel is about 4 inches. The lamp emits such energy for up to about 1/1,000 second.

The lamp is about 8 inches long and 10 mm in diameter and housed in an elongated housing of essentially square cross section that is black inside. The UV light is emitted from a 1-inch diameter exit port about 1 inch in front of the lamp surface and mid-way to the length of the lamp tube. The power supply is 400 volts DC charging a bank of 10 capacitors in parallel, each of 100 microfarad rating. The housing acts poorly as a reflector; it is estimated that about one-half to 1 joule of near-UV energy per flash is the near-UV light output through such exit port.

A superficial top surface cure results on the paint of the panel struck by the flash of UV energy. This area of the paint is nontacky to the touch, but examination shows that the film is soft and decidedly underpolymerized slightly below its surface and further to its bottom. The thus-treated panel then is passed, paint side-up, by conveyor successively under a pair of like commercial mercury vapor lamps in parallel array and designed for emitting far-UV energy. These lamps are Hanovia model No. 652-OA431 UV lamps having a 4,000-watt demand, and they are 20 inches long. Each is equipped with an efficient reflector.

The conveyor travel is normal to the axis of the lamps and 6 inches below the lamps at the rate of about 100 feet per minute. Air is the atmosphere between lamps and panel. The result is a fully cured (substantially completely polymerized throughout) paint film having excellent adhesion, gloss, and resistance for its type.

Flash Lamp to Initiate Curing

C. Sander; U.S. Patent 3,930,064; December 30, 1975 found that a coating containing ethylenic unsaturation, i.e., double bonds between carbon atoms such as are contained in a varnish which dries by oxidation, may be cured most effectively without the use of added photosensitizers by exposing the coating to at least one flash of ultraviolet light to initiate reaction in the coating. The energy of the flash of light should produce at least 4.5 milliwatt-seconds per cm^2 of the surface of the coating. Such a flash produces at least about 0.03% of radiation having an effective wavelength of 197.4 nm. It is also advantageous that the flash produce at least about 0.004% of radiation having an effective wavelength of 184.9 nm and at least about 0.001% of radiation having an effective wavelength of 389.0 nm. It must be noted that these values relate to the radiation produced and not to the energy used in a lamp to create the radiation. These results are not obtained by conventional ultraviolet lamps.

Another feature of the process consists in employing a flash of ultraviolet light of a wave range at least a part of which has a wavelength which is effective for curing the coating material and lies within the resonant range of the radical of the material. This effective wavelength does not have to amount to a large component of the entire wave range of the rays of each flash but only a small component of this range will suffice which lies within the resonant range of the radical of the material to be cured. However, the larger this component is made, the shorter will be the length of time which is required for fully curing and hardening the material.

The means employed for producing the flashes is a flash lamp supplied with a current per flash having a density of more than 4,000 amp/cm^2 of the cross section of the light arc which is produced between the electrodes in the flash lamp and an output of less than 100 watt-seconds, the flash lasting for a period of less than 50 microseconds, each of the flashes being separated from the next flash for a period of time having a length equal to more than 100 times the length of time of one of the flashes. Details of the apparatus are included in the patent.

IMAGING COMPOSITIONS AND PROCESSES

COMPOSITIONS FOR GENERAL USE

Polymers Containing Stilbene Moiety

J.G. Pacifici; U.S. Patent 3,879,356; April 22, 1975; assigned to Eastman Kodak Company provides a class of light-sensitive polymers which contain a light-sensitive moiety selected from the group consisting of stilbenebenzimidazole, stilbenebenzoxazole, stilbenebenzotriazole and stilbenebenzothiazole attached to the polymer backbone. The point of attachment of the light-sensitive moiety to the polymer backbone is the carboxyl group which is attached to the stilbene nucleus. When the polymer backbone is derived from a hydroxyl-containing polymer, the light sensitive moiety is attached through a carbonyloxy linkage. When the polymer backbone is derived from a polymer containing reactive amino groups, the attachment of the light-sensitive moiety is through an amido linkage.

These light-sensitive polymers are prepared by the reaction of the hydroxy or amino groups on the polymer backbone with an acid halide of one light-sensitive moiety or the acid halide of a mixture of light-sensitive moieties. The polymers of the process are useful in a variety of photographic applications to prepare photomechanical images such as lithographic printing plates and photoresists.

Example 1: A mixture of 5.0 grams polyvinyl alcohol (Vinol 523, 12% acetyl) and 100 ml of dry pyridine is heated on a steam bath overnight. After cooling, 3.7 grams (0.01 mol) of 4'-(2-benzoxazolyl)-4-stilbenecarbonyl chloride is added to the reaction flask and the mixture heated for 12 hours over a waterbath at 50°C. To this mixture is then added 12.5 grams (0.096 mol) of benzoyl chloride and the mixture heated for an additional 3 hours. The mixture is cooled, 600 ml of acetone added, and the soluble product filtered. The acetone solution is poured into water and the precipitated polymer isolated by filtration. After air drying for 24 hours, the polymer is dissolved in chloroform and precipitated from methyl alcohol. The yield of dried product is 14.5 grams. Fluorescence analysis confirmed the presence of the stilbenebenzoxazole moiety.

Example 2: A mixture of 5 grams Eponol 55B-40 (a phenoxy-type resin from bisphenol A and epichlorohydrin), 75 ml of dry pyridine and 75 ml of methylene chloride is heated gently for 1 hour at 50°C. To this mixture is added 5.0 grams (0.013 mol) of 4'-(2-benzoxazolyl)-4-stilbenecarbonyl chloride and the mixture is heated for 24 hours at 50°C. The solution is diluted with 200 ml of methylene chloride and the product precipitated by adding the solution to methanol. The precipitated product is collected by filtration and air dried overnight.

Example 3: Films (0.5 mil) of the light-sensitive polymers prepared according to Examples 1 and 2 are cast on aluminum supports from appropriate solvents with a Garner Film Casting Knife. For the polymer prepared in Example 1 acetone/benzene (50/50) is used as the solvent and for the polymer prepared in Example 2 dichloroethane is used as the solvent. Samples are exposed through a negative mask to a 250 watt medium pressure Hanovia mercury lamp at a distance of 15 cm. The exposure time for all samples is 1 minute and the samples are developed for 5 minutes in the appropriate solvent. This exposure time is sufficient to render the exposed portions insoluble in the developer solution.

Dye Crosslinking Agent

D.W. *Heseltine, P.W. Jenkins and J.D. Mee; U.S. Patent Reissue 27,922; Feb. 19, 1974; assigned to Eastman Kodak Company* describe crosslinkable compositions comprising a polymer having hardenable groups and a dye crosslinking agent capable of being activated by heat, pressure or radiation having one of the following structures:

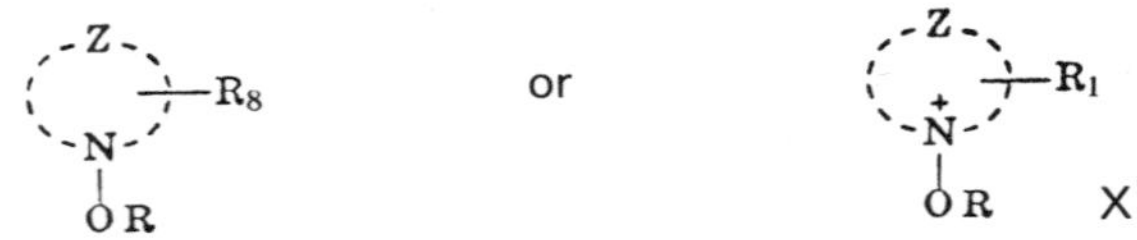

Z ring with N, substituent R_8; N–OR or Z ring with $\overset{+}{N}$, substituent R_1; N–OR X^-

In the formulas R_1 can be any of the following: (a) a methine linkage terminated by a heterocyclic nucleus of the type contained in cyanine dyes; the methine linkage can be substituted or unsubstituted; (b) an alkyl radical preferably containing one to eight carbon atoms including a substituted alkyl radical; (c) an aryl radical including a substituted aryl radical; (d) a hydrogen atom; (e) an acyl radical; (f) an anilinovinyl radical and (g) a styryl radical including substituted styryl radicals.

R can be either (a) an alkyl radical preferably having 1 to 8 carbon atoms; or (b) an acyl radical. Z represents the atoms necessary to complete a five to six membered heterocyclic nucleus including a substituted heterocyclic nucleus which nucleus can contain at least one additional hetero atom such as oxygen, sulfur, selenium or nitrogen.

R_8 can be either of the following: (a) a methine linkage terminated by a heterocyclic nucleus of the type contained in merocyanine dyes; the methine linkage can be substituted or unsubstituted; or (b) an allylidene radical including a substituted allylidene radical and X represents an acid anion.

When these compounds are exposed to electromagnetic radiation, they are decomposed by a heterolytic cleavage of the nitrogen-oxygen bond to produce a RO^+ ion, a dye base and an acid anion. The RO^+ ion may be decomposed even further producing an aldehyde and H^+ ion. The aldehyde is itself a crosslinking agent and causes crosslinking to occur in those areas of polymer with which it is in contact.

These compositions can be crosslinked by exposure to any form of light or other energy source simply by selecting a crosslinking agent which is sensitive to the particular form of energy employed. For example, the dye crosslinking agents can be made to have maximum light absorption over wide portions of the visible spectrum facilitating several types of exposing means. Mixtures of such dyes can also be used to facilitate imagewise exposing.

Example 1: A composition containing gelatin and 1-methoxy-2-methylpyridinium p-toluene-sulfonate in an amount sufficient to produce 3.0% aldehyde by weight of the total composition is coated on a polyethylene terephthalate film support at a thickness of 6 μ. The element is exposed to a transparency having opaque and transparent areas with a 140 watt ultraviolet lamp placed 3 inches away from the surface of element causing crosslinking to occur in the exposed areas. After exposure, the element is washed with water. The unexposed areas wash off leaving only an image formed by the exposed hardened areas.

Example 2: This example illustrates the preparation of a lithographic plate. 1-methoxy-2-methylpyridinium p-toluene-sulfonate is mixed with gelatin at a level calculated to produce theoretically 6% of formaldehyde with respect to dry weight of gelatin. The composition is coated on an anodized aluminum sheet at a thickness of 0.002 in. The dried coating is exposed for 15 minutes to the light from a 140 watt Hanovia ultraviolet lamp through a definition chart. The exposed plate is washed with 85°F water for 10 minutes and dried. The sheet is placed on a lithographic printing press and a good press copy is obtained.

Benzoate- and Furoate-Esterified Polyvinyl Alcohol

J.E. Apellaniz; U.S. Patent 3,881,935; May 6, 1975; assigned to Powers Chemco, Inc. provides a light-sensitive organic-solvent-soluble film-forming polymer capable of forming a continuous coating on a base which comprises recurring units of benzoate- and furoate-esterified polyvinyl alcohol having the structure:

```
—[ CH2 —— CH ——— CH2 —— CH ]—
          |                |
          O                O
          |                |
          C = O            C = O
          |          R     |
    (CH=CH)m—C6H4(R)    (CH = CH)n—(furan ring)—R'
```

In the structure R and R' are each H, alkyl or alkenyl of up to 12 carbon atoms, aryl, alkaryl or aralkyl of from 6 to 9 carbon atoms, halogen (F, Cl, Br, I), nitro, cyano, hydroxy, acetylamine, amino, alkoxy, carboalkoxy, alkylthio, mono- or dialkylamine, N-alkyl-carbamyl, N,N,-dialkylcarbamyl, alkylsulfonyl, the alkyl groups containing from 1 to 4 carbon atoms, trifluoromethyl, trifluoromethoxy, methoxymethyl, carbamyl, alkanoyloxy containing up to 4 carbon atoms, phenyl, p-chlorophenyl, p-methylphenyl or p-aminophenyl; m is a whole number from 0 to 1; and n is a whole number of 0 to 1. The polyvinyl alcohol has a molecular weight of from 14,000 to 115,000 and less than 59% of the possible -OH groups esterified with the benzoate moiety. The polymers are effectively utilized in the preparation of photographic resist materials and printing plates for lithography.

Example 1: Preparation of Polyvinyl Cinnamate Half-Ester — A mixture of polyvinyl alcohol of molecular weight 86,000 (3.0 mols) and N-methyl-2-pyrrolidone (2 l) is heated until a clear solution results.

While maintaining a temperature of about 55°C, cinnamoyl chloride (1.5 mols) is added dropwise over a period of 4 hours. The reaction is then diluted with acetone (6,000 ml) and poured into a large amount of water. The cinnamated PVA forms as a white precipitate which is filtered, washed several times (3) with water and finally air-dried. A volumetric analysis (saponification) shows the product to contain 42 to 45% cinnamate esterification.

Preparation of Polyvinylcinnamate Furoate — Polyvinyl cinnamate half-ester (3 mols) as prepared by the above procedure is dissolved in pyridine (1.5 l). To this solution is added 2-furoyl chloride (1.5 mols) and after completion of addition is maintained at 75°C for 3 to 4 hours. The mixed ester product is precipitated as described in the above procedure. After air drying, the yield of product is 85%.

Example 2: Photochemical Insolubilization — Polyvinylcinnamate furoate (0.75 g) as prepared by the procedure of Example 1 is dissolved in chlorobenzene (10 ml) containing benzanthracene-7,12-dione (0.056 g). This solution is coated on a copper plate such that after evaporation of the solvent, a layer of about 5 microns in thickness remained.

A negative film transparency is laid upon the layer and exposed for 200 seconds to an 8,000 watt pulsed xenon lamp at a distance of 36 inches. The unexposed portions of the layer are removed by washing with a xylol-methyl Cellosolve acetate mixture to yield a relief image.

Acrylic Esters of Tetraalcohols

K.-W. Klüpfel, H. Steppan and M. Hasenjäger; U.S. Patent 3,882,168; May 6, 1975; assigned to Hoechst AG, Germany provide photopolymerizable compounds of the general formula:

$$
\begin{array}{ccccccc}
 & & & Z & & & \\
R_1OCH_2 & \diagdown & & \diagup\ \ \diagdown & & \diagup & CH_2OR_3 \\
 & & C & & C & & \\
R_2OCH_2 & \diagup & | & & | & \diagdown & CH_2OR_4 \\
 & & CH_2 & & CH_2 & & \\
 & & & \diagdown\ X\ \diagup & & &
\end{array}
$$

where X is a single bond, an oxygen atom, or a CH_2 group, Z is one of the groups –CO– and –CHOH–, and R_1, R_2, R_3, and R_4 are the same or different and are hydrogen atoms, or acryloyl or methacryloyl groups; Z is a carbonyl group, if X is a single bond, and, on an average, at least one of R_1, R_2, R_3, and R_4 is an acryloyl group.

The process further provides a photopolymerizable copying composition which, as the primary constituents, contains at least one binder, at least one polymerizable compound and at least one photoinitiator. As the polymerizable compound, it contains at least one compound of the general formula as defined above.

Also in the case of a high degree of esterification, the compounds of the process show practically no crystallization tendency. The acrylic esters are obtained by acid azeotropic esterification of the corresponding tetra- or pentaalcohols.

Example: A coating solution containing:

1.4 parts by weight of the acrylic acid-esterified 2,2,6,6-tetrahydroxymethylcyclohexanol (saponification number 480.6),
0.5 part by weight of an acrylic resin having an average molecular weight of 36,000 (Elvacite 2008),
0.9 part by weight of a copolymer of methyl methacrylate and methacrylic acid, having an average molecular weight of 40,000 and an acid number of 90 to 105.
0.05 part by weight of 2,3-di(4-methoxyphenyl)-6-methoxyquinoxaline,
0.2 part by weight of 1,6-dihydroxyethoxyhexane,
0.02 part by weight of Supranol Blue GI and
15.0 parts by weight of ethylene glycol monoethyl ether.

This solution is whirl-coated onto electrolytically roughened 0.1 mm thick aluminum (Rotablatt). The layer is then dried for 2 minutes at 100°C in a drying cabinet and then has a thickness corresponding to a weight of about 5 g/m^2.

In a second process step, a top coating is prepared from a solution of:

15.0 parts by weight of polyvinyl alcohol (Elvanol 52-22),
0.25 part by weight of sodium alkylbenzenesulfonate as wetting agent in,
485.0 parts by weight of water.

This solution is applied in a thickness of 0.1 to 1 g/m^2 to the photopolymer layer.

The photopolymer plate obtained is exposed for one minute under a negative to a xenon point light lamp (Xenokop) at a distance of 80 cm between the lamp and the printing frame. After exposure, the nonimage areas are removed with an aqueous alkaline developer. The plate is then briefly wiped over with dilute 1% aqueous phosphoric acid.

A positive printing form is obtained which readily accepts ink, is free from scumming, and yields long printing runs. Similar results are obtained by using the same sensitizing solution in which, however, the acrylic ester of the tetrahydroxymethylcyclohexanol is replaced by the same quantity of a monomer obtained by esterification of 2,2,5,5-tetrahydroxymethylcyclopentanone with a mixture of equimolar portions of acrylic acid and methacrylic acid (saponification number 517).

Alpha-Substituted Maleimido Groups

K. Ichimura and H. Ochi; U.S. Patent 3,920,618; November 18, 1975; assigned to Director-General of the Agency of Industrial Science and Technology, Japan describe photopolymers having as photocrosslinking radicals α-substituted maleimido groups of the formula:

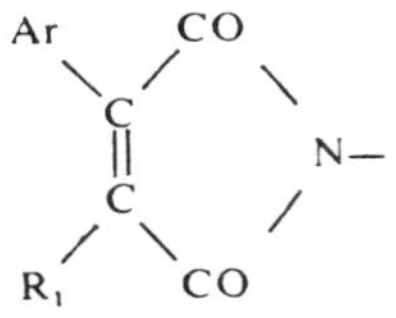

```
Ar      CO
  \    /   \
    C       \
    ||       N—
    C       /
  /    \   /
R1      CO
```

wherein Ar stands for a cyclic group of aromaticity and R_1 for a hydrogen atom or cyano group. These photopolymers can be produced by introducing the α-substituted maleimido groups into a backbone molecular chain of a linear polymer or copolymer according to a known reaction, such as amidation, imidation, esterification, quaternization or nuclear substitution.

The photosensitive resins of this process show good sensitivity at a rate of substitution as low as 10% or less and attain insolubilization even at a rate of substitution of 1% or less.

As the photosensitive resins of this process permit introduction of the specific photosensitive radicals at a high rate of substitution up to 70 to 80%, this process makes it possible to produce photosensitive resins of any desired rate of substitution. Thus, photosensitive elements of a desired sensitivity can be obtained. The photosensitive radicals in the resins of this process show a maximum absorption at 340 to 360 μ but show no substantial change to light having a wave length of 400 μ or more.

Accordingly, no special darkroom is required for preparation and handling of the photosensitive resins of this process. Moreover, the reaction efficiency of the photosensitive resins is scarcely affected by oxygen, thus bringing about the feature that the photosensitive resins or elements can be handled in the atmosphere.

The photosensitive resins of this process are insolubilized in the absence of any sensitizer by irradiation of light for a short period of time, say a few seconds, and so are suited as photoresists, PS plates for printing use, copying materials, memory elements, etc.

Example: Into an ampule were charged 2.0 grams of p-aminostyrene, 4.0 grams of styrene and 0.12 gram of azobisisobutyronitrile. The ampule was evacuated, sealed and heated at 60°C for 23 hours to effect reaction. The reaction product was dissolved in benzene and poured into cyclohexane to precipitate a slightly brown polymer in an almost quantitative yield.

In 5 ml of dimethylformamide was dissolved 0.5 gram of this polymer and the solution was filtered to remove any insoluble matter. To the filtrate was then added 0.25 gram of phenylmaleic anhydride and the resulting mixture was stirred for 5 minutes at room temperature to obtain a slightly yellow liquid. By IR-absorption spectrometry, this reaction product was observed to have amidocarboxylic acid residues. A solution of the reaction product was coated on an aluminum plate having been subjected to a surface treatment and dried at 50°C to obtain a colorless to light yellow, transparent film showing good adhesiveness to the support. This film was exposed for 5 to 10 seconds to the light from a 450 w ultra-high-pressure mercury lamp through an original image, dipped into dimethylformamide for 30 seconds and washed with acetone to obtain a sharp reproduced image.

Polymers with Urethane Groups Crosslinkable by Vinyl Polymerization

H.J. Rosenkranz, H. Rudolph, E. Wolff and H. von Rintelen; U.S. Patent 3,928,299; December 23, 1975; assigned to Bayer AG, Germany provide polymers which contain urethane groups and crosslinkable vinyl or vinylidene groups suitable as photocrosslinking layers and molded products which are free from tackiness and do not show any friability in the presence of atmospheric oxygen.

The organic homopolymer or random copolymer has a number average molecular weight, determined by the osmometric method, of greater than 1,000 which contains per molecule y recurring units of the formula

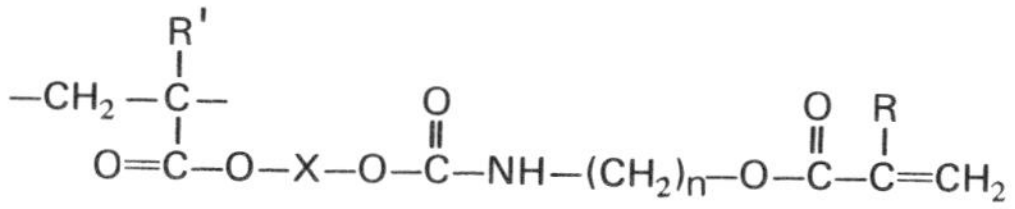

$$
\begin{array}{l}
\quad\ \ R' \\
\quad\ \ | \\
-CH_2-C- \\
\quad\ \ | \\
O{=}C-O-X-O-\overset{\displaystyle O}{\overset{\|}{C}}-NH-(CH_2)_n-O-\overset{\displaystyle O}{\overset{\|}{C}}-\overset{\displaystyle R}{\overset{|}{C}}{=}CH_2
\end{array}
$$

where R and R' are hydrogen or alkyl having 1 to 4 carbon atoms, n is an integer from 1 to 6, y is an integer from 2 to 500 and X is an alkylene group having 1 to 9 carbon atoms or an alkylene group having 2 to 9 carbon atoms interrupted by from one to two —O— groups.

Example: Preparation of the Polymer — 400 ml of tertiary butanol were heated under reflux under nitrogen in a 2-liter three-necked flask equipped with stirrer, reflux condenser and dropping funnel. A mixture of 288 grams of propyl methacrylate, 150 grams of methyl methacrylate, 150 grams of methyl acrylate, 100 grams of tertiary butanol and 1.5 grams of benzoyl peroxide was added dropwise in the course of 5 hours. When all the mixture had been added, another 0.5 gram of benzoyl peroxide in tertiary butanol was added and the mixture was kept under reflux for 4 hours. After removal of the tertiary butanol by evaporation, which was carried out partly under vacuum, the polymer was freed from residual solvent by drying in a vacuum drying cupboard at 15 mm Hg/50°C. 580 grams of a colorless solid resin were obtained.

480 grams of the polymer were dissolved in methylene chloride to prepare a solution having a concentration of 30% by weight, and 194 grams of isocyanatoethyl methacrylate were added to the solution at room temperature with stirring in the course of 30 minutes. The mixture was then kept at room temperature for another 48 hours.

Light-Sensitive Material —The solution prepared as described above was sensitized with 3% by weight, based on the dry film-forming polymer, of 2-chloromethylanthraquinone and colored by the addition of 0.5% by weight of Sudan Blue. This solution was used for coating an aluminum foil by means of a whirler and the coating was dried in the usual manner. The light-sensitive layer had a thickness of 28 to 30 μ after drying. It was then laminated with a polyethylene foil 30 μ in thickness to protect it against atmospheric oxygen which acts as a polymerization inhibitor.

Processing — The material described above was exposed through an original for 4 minutes in an exposure apparatus. This exposure corresponds approximately to 2 minutes exposure to a carbon arc lamp (42 v, 30 amp) at a distance of 45 cm. After removal of the protective foil, the layer may be developed in a mixture of ethyl acetate and trichloroethylene. A sharp positive relief of the original is obtained.

Beta-Vinyloxyethyl 5-Furyl-2,4-Pentadienate

Y. Maekawa, M. Satomura, A. Umehara; U.S. Patent 3,931,248; January 6, 1976; assigned to Fuji Photo Film Co., Ltd., Japan provide a high polymer compound containing therein the reactive group represented by the following general formula:

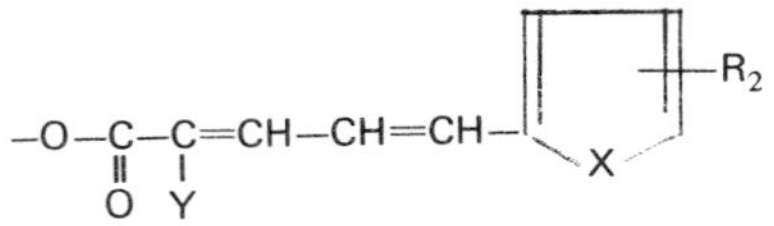

$$-O-\underset{\underset{O}{\|}}{C}-\underset{\underset{Y}{|}}{C}=CH-CH=CH-\left[\text{ring containing } X\right]-R_2$$

where X represents O or S; Y represents a hydrogen atom or a cyano group; and R_2 represents a hydrogen atom, an alkyl group having not more than 4 carbon atoms, a halogen atom or a nitro group.

The high polymer compound has greater light sensitivity as compared with polyvinyl cinnamate, and is useful particularly for a light-sensitive composition.

Example 1: Synthesis of β-Vinyloxyethyl 5-Furyl-2,4-Pentadienate — 30 grams of potassium 5-furyl-2,4-pentadienate, 200 ml of β-chloroethyl vinyl ether and 1.0 gram of trimethylbenzylammonium chloride were placed, together with a small amount of a polymerization inhibitor (0.5 gram of phenylnaphthylamine), in a 500 ml three-necked flask equipped with a condenser and the reaction was conducted at 110° to 120°C for 5 hours under vigorous stirring.

The potassium chloride produced was filtered off and washed with 20 ml of chloroethyl vinyl ether. This washing and the mother liquor were combined and the chloroethyl vinyl ether was distilled off under reduced pressure.

Then, the residue was rectified under reduced pressure to obtain the product as a pale yellow solid. Upon recrystallization from n-hexane, 31 grams of white crystals was obtained. Yield: 90%, MP 63° to 64°C.

Polymerization — In a glass polymerization vessel purged with dry argon gas were placed 1.0 gram of β-vinyloxyethyl 5-furyl-2,4-pentadienate and 4.0 ml of methylene chloride and, at 0°C, a 4.3×10^{-4} mol/ml methylene chloride solution of boron trifluoride etherate was added thereto as a polymerization catalyst in an amount of 4 mol % based on the monomer.

After maintaining the temperature at 0°C for 1 hour, the contents of the polymerization vessel were poured, under stirring, into 150 ml of methanol containing a small amount of ammonia. Thus, the high polymer compound was obtained as a white solid, which was soluble in ordinary organic solvents such as benzene, tetrahydrofuran, methyl ethyl ketone, etc.

The yield of the high polymer compound after drying at room temperature (about 20° to 30°C) in vacuo was 0.18 gram. The intrinsic viscosity, (η), measured in tetrahydrofuran at 30°C was 1.67 (dl/g).

The compound of this process finds application as a coating material, a photographic copy, a printing plate, a resist, a relief image and the like.

Example 2: A 5% (by weight) methyl ethyl ketone solution was prepared using the high polymer compound obtained in Example 1. To this was added a sensitizing agent 5-nitroacenaphthene in an amount of 5% (by weight) based on the high polymer compound and stirred to completely dissolve.

The resulting solution was applied to a surface-processed aluminum plate under a safe light using a No. 26 coating rod and was allowed to dry at room temperature followed by maintenance at 70°C for 5 minutes to dry it completely. The thus obtained light-sensitive plate was irradiated through a line image original superposed thereon using a 450 w high-pressure mercury lamp spaced at a distance of 28 cm as a light source. Upon development with methyl ethyl ketone, the exposed area was found to be rendered insoluble as a result of crosslinking.

Alternatively, when the light-sensitive plate was irradiated for 1 minute through a step wedge (optical wedge) of a step difference of 0.11 to 0.16 superposed thereon using a 450 w high-pressure mercury lamp spaced at a distance of 28 centimeters and processed with methyl ethyl ketone, photoinsolubilization was found to have occurred up to the 12th step.

In order to obtain the same effect using a commercially available polyvinyl alcohol cinnamic acid ester containing a sensitizer [polyvinyl alcohol (molecular weight about 2,000) esterified with cinnamoyl chloride to about 90 mol % esterification in the presence of pyridine; 2-benzoylmethylene-N-methyl-β-naphthothiazoline, 8 to 10% by weight of the polymer, as a sensitizer, with the polymer and sensitizer being dissolved in an ethoxyethanol and toluene mixture, at 10 weight %], exposure for about 10 minutes was necessary.

Styrene-Allyl Alcohol Copolymer Based Polythiol and Polyene Compositions

C.R. Morgan; U.S. Patent 3,904,499; September 9, 1975; assigned to W.R. Grace & Co. provides solid polythiols which are mercaptoester derivatives of styrene-allyl alcohol copolymers. These solid polythiols are readily prepared by esterifying a styrene-allyl alcohol copolymer with a mercaptocarboxylic acid, e.g., β-mercaptopropionic acid. The solid styrene-allyl alcohol based polythiols may be admixed with liquid polyenes thereby forming solid polyene-polythiol polymeric systems which are curable, particularly photocurable, in the solid state.

Upon exposing the solid, curable polyene-polythiol compositions to a free radical generator, e.g., UV light, solid crosslinked and chemically resistant polythioether products are formed which are particularly useful as coatings, photoresists, printing plates, etc.

Operable styrene-allyl alcohol copolymers are those containing from about 30 to 94% by weight of the styrene monomer, and preferably 60 to 85% by weight and correspondingly, from about 70 to 6% by weight of the ethylenically unsaturated alcohol, and preferably from about 40 to 15% on the same basis. In general, styrene-allyl alcohol copolymers have from about 1.8 to 10% hydroxyl groups by weight, preferably 4 to 8%.

The styrene monomer moiety of the copolymer may be styrene or a ring-substituted styrene in which the substituents are 1 to 4 carbon atom alkyl groups or chlorine atoms. The ethylenically unsaturated alcohol moiety may be ally alcohol, methallyl alcohol, or a mixture thereof. The polythiol component of the solid curable composition is solid mercaptoester having at least two thiol groups per molecule. The liquid polyene contains at least 2 reactive unsaturated carbon to carbon bonds and has a molecular weight in the range of about 50 to 20,000.

Example 1: Formation of Solid Polythiols — 220 grams of a copolymer of styrene-allyl alcohol having an equivalent weight of about 220 and a hydroxyl content of about 7.7% and available from Monsanto Company as RJ 101, and 106 grams of β-mercaptopropionic acid along with 400 ml of benzene as a solvent and 2.0 grams of p-toluenesulfonic acid as a catalyst were charged to a resin kettle equipped with a stirrer, condenser, Dean-Stark trap, thermometer and gas inlet and outlet. The mixture was heated to reflux and the benzene-water azeotrope was collected. The amount of water obtained was about 18 milliliters.

The reaction mixture was then vacuum-stripped to remove the benzene. The mixture was then dried in a vacuum oven at 40°C resulting in a white rubbery solid polythiol having a styrene-allyl alcohol copolymer based polymeric backbone which had a mercaptan content of 2.65 meq/g. This polythiol will be referred to as Polythiol A.

Example 2: Formation of Polyene Prepolymers — 2.0 mols of trimethylolpropane diallyl ether and 0.2 gram of dibutyltin dilaurate as a catalyst was charged to a resin kettle maintained under nitrogen and equipped with a stirrer, thermometer, dropping funnel and a glass inlet and outlet. 1.0 mol of tolylene diisocyanate was added slowly with stirring and the reaction temperature was maintained at 70°C by means of a water bath for the flask. After the addition of the tolylene diisocyanate, the reaction was continued for about 1 hour at 70°C until the NCO content was substantially zero. The thus formed allyl terminated liquid prepolymer will be referred to as Polyene A.

Example 3: Curing Process — To a solution containing 37.0 grams of solid Polythiol A from Example 1 and 58.0 grams of 1,2-dichloroethane were added 7.5 grams of liquid Polyene A from Example 2, 0.44 gram of dibenzosuberone and 0.016 gram of phosphorous acid. The thus formed solution was applied uniformly onto a 5 mil thick polyethylene terephthalate, i.e., Mylar, film in a layer of approximately 1.0 mil thickness by means of a drawbar.

The dichloroethane was allowed to evaporate leaving a solid photocurable coating of the admixture on the support film. Thereafter the solid photocurable coating on the Mylar film was brought in contact with the surface of the copper cladding of a clean copper clad epoxy-glass printed circuit board blank. Heat (60°C) and pressure are applied to make the laminate. A negative image-bearing transparency of a printed circuit was placed in contact with and over the Mylar film and the solid photocurable coating was exposed through the transparency and UV-transparent polyethylene terephthalate film to UV radiation from an 8,000 watt Ascorlux pulsed xenon arc lamp at a surface intensity of 3,600 microwatts/cm^2 for about 5 minutes. The major spectral lines of this lamp are all above 3,000 A. The negative transparency was removed and the Mylar film was stripped off. The coating was washed in 1,1,1-trichloroethane to remove the unexposed, uncured portion thereof, thus exposing the copper thereunder.

The image coated circuit board was then etched by spraying with a ferric chloride solution (42° Baumé) for about 30 minutes at 40°C to remove the exposed copper, followed by a water wash. The cured photoresist coating which was not affected by the etching solution was left on the etched printed circuit board as a protective cover for the desired electrical circuit thereunder.

In related work *C.R. Morgan; U.S. Patent 3,936,530; February 3, 1976; assigned to W.R. Grace & Co.* provides a method for preparing imaged surfaces which includes coating the surface of a substrate with a solid curable composition containing styrene-allyl alcohol copolymer based polyene and polythiol components, curing the composition by exposing selected areas thereof to a free radical generating source, e.g., actinic radiation and removing, e.g., by dissolving, the uncured, unexposed areas of the curable composition to bare the underlying substrate. The solid polyene component is a reaction product of a copolymer of styrene-allyl alcohol and an unsaturated isocyanate, e.g., allyl isocyanate or an unsaturated acid, e.g., acrylic acid. The solid polythiol is a reaction product of a copolymer of styrene-allyl alcohol and a mercapto carboxylic acid.

Example 1: Preparation of Solid Polyenes — Into a one-liter flask maintained under a nitrogen atmosphere and equipped with a stirrer, thermometer, condenser and a gas inlet and outlet, was added 348 grams (2.0 mols) of toluene diisocyanate (a 80/20 mixture of the 2,4 and 2,6 isomers). 116 grams (2.0 mols) of allyl alcohol was slowly added over a period of 2 hours to the reaction vessel with stirring during which time the exotherm and reaction temperature were maintained below about 70°C.

After the addition of the allyl alcohol was completed the reaction was continued for about 15 hours at about 25°C. The thus formed liquid monoallyl urethane had an NCO content of 4.13 meq/g and an unsaturation of 4.5 mmol/g. This product will be referred to as ene-isocyanate A.

110 grams of a copolymer of styrene-allyl alcohol (RJ 101), was dissolved in 300 ml of benzene in a 1-liter reaction flask maintained under a nitrogen atmosphere and equipped with a stirrer, condenser, thermometer and a gas inlet and outlet. 0.6 gram of dibutyltin dilaurate as a catalyst was added to the reaction flask followed by dropwise addition over a period of ½ hour of 116 grams of the ene-isocyanate A prepared above.

The reaction was allowed to continue for about 15 hours while maintaining the temperature at about 70°C. Thereafter, the reaction mixture was cooled to room temperature and the solvent was removed under vacuum. The resulting solid polyene having a styrene-allyl alcohol based polymeric backbone (190 g) had an unsaturation of 2.2 mmol/g and a melting point of 85° to 105°C. This polyene will be referred to as Polyene A.

Example 2: Curing Processes — An admixture of 3.7 grams of solid Polyene A from Example 1, 4.5 grams of solid Polythiol A from Example 1 of U.S. Patent 3,904,499, above, and 0.2 gram benzophenone were dissolved in acetone. The solution was applied uniformly to the copper surface of a circuit board comprising a 0.001 inch thick copper cladding on a 0.050 inch epoxy-glass substrate.

The acetone was allowed to evaporate, leaving about a 1 mil solid photocurable coating of the admixture on the copper. A negative image-bearing transparency of a printed circuit was placed in contact with and over the coating, and the photocurable coating was exposed through the transparency to UV radiation from an 8,000 watt Ascorlux pulsed xenon arc lamp at a surface intensity of 3,800 microwatts/cm^2 for about 100 seconds. The major spectral lines of this lamp are all above 3000 A.

The negative transparency was removed and the coating was washed in acetone to remove the unexposed, uncured portion thereof, thus exposing the copper thereunder. The imaged circuit board was then etched in an aqueous 30% by weight ammonium persulfate solution for about 5 minutes at 120°F to remove the exposed copper, followed by a water wash. The cured photoresist coating was then removed in methylene chloride solution, thus revealing the desired copper electrical circuit.

In related work *C.R. Morgan; U.S. Patent 3,925,320; December 9, 1975; assigned to W.R. Grace & Co.* provides styrene-allyl alcohol copolymer based solid polyene compositions which when mixed with liquid polythiols can form solid curable polyene-polythiol systems. These solid polyenes, containing at least two reactive carbon-to-carbon unsaturated bonds, are urethane or ester derivatives of styrene-allyl alcohol copolymers.

The solid polyenes are prepared by treating the hydroxyl groups of a styrene-allyl alcohol copolymer with a reactive unsaturated isocyanate, e.g., allyl isocyanate or a reactive unsaturated carboxylic acid, e.g., acrylic acid. Upon exposure to a free radical generator, e.g., actinic radiation, the solid polyene-polythiol compositions cure to solid, insoluble, chemically resistant, crosslinked polythioether products. Since the solid polyene-liquid polythiol composition can be cured in a solid state, such a curable system finds particular use in preparation of imaged surfaces.

Example: Curing Process — An admixture of 2.0 grams of solid Polyene A from Example 1, of U.S. Patent 3,936,530, above, 0.6 gram of pentaerythritol tetrakis(β-mercaptopropionate), a liquid polythiol available from Carlisle Chemical Co. as Q-43 and 0.15 gram of benzophenone was dissolved in 20 grams of methyl ethyl ketone. The solution was spin coated to the copper surface of a clean copper clad epoxy-glass printed circuit board blank. The methyl ethyl ketone was allowed to evaporate leaving a 1.0 mil solid photocurable coating of the admixture on the copper.

A negative image-bearing transparency of a printed circuit was placed in contact with the coating, and the photocurable coating was exposed through the transparency to UV radiation from 8,000 watt Ascorlux pulsed xenon arc lamp at a surface intensity of 4,000 microwatts/cm^2 for about 2 minutes. The major spectral lines of this lamp are all above 3000 A. The negative transparency was removed and the coating was washed in acetone to remove the unexposed, uncured portion thereof, thus exposing the copper thereunder. The imaged circuit board was then etched by spraying it with a ferric chloride solution (42° Baumé) for about 5 minutes at 50°C to remove the exposed copper, followed by a water wash. The cured photoresist coating which was not affected by the etching solution was left on the etched printed circuit board as a protective cover for the desired electrical circuit thereunder.

Vinyl Crosslinking Without Protective Foil

H.-J. Rosenkranz, H. Rudolph, E. Wolff and H. von Rintelen; U.S. Patent Application (Published) B 530,873; February 17, 1976; assigned to Agfa-Gevaert AG, Germany describe a process for the production of photographic relief images or photoresists comprising the steps of imagewise exposing a supported light-sensitive material with a photocrosslinkable polymer in the layer, washing away the unexposed areas of the photosensitive layer to form a relief image of photocrosslinked polymeric material. The improvement is that the photocrosslinkable polymer compound is a photopolymer or recurrent units of the general formula

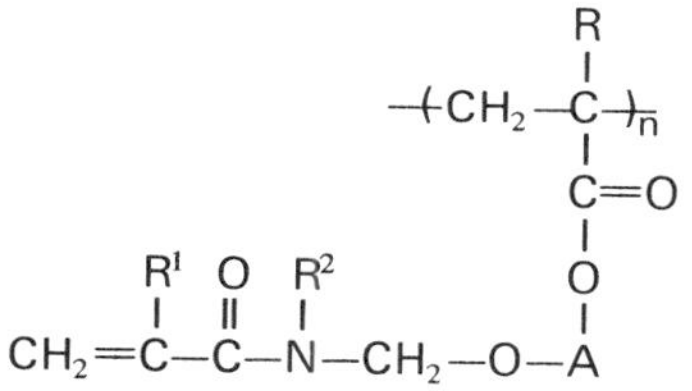

$$-\!\!\left(CH_2-\underset{\substack{|\\C=O\\|\\O\\|\\CH_2=\overset{R^1}{\overset{|}{C}}-\overset{O}{\overset{\|}{C}}-\overset{R^2}{\overset{|}{N}}-CH_2-O-A}}{\overset{R}{\overset{|}{C}}}\right)_{\!n}$$

where R stands for hydrogen or alkyl having up to 4 carbon atoms and R^1 represents hydrogen or alkyl having up to 4 carbon atoms, R^2 stands for hydrogen or alkyl having up to 8 carbon atoms, and A represents a divalent aliphatic bridging member having 2 to 4 carbon atoms or a bridging member of the formula $-(CH_2-CH_2-O)_m-CH_2-CH_2-$ where m represents an integer of 1 to 3 and n is 2 or more.

In contrast to the light-sensitive polymer hitherto used which can be crosslinked by vinyl polymerization, the layers according to the process make it possible to dispense with the use of a protective foil and with working in an oxygen-free atmosphere.

Example: Production of the Polymer — 868 grams ethyl acetate were heated in a nitrogen atmosphere under reflux in a three-neck flask of two liters capacity fitted with stirrer, reflux condenser and dropping funnel. A mixture of 288 grams methacrylic acid hydroxypropyl ester, 150 grams methyl methacrylate, 150 grams ethyl acrylate and 9 grams azoisobutyric acid dinitrile were added dropwise within three hours.

When the dropwise addition was completed, there was added one gram azoisobutyric acid dinitrile in 10 grams ethyl acetate, and heating under reflux was continued for a further 5 hours. The ethyl acetate was partially distilled off, and the highly viscous polymer solution was concentrated by evaporation in a vacuum drying cabinet at 15 mm/50°C for 24 hours to give a solid colorless resin (580 g).

234 grams of this resin, 800 mg p-toluenesulfonic acid and 60 mg hydroquinone were dissolved in 300 grams glycol acetate monomethyl ether. 80 grams N-methylolacrylamide dissolved in 200 grams methanol were added to this solution at room temperature, and this mixture was stirred at room temperature in a water jet vacuum for 3 hours, methanol and the liberated reaction water being thus evaporated off.

Light-Sensitive Material — A solution of the polymer prepared in this way is diluted with glycol acetate monomethyl ether to 30% by weight and sensitized with 2% by weight, referred to the dry film-forming polymer, of 2-tert-butyl-anthraquinone. A clean copper sheet was coated with this solution by means of a whirling coater at 200 rpm and dried at room temperature for 8 hours. The dried layer had a thickness of 25 μ.

Processing — The above layer is exposed for 4 minutes in a Chem-Cut vacuum frame through a 0.15 grey step wedge. Development with ethyl acetate-isopropanol gives a sharp relief image of 9 steps of the test wedge.

On account of its good thermoplastic properties, the product prepared and sensitized as described above, after coating and drying, is especially suitable for transfer from a carrier foil (e.g., Hostaphan RHH 30) to another support by means of hot rollers at 120°C. After imagewise exposure through a line pattern and subsequent development, a sharp relief image of the pattern is obtained. Moreover, the relief image can also be transferred, after exposure and development, in exactly the same manner.

Beta-Thienylacrylic Acid Ester or Amide

Photosensitive resins comprising high molecular weight compounds containing a β-thienylacrylic acid ester or amide group are provided by *M. Satomura; U.S. Patent 3,945,831; March 23, 1976; assigned to Fuji Photo Film Co., Ltd., Japan.* These resins can be used for formation of images, optionally, in combination with at least one sensitizer. These resins have extremely high photosensitivity in comparison with conventional photosensitive high molecular weight compounds. The methods of preparing the polymer and the thienylacrylic acid ester or amide functional group containing monomers are disclosed.

Example 1: Synthesis of High Molecular Weight Compound — 2.0 grams of β-vinyloxyethylthienylacrylate and 8.0 ml of methylene chloride were charged into a polymerization ampoule purged with argon gas and, at −78°C, 4 mol percent, based on the monomer, of a methylene chloride solution containing 4.3 x 10^{-4} mol per milliliter of boron trifluoride etherate was added thereto. After maintaining the mixture at −78°C for about one hour, the contents of the ampoule were poured, under stirring, into 100 ml of methanol containing a small amount of aqueous ammonia. Thus, a white high molecular weight compound precipitate was formed in a yield of about 90%, $[\eta]$ 0.21 (at 30°C in tetrahydrofuran).

Example 2: 95% by weight methyl ethyl ketone solution of high molecular weight compound synthesized in Example 1 containing 0.25% of Michler's ketone was prepared. This solution was applied to an anodized aluminum plate in a dry thickness of 2 to 3 microns and dried. The plate was exposed to light through a transparent original for 1 minute using a 450 w high-pressure mercury lamp at a distance of 30 cm from the plate. Then the plate was treated with dimethylformamide for development to dissolve off the unexposed areas and then the plate was immersed in water. After treating with an oil-soluble dye, the surface exposed was colored and these areas were sharply distinguished from the plate surface.

Pullulan Compositions

T. Sano, Y. Uemura and A. Furuta; U.S. Patent 3,960,685; June 1, 1976; assigned to Sumitomo Chemical Company, Ltd., and Hayashibara Biochemical Laboratories, Inc., Japan describe a photosensitive resin composition comprising 30 to 90 parts by weight of at least one member selected from the group consisting of pullulan, an acetylation product of pullulan, an ester of pullulan with an unsaturated organic acid and a cinnamoylated ester of pullulan, 70 to 10 parts by weight of a photopolymerizable monomer, 0.5 to 5 parts by weight of a photosensitizer and 0.1 to 1 part by weight of a thermal polymerization inhibitor per 100 parts by weight of the polymerizable monomer.

Pullulan having a molecular weight of 10,000 to 4,000,000, preferably 20,000 to 200,000, is used. The pullulan may be used as it is, or, for improvement in compatibility or crosslinkability, may be modified by acetylation or the like esterification reaction, utilizing the hydroxyl groups in the pullulan.

The photopolymerizable monomer, which is used in the case where pullulan is used as it is, is a monomer high in compatibility with pullulan such as, for example, acrylamide, acrylic acid, methacrylic acid, 2-hydroxyethyl methacrylate or N-vinylpyrrolidone. Alternatively, a crosslinkable monomer such as nonaethylene glycol dimethacrylate or the like may also be used. In this case, the curing of resin at the exposed portion is further promoted to give a clearer image. The mixing proportions of pullulan and the abovementioned monomer are 30 to 90 and 70 to 10 parts by weight, respectively.

In case pullulan has been enhanced in compatibility by acetylation, a photosensitive resin composition may be prepared by mixing the pullulan with a monomer incompatible therewith, e.g., a monomer high in crosslinking efficiency such as ethylene glycol dimethacrylate or triethylene glycol dimethacrylate, or a water-insoluble monomer such as vinyl acetate, methyl methacrylate, glycidyl methacrylate, ethyl acrylate, cyclohexyl acrylate, and the like.

In this case, the proportion of the monomer is 10 to 200 parts by weight per 100 parts by weight of the acetylated pullulan. In case the acetylation is effected to a substitution degree of less than 1.2, the pullulan is not only improved in compatibility with the monomer but also can give a resin capable of forming a water-developable relief.

Example 1: Ten grams of pullulan having a molecular weight of 150,000 was dissolved in 40 grams of water. To the resulting aqueous solution was added 21.4 grams of an aqueous solution formed by dissolving 5 grams of acrylamide in 20 grams of water. This solution was incorporated with 0.0432 gram of sodium anthraquinone-α-sulfonate and 0.00432 gram of hydroquinone to prepare a photosensitive resin composition.

The composition was cast on an iron plate coated with a paint and then air-dried overnight to obtain a photosensitive resin plate having a thickness of 0.6 mm. The resin plate was completely transparent. This resin plate was brought into close contact with a negative film, exposed for 10 minutes to a 3 kw high pressure mercury lamp at a distance of 75 cm and then washed for one minute in running water to obtain a relief image having a thickness of 0.6 mm.

Example 2: Example 1 was repeated, except that the acrylamide was replaced by acrylic acid, whereby a relief image corresponding to the negative was obtained.

Example 3: A mixture comprising 50 grams of pullulan anhydride having a molecular weight of 320,000 and 100 grams of pyridine was dissolved in 500 grams of dimethylformamide. Into the resulting solution, 30 grams of acetic anhydride was dropped over a period of 1 hour with stirring at 65°C. The resulting mixture was reacted at that temperature for an additional hour and then cooled. Subsequently, ethanol was added to the reaction mixture to deposit a precipitate. The precipitate was recovered by filtration, dissolved in water, deposited again by addition of ethanol, and then washed and dried to obtain 65 grams of a polymer having a hydroxyl group substitution degree of 0.6.

A solution of 70 grams of the above-mentioned polymer in 350 grams of dimethylformamide was incorporated with 30 grams of triethylene glycol dimethacrylate, 0.3 gram of sodium anthraquinone-α-sulfonate and 0.03 gram of hydroquinone to prepare a photosensitive resin composition. This composition was treated in the same manner as in Example 1 to obtain a clear relief image.

Epoxy-Acrylate Compositions

K. Tsukada, A. Isobe, N. Hayashi, M. Abo and K. Ogawa; U.S. Patent 3,989,610; November 2, 1976; assigned to Hitachi Chemical Company, Ltd., Japan describe a photosensitive resin composition which consists essentially of

(a) 10 to 90% by weight of a photopolymerizable unsaturated compound containing at least two terminal ethylene groups,

(b) 0.1 to 15% by weight of a sensitizer capable of initiating polymerization of the above unsaturated compound upon irradiation with active rays,

(c) 5 to 80% by weight of a compound having at least two epoxy groups and

(d) A compound selected from the group consisting of 0.1 to 20% by weight dicyandiamide, 1 to 30% by weight p,p'-diaminodiphenyl compounds, and 0.3 to 1.5 mols per mol of epoxy group of a polycarboxylic acid having at least two carboxyl groups, a polycarboxylic anhydride or mixtures thereof.

Example: A photosensitive resin of the following composition was prepared.

	Parts by Weight
Methyl methacrylate-methacrylic acid copolymer (98/2 weight ratio)	40
Pentaerythritol triacrylate	30
Epoxy resin (ECN-1280)	25
Dicyandiamide	1.5
Benzophenone	2.7
Michler's ketone	0.3
p-Methoxyphenol	0.6
Methyl ethyl ketone	200

The composition was applied to a copper-clad laminate, and dried at room temperature for 10 minutes and at 80°C for 10 minutes to form a sensitive layer having a thickness of 20 μ. The resulting sensitive layer is covered with transparent polyethylene terephthalate film having a thickness of 25 μ, and then exposed to rays through a negative mask from a 3 kw super-high-pressure mercury lamp at an intensity of 4,000 $\mu w/cm^2$ for 60 seconds, immediately after which the layer was heated at 80°C for 5 minutes and then cooled.

The polyethylene terephthalate film was thereafter peeled off from the layer, and the layer was subjected to spray-development with 1,1,1-trichloroethane for one minute. The thus developed layer was subjected to heat-treatment at 150°C for 2 hours to obtain a protective film in the precision form of image corresponding to the negative mask.

This protective film did not change at all even when dipped for 10 hours in each of methyl ethyl ketone, acetone, chloroform, Triclene, methanol, isopropanol, 50% aqueous sulfuric acid solution, toluene, benzene and xylene, and no cracking or whitening of the protective film was found even after the film was further dipped in an aqueous sodium hydroxide solution of a pH of 12 at 70°C for 100 hours, and no separation of the film from the copper foil was seen.

After the above tests on solvent resistance and chemical resistance, the film was dipped in a solder bath at a temperature of 260° to 270°C for two minutes to find no change in the film. From these results, it was seen that the protective film obtained in this example could sufficiently be used as a resist for etching, plating or strongly alkaline nonelectrolytic chemical plating and as a solder resist.

PRINTING PLATE COMPOSITIONS

Polyamide with Pendant Sulfonate Salt Groups

A. Koshimo, K. Tomita, K. Takagi, M. Mitsui and K. Isikawa; U.S. Patent 3,884,702; May 20, 1975; assigned to Unitika, Ltd., Japan developed a photosensitive polyamide composition capable of producing a printing plate which has good storage stability for prolonged periods of time and good resolving power, which can be easily developed with water and which can be produced at low cost.

According to this process, there is provided a photosensitive polyamide composition comprising at least one polyamide having pendant sulfonate salt groups, at least one compound having at least two polymerizable ethylenic double bonds and at least one photoinitiator. The polyamide can be prepared by reacting a polyfunctional compound having a sulfonate salt group with a polyamide-forming reactant which generally is free of sulfonate salt groups.

Example: An autoclave was charged with 17 grams of ε-caprolactam, 2 grams of hexamethylenediamine, 10 grams of 2-hydroxy-4-phenoxy-6-[p-(sodium sulfo)-aminophenyl]-s-triazine and 1 gram of ε-aminocaproic acid, and after purging the reactor with nitrogen, the contents were reacted for 5 hours at 250°C and 30 atmospheres. The polyamide obtained was dissolved in a 50% aqueous solution of methanol, and the terminal amino groups were titrated with an aqueous solution of p-toluenesulfonic acid using thymol blue as an indicator. On the other hand, the terminal carboxyl groups were titrated with an aqueous solution of potassium hydroxide using phenolphthalein as an indicator. The molecular weight of the polyamide, calculated from the terminal group contents, was 4,200. (The molecular weight will be calculated in the same way hereinafter.)

Twenty grams of this polyamide was dissolved in 200 ml of 50% aqueous solution of methanol, and further, 3 grams of N,N'-methylene bisacrylamide, 0.6 gram of benzophenone and 0.06 gram of methylene blue were added thereto to yield an homogeneous solution. The solvent (50% aqueous solution of methanol) was then removed to give a homogeneous solid composition. This composition was cut into granular form and dried for one day at room temperature. The dried product was then pressed at 100°C and 100 kg/cm^2 to form a sheet-like material having a thickness of 1 mm.

The sheet-like material was bonded with an epoxy adhesive to a chromium plated plate, and exposed to photo-irradiation by a 400 w mercury lamp (60 cm from the plate) through a negative transparency for 5 minutes. After exposure, the laminated material was washed with water for 20 minutes, and there was obtained a relief having a very sharp image.

The above procedure was repeated except that instead of the 2-hydroxy-4-phenoxy-6-[p-(sodium sulfo)-aminophenyl]-s-triazine, the corresponding potassium or tetramethyl ammonium salt was used in preparing the polyamide. There was obtained a relief having a sharp image.

Bis-Acrylic Derivatives Compatible with Polyamides

M. Hasegawa, F. Nakanishi, H. Nakanishi, S. Watanabe, Y. Suzuki, H. Nakane, T. Aoyama, Y. Komatsubara, S. Nojima and H. Tagami; U.S. Patent 3,890,150; June 17, 1975; assigned to Agency of Industrial Science & Technology and Tokyo Ohka Kogyo Co., Ltd., Japan provide a class of ethylenically unsaturated compounds possessing excellent compatibility with linear polyamides and a planographic relief printing plate manufactured from this photosensitive composition.

There is provided a group of aromatic bis-acrylic derivatives of the general formula:

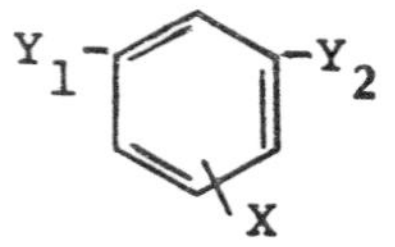

$$Y_1-C_6H_3(X)-Y_2$$

where Y_1 and Y_2 each represent a

$$CH_2{=}\underset{R}{\underset{|}{C}}-\underset{O}{\underset{\|}{C}}-O- \quad \text{or} \quad CH_2{=}\underset{R}{\underset{|}{C}}-\underset{O}{\underset{\|}{C}}-NH-$$

group, R is a hydrogen atom or a lower alkyl group, and X is a hydrogen atom or a lower alkyl group.

In the formula Y_1 and Y_2 may be the same or different. In a preferred embodiment, each of R and X is a hydrogen atom or methyl group. Ideally, both R and X are hydrogen atoms.

In the photosensitive composition of this process, the mixing ratio of the linear polyamide to the aromatic bis-acrylic derivative is usually 100:1 to 1:2, preferably 10:1 to 1:1, by weight.

Example 1: In a 500 ml 4-necked flask equipped with a stirrer, dropping funnel and thermometer are placed 45.9 grams (0.507 mol) of acryloyl chloride and 100 ml of dry benzene. The inner temperature is maintained at 5±3°C and then a solution of 22.0 grams (0.2 mol) of resorcinol in 100 ml of distilled water and a solution of 75.0 grams (0.543 mol) of anhydrous potassium carbonate in 150 ml of distilled water are simultaneously added dropwise over 30 minutes. After completion of the addition, the mixture is stirred for another 30 minutes at the same temperature.

The reaction mixture is then allowed to stand to form two separate layers and the benzene layer is taken up in a separating funnel and washed with a dilute aqueous solution of caustic soda to remove unreacted resorcinol and then with water until the pH value of the washing liquid becomes 6 to 7.

The benzene layer is then dried over anhydrous magnesium sulfate and concentrated under reduced pressure to yield 40.0 grams (yield: 91.7%) of crude 1,3-bis(acryloyloxy)benzene. The pure end product is obtained by distilling this crude product in the presence of pyrogallol (polymerization inhibitor) under reduced pressure and collecting a fraction distilled at 125° to 126°C at 1.8×10^{-1} mm Hg. Elementary analysis as $C_{12}H_{10}O_4$—Calculated: C 66.10%, H 4.60%. Found: C 65.52%, H 4.55%.

Example 2: One hundred parts of an alcohol-soluble copolymeric polyamide (Ultramid-1c), 100 parts of 1,3-bis(acryloyloxy)benzene and 2 parts of benzophenone are added to 700 ml of methanol. The mixture is stirred at about 50°C until the solid substances are entirely dissolved and then the solution is concentrated under reduced pressure.

A significantly viscous liquid thus obtained is spread over a glass plate which has been cleaned and the plate is dried under heating at about 50°C. The resulting sheet is bonded to an aluminum sheet by the aid of an adhesive of the phenol resin series and then exposed at a distance of 6 cm from a chemical lamp (Model FL 20 BL, Tokyo Shibaura Denki KK) to light for 15 minutes through a negative image plate. Methanol maintained at 35°C is then sprayed onto the exposed sheet for 5 minutes to effect development whereby a positive image in the form of figure and half tones is formed in conformity with lightened areas of the negative image plate. This apparently shows that the exposed areas become insoluble in methanol.

When an unexposed sheet of the above composition was wrapped with aluminum foil and stored for 30 days in a cool place, no change was observed in connection with the appearance of surface and transparency. When exposure and development treatments were carried out in a manner as described above using such sheet, a clear positive image was obtained likewise.

Polymer Containing Quinone Diazide Moiety

H. Iwama, N. Kita, K. Yumiki, H. Kawada and H. Yamaguchi; U.S. Patent 3,902,906; September 2, 1975; assigned to Konishiroku Photo Industry Co., Ltd., Japan describe an improved photosensitive material which comprises an addition polymer having a molecular weight of at least 5,000 as a photosensitive component. The polymer comprises the following repeating monomer units (1) and (2).

(1)

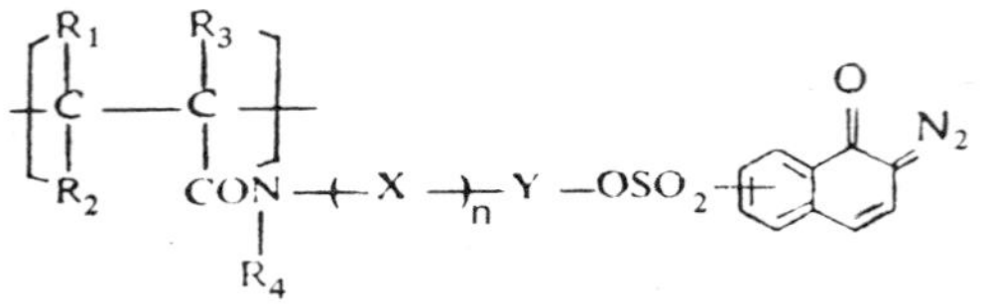

$$\left[-C(R_1)(R_2)-C(R_3)\left(CON(R_4)-(X)_n-Y-OSO_2-\text{(naphthoquinone diazide, } =O,\ =N_2)\right)- \right]$$

(2)

$$\left(-C(R_1)(R_2)-C(R_3)\left(CON(R_4)-(X)_n-Y-OH\right)- \right)$$

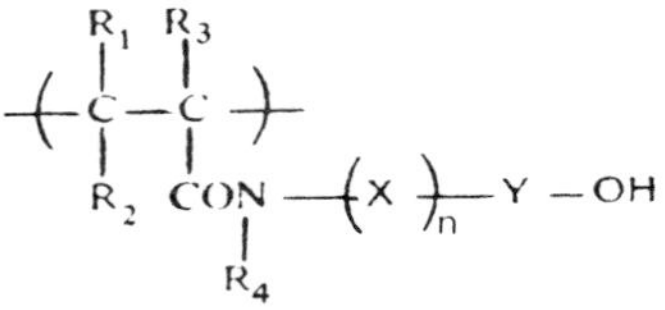

where R_1 and R_2 are individually hydrogen, alkyl or carboxylic acid; R_3 is hydrogen, halogen or alkyl; R_4 is hydrogen, alkyl, phenyl, or aralkyl; X is a divalent organic group selected from

$-CH_2-$, $-(CH_2)_3$, $-CH_2CH_2CO-$,

$-CH_2CH_2N(C_2H_5)-$, $C_6H_5-NH-SO_2-$, and $-C_6H_4-SO_2-$;

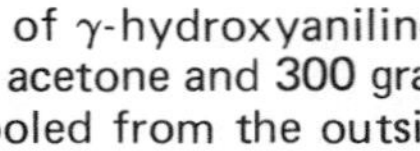

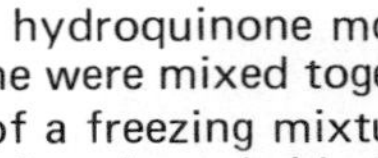

Y is unsubstituted phenylene or naphthylene, or phenylene or naphthylene having at least one of alkyl, nitro, carboxylic acid, cyano, sulfonic acid, hydroxyl, acyl, alkoxy or halogen; and n is 0 or 1.

Example 1: 400 grams of γ-hydroxyaniline, 4 grams of hydroquinone monomethyl ether, 4 liters of acetone and 300 grams of pyridine were mixed together, and the mixture was cooled from the outside with use of a freezing mixture. When the inside temperature reached -10°C, 420 grams of methacryl chloride was added dropwise to the above mixture with stirring.

The dropping rate was adjusted so that the reaction temperature was maintained below 0°C. After completion of the dropwise addition, the mixture was stirred at 0° to 3°C for about 2 hours, and then it was further stirred at 25°C for two hours. Then, the resulting reaction mixture was condensed until its volume was reduced to about one-third, and the concentrate was poured into 10 liters of dilute hydrochloric acid (pH 1.0). The resulting precipitate was suction filtered to obtain a white solid. This white solid was dissolved under heating in 2 liters of methanol, and 2 liters of 5% sodium carbonate was added to the solution and the mixture was stirred at 40°C for 30 minutes. The resulting dark red solution was poured into 8 liters of a 5% aqueous solution of hydrochloric acid to form a large quantity of a precipitate, which was recovered by suction filtration and dried to obtain a light pink solid.

Recrystallization of this solid from a mixed solvent of ethanol and water gave 450 grams of white crystals of p-hydroxymethacryl anilide. 124 grams of so obtained p-hydroxymethacryl anilide and 164 mg of α,α'-azobisisobutyronitrile were dissolved in a 1:1 mixed solvent of acetone and methanol, and heated at 65°C for 30 minutes in a sealed tube, the inside atmosphere of which had been substituted by nitrogen gas, to thereby form a polymer solution.

This polymer solution was diluted with 150 ml of methanol, the dilution was poured into water, and the resulting white precipitate was recovered by filtration and dried to obtain 101 grams of a white polymer (A). The molecular weight of this polymer (A) was about 250,000 as measured according to the osmotic pressure method. 20 grams of this white polymer (A) and 15.2 grams of o-naphthoquinone-diazide-5-sulfonyl chloride were dissolved in a mixed solvent of 400 ml of γ-butyrolactone and 10 ml of water. 100 ml of a 5% solution of sodium carbonate was added dropwise to the solution at 40°C under stirring. After completion of the dropwise addition, the reaction liquor was stirred for 30 minutes and poured into 2 liters of dilute hydrochloric acid (pH 1 to 3) to form a precipitate. The resulting yellow precipitate was recovered by filtration, washed with methanol and then with ethyl acetate sufficiently, and dried to obtain 27 grams of the compound shown below.

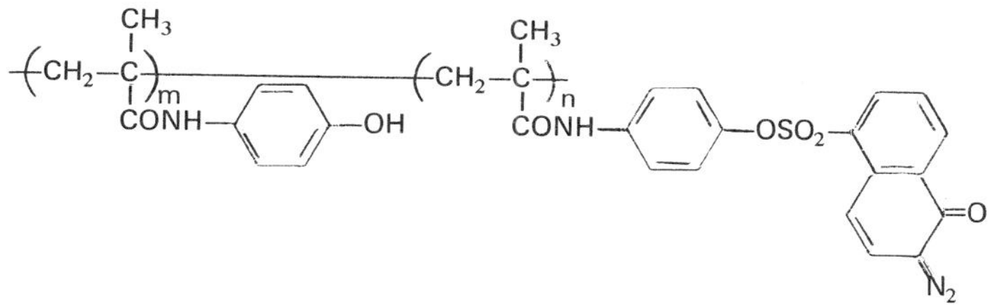

The compound had an average molecular weight of 310,000 and a mol ratio of m:n of 70:30.

Example 2: 40 grams of the compound obtained in Example 1 was dissolved in 800 ml of a 1:1 mixed solvent of methyl Cellosolve and cyclohexanone, and the solution was coated on a brush-grained aluminum plate by means of a whirler and dried. Then a positive photo original having lines and screen was applied to the face of the so obtained photosensitive plate, and light exposure was conducted for 1 minute by means of a mercury lamp of 3 kw disposed 1 meter away from the photosensitive plate. Then, the plate was immersed in a 3% aqueous solution of sodium metasilicate for 1 minute and the exposed surface was rubbed lightly with absorbent cotton.

Thus, the coating of the photosensitive material was removed at the light-exposed area and a positive relief image excellent in ink receptivity was obtained. When the relief image was given a water-retaining property by dampening solution and was used for printing by an offset printing machine, a great number of prints having a good image can be obtained.

N-Allyl-Maleimide Units

The process of *W. Frass; U.S. Patent 3,905,820; September 16, 1975; assigned to Hoechst AG, Germany* relates to a light-sensitive copying composition which contains from 65 to 99% by weight of at least one copolymer containing from 10 to 55 mol percent of N-allyl-maleimide units, from 0 to 15 mol percent of maleic acid units and from 45 to 80 mol percent of units of the formula

$$-CH_2CXZ-$$

wherein X is selected from the group consisting of a hydrogen atom, a halogen atom, an alkyl or alkoxy group containing from 1 to 3 carbon atoms, an aryl group or an acyloxy group, and Z is selected from the group consisting of a hydrogen atom, a halogen atom or an alkyl group containing 1 to 2 carbon atoms, from 0 to 35% by weight of at least one photopolymerizable compound having at least two vinyl or vinylidene groups in the molecule and which boils above 100°C under standard pressure, and from 1 to 8% by weight of at least one photoinitiator.

Example: A coating solution comprising 4.0 grams of the copolymer described below and 0.2 gram of Michler's ketone in 46.0 grams of methyl ethyl ketone is filtered and applied with a whirler-coating apparatus, at 100 revolutions per minute, to a commercially available electrochemically roughened aluminum foil. The resulting plate is dried for 10 minutes at 50°C in a drying cabinet and is exposed under a negative original for 6 minutes in the vacuum frame of an 8,000 w xenon exposure apparatus (72 cm distance from the light source).

The exposed plate is developed by one minute's immersion in methyl ethyl ketone. After rendering the plate hydrophilic with a solution comprising 80 parts by volume of gum arabic (14°Bé), 12 parts by volume of phosphoric acid (85%), 0.2 part by volume of hydrofluoric acid (50%), 0.5 part by volume of hydrogen peroxide (30%), and 7.3 parts by volume of water and after inking with a fatty ink, several tens of thousands of prints may be obtained from the plate on an offset printing machine.

The copolymer used is manufactured as follows: 126 grams (1 mol) of ethylene/maleic anhydride copolymer in a 1:1 molar ratio are suspended in 750 grams of glacial acetic acid. 43 grams (0.75 mol) of allylamine are added dropwise and the reaction mixture is heated to 112°C for 2 hours. After this time, 0.5 gram of p-methoxyphenol is added, the mixture is cooled and the product is precipitated in water. The coarsely granular precipitate is filtered off, washed with water and twice reprecipitated by dissolving it in acetone and pouring it into water acidified with hydrochloric acid. After drying, the yield is 137 grams, 88% of theoretical yield. Analysis for 74% conversion of ethylene/maleic acid/N-allylimide copolymer ($C_9H_{11}NO_2$)—Found: 6.3% N, iodine number 116.8, acid number 163.5. Calculated: 6.5% N, iodine number 118, acid number 179.

Polyurethane Using 1-Butene-3,4-diol

P. Richter, H. Stutz, L. Metzinger, O. Volkert, H.-U. Werther and A. Wigger; U.S. Patent 3,948,665; April 6, 1976; assigned to BASF AG, Germany found that relief plates which are particularly suitable for flexographic printing can be produced by imagewise exposure of a layer A of a photocrosslinkable material, containing 0.01 to 10% by weight of photoinitiator and, if desired, further conventional additives, on a base and subsequent washing out of the unexposed areas of the layer A with a developer. The photocrosslinkable material of layer A consists essentially of (a) 70 to 100% by weight of a noncrystallizing polyurethane elastomer (A11) which is soluble in the developer and has been produced by reaction of an aliphatic saturated polyester glycol having a molecular weight of about 400 to 4,000 (PG) with an aliphatic diisocyanate (DI) and at least one low molecular weight aliphatic diol (OL), of which 25 to 100 mol percent is 1-butene-3,4-diol, as chain extender, the molar ratio of polyester glycol to diisocyanate to low molecular weight diol being about 1:3 to 6:2 to 5 and the NCO/OH molar ratio always being about 0.9 to 1.0, or of a mixture of the polyurethane elastomer (A11) with up to 30% of its weight of another compatible light-transmitting polymer (A12), and (b) 0 to 30% by weight of a photocrosslinkable monomer with C—C multiple bonds which is substantially compatible with the polyurethane elastomer and layer A having Shore A hardness of 30° to 90° in the photocrosslinked state.

Laminates having the photocrosslinkable material in layer A are particularly suitable for the production of flexographic printing plates when layer A is firmly bonded to a nonphotocrosslinkable layer Z based on a polyurethane elastomer which is not soluble in the developer solution and which has a Shore A hardness of from 15° to 70° which is at least 20° less than the Shore A hardness of the layer A in the photocrosslinked state.

Example 1: A light-sensitive polyurethane is prepared in the following manner from the components listed below. The components are 218 parts (0.2 mol) of a polyester prepared from adipic acid, ethylene glycol and 1,4-butanediol and having an OH number of 103, 131.9 parts (0.785 mol) of hexamethylenediisocyanate, 26.42 parts (0.3 mol) of 1-butene-3,4-diol, and 26.42 parts (0.3 mol) of 2-butene-1,4-diol.

The polyester is melted in a reaction vessel provided with a stirrer and a vacuum connection and is dehydrated for 30 minutes at 120°C under a vacuum of 20 mm Hg. After adding the diisocyanate, the mixture is stirred for 2 hours at 80°C. The chain extender and 0.01% of dibutyltin dilaurate are stirred, at 80°C, into the melt of the isocyanate prepolymer thus produced and the melt is then poured into molds and is maintained at 80°C for a further 16 hours. A Shore A hardness of the polyurethane obtained is 60.

Example 2: A light-sensitive mixture is prepared from 178.4 parts of the polyurethane of Example 1, 20.0 parts of triethylene glycol diacrylate which is stabilized with 0.2% of hydroquinone, 1.6 parts of benzoin methyl ether and 0.02 part of New Fuchsin (Color Index 42520) and 130 parts of tetrahydrofuran. The solution is applied with a doctor blade to an elastomeric polyurethane film of 2.1 mm in thickness and having Shore A hardness of 40, which is insoluble in methyl ethyl ketone and tetrahydrofuran, and the excess removed with it, so that after evaporation of the solvent a layer 0.6 mm thick remains on the layer Z. The bond between the two layers is excellent. The Shore A hardness of a sample of the light-sensitive layer exposed is 77.

For the imagewise exposure of the upper layer applied to the intermediate layer Z of elastomeric polyurethane, the system is exposed for 15 minutes using a commercial flat-plate exposure unit. For reinforcement, the noncovered side of the layer Z is glued with an adhesive based on polychloroprene to a 0.1 mm thick polyethylene terephthalate film. After exposure, the printing plate is washed out for 8 minutes in a spray washer using methyl ethyl ketone. After a further 5 minutes' exposure and after drying for 1 hour at 80°C, the relief printing plate is ready for use.

Vinyl Urethane Monomers from Xylylene Diisocyanate

N. Miyata and H. Nakayama; U.S. Patents 3,907,865, September 23, 1975 and 3,954,584, May 4, 1976; both assigned to Kansai Paint Co., Ltd., Japan provide a photopolymerizable vinyl urethane monomer and a composition thereof useful for the preparation of printing plates and reliefs. The vinyl urethane monomer has the following chemical formula

$$CH_2{=}\underset{R}{C}-\underset{O}{\overset{}{C}}-O-A-\underset{O}{C}-NH-CH_2-C_6H_4-CH_2-NH\underset{O}{C}-O-A-\underset{O}{C}-\underset{R}{C}{=}CH_2$$

or

$$\left[CH_2{=}\underset{R}{C}-\underset{O}{C}-O-A-\underset{O}{C}-NH-CH_2-C_6H_4-CH_2-NH\underset{O}{C}-O\right]_n-X$$

in which R represents a hydrogen atom or a methyl group, X represents an alkylene or alkane-triyl to -hexayl group or polyalkyleneoxy group having a number average molecular weight not exceeding 2,000 or 3,000 respectively, and n represents a corresponding number to the valency of the group X selected from 2 to 6, and A represents an alkyleneoxy group or polyalkyleneoxy group of the formula:

$$-\!\left(CH_2-\underset{R}{CH}-O\right)_m\!-$$

where R is a hydrogen atom or a methyl group and m is a positive integer of 1 to 11. The composition comprises the vinyl urethane monomer, ethylenically unsaturated liquid monomers and a photosensitizer.

The photopolymerizable vinyl urethane monomer is prepared from xylylene diisocyanate, acrylic esters or methacrylic esters having one hydroxyl group and di-, tri-, tetra-, penta- or hexahydric alcohols through the reaction of the isocyanate groups with the hydroxyl groups to form urethane linkages.

Vinyl urethane monomers prepared from xylylene diisocyanate show extremely high photocurability. Fluid compositions consisting of such vinyl urethane monomers, photosensitizer and, if necessary, other ethylenically unsaturated monomers can cure to a hard mass having an adequate mechanical property enough for use as a printing plate by irradiation of actinic rays in a time not exceeding 10 min.

The unirradiated portion can be washed off completely with developing liquid such as water, aqueous solution of surfactants, organic solvents or mixtures thereof. The cured mass derived from xylylene diisocyanate does not show yellowing in course of time as is the case when tolylene diisocyanate is used.

Example: 188 grams (1 mol) of xylylene diisocyanate m-isomer/p-isomer (65-75/35-25) was added to 260 grams (2 mols) of 2-hydroxyethyl methacrylate containing 90 mg (200 ppm) of 2,6-di-tert-butyl-4-methyl phenol in a reaction flask. The temperature of the reaction mixture was held at 80°C for 6 hours under air atmosphere. Then the reaction mixture was cooled to room temperature to obtain the reaction product as a white and sticky solid. The isocyanate value of the product was below 5.

75 grams of the reaction product was mixed well with 25 grams of 2-hydroxyethyl methacrylate and to the mixture was added 1 gram of benzoin ethyl ether to obtain a photocurable liquid composition. The composition was applied to a primed steel plate to form a layer of 0.6 mm in thickness. The layer was then irradiated with light using a low-pressure mercury lamp at a 10 cm distance for 3 minutes. The irradiated layer was hard, tough and adherent to the steel substrate.

Polyester-Based Polyurethanes as Binding Agent

The process of *F.P. Recchia and T.M. Shah; U.S. Patent 3,912,516; October 14, 1975; assigned to The Upjohn Company* comprises a photopolymerizable element for use in preparing relief printing plates comprising:

- (a) An addition polymerizable ethylenically unsaturated compound having a boiling point above 100°C at normal atmospheric pressure;
- (b) A free radical generating addition polymerization initiator activatable by actinic radiation;
- (c) A compatible polyurethane binding agent comprising the product of reaction of
  - (1) 4,4'-methylenebis(phenyl isocyanate)
  - (2) A polycaprolactone diol having a molecular weight in the range of about 1,000 to 2,500; and
  - (3) A mixture of at least two different aliphatic diols of from 2 to 6 carbon atoms, inclusive.

The ratio of equivalents of polycaprolactone diol (2) to total equivalents of the aliphatic diols (3) are within the range of 1:1.5 to 1:7, and the ratio of equivalents of isocyanate (1) to total equivalents of polyols (2) and (3) are within the range of 0.94:1 to 0.98:1.

The improved properties of the photopolymerizable elements of the process are, in large measure, attributable to the use of the particular polyester-based polyurethanes which are employed as the compatible binding agent. The amount of addition polymerizable compounds is advantageously within the range of about 5 to about 20% by weight based on total weight of the composition.

Example: (a) Preparation of Photosensitive Polyurethane — A blend was prepared of 25 grams of benzophenone, 50 grams of trimethylolpropane trimethacrylate, 347 grams (0.35 equivalent) of a polycaprolactonediol (equivalent weight, 989; prepared from ϵ-caprolactone and 1,4-butanediol using the procedure described in U.S. Patent 2,933,478), 15.79 grams (0.35 equivalents) of 1,4-butanediol and 18.8 grams (0.28 equivalent) of dipropylene glycol.

The polycaprolactonediol had been dried and degassed previously by heating at 100°C in vacuo for 2 hours. The 1,4-butanediol and dipropylene glycol had previously been dried over molecular sieves. The resulting blend of components was stirred vigorously while 6 drops (0.125 gram) of a 50% w/w solution of stannous octoate in diethyl phthalate was added using a pipette. The mixture so obtained was stirred vigorously for approximately 30 seconds before adding, in one batch, 118.37 grams (0.94 equivalent) of 4,4'-methylenebis(phenyl isocyanate) in the form of a liquid at 45°C. The vigorous stirring was continued for a further 10 seconds and the reaction mixture was then poured on to a Teflon-lined shallow aluminum pan. The cast elastomer (NCO/OH index = 0.96) gelled in about 30 seconds after pour and was tack free in about 15 minutes.

(b) Molding of Photosensitive Element — A 4.75 x 4.75 x 0.062 inch sheet was prepared by compression molding a portion of the photosensitive polyurethane prepared as described above. A 50-gram charge of the material was placed in an appropriately configured two-piece aluminum mold having Teflon-coated inner surfaces. The mold was housed in a bench-type hydraulic press capable of exerting a maximum 50,000-pound force on a 3⅝-inch ram. The mold was heated to 230°F and the polymer was held under moderate pressure until flow began. The pressure on the mold was then increased to 10,000 pounds and held there for 2 minutes before increasing the pressure to 40,000 pounds.

After the polymer had been exposed to the latter pressure for 30 seconds the mold was cooled to room temperature using cold water while the pressure was maintained at the same level. Cooling time was about 16 minutes. The finished plate was demolded and found to be freely flexible and free from bubbles or other flaws. The plate had a Shore A hardness of 48.

(c) Preparation of Flexographic Printing Plate — The above plate was covered with a photographic negative showing a weather map and a vacuum frame enclosed in a Mylar sheet was placed over the negative and plate. A vacuum (25 torrs) was applied to the set up to ensure intimate contact between the negative and plate. The plate was then exposed, via the negative, to radiation from a scanning medium pressure tubular mercury vapor lamp (range 260 to 420 nm with a peak at 365 nm). The exposure time was 90 seconds. The negative was removed and the opposite side of the plate, i.e., the side which had not been exposed imagewise, was exposed totally to the same source of radiation for a period of 25 seconds.

The resulting plate was then subjected to solvent etching by clamping the plate, with the imaged side exposed, to a rotating drum in an etching unit equipped with four solvent spray nozzles and a recirculating solvent system. The plate was exposed to spray by methyl ethyl ketone at a pressure of 9 psi for 180 seconds. At the end of this time the plate was dried in a forced air oven at 70°C for 5 minutes, and found to have a depth of etch of 13 mils. The plate showed no tendency to curl or distort and the top surfaces of the etched image were planar and showed no evidence of cupping.

Degradable Polyaldehydes

W.W. Limburg and M. Stolka; U.S. Patent 3,915,704; October 28, 1975; assigned to Xerox Corporation provide a method for the selective degradation of a composition comprising at least one acid degradable polymer of the formula:

$$\left[\begin{array}{c} H \\ | \\ -C-O- \\ | \\ R \end{array} \right]_n$$

wherein R is an aliphatic hydrocarbon radical of 1 to 6 carbon atoms, a chlorinated aliphatic hydrocarbon radical of 1 to 6 carbon atoms, or a nitrile substituted aliphatic hydrocarbon radical of 1 to 5 carbon atoms; and n is at least 50.

In this method, a thin film of a composition comprising at least one degradable polymer of the above formula, at least one polymer capable of undergoing photo-induced dehydrohalogenation (hereinafter referred to as acid generating polymer) and an organic electron acceptor are subjected to selective illumination with ultraviolet light for a brief interval whereupon a visible image is formed within the thin film. The relative weight ratio of acid degradable polymer to acid generating polymer in the film can range from about 99:1 to about 1:99 and preferably from about 75:25 to about 50:50. The concentration of electron acceptor will generally range from 0.1 to 5 weight percent based upon acid generating polymer concentration.

In the preferred embodiments, the acid degradable polymer is poly(acetaldehyde); and the acid generating polymer is poly(vinylbromide). It is also preferable that such films be illuminated with ultraviolet light at a wavelength of less than about 3000 A.

The acid generating polymer of the composition can be any one or a combination of the polyvinylhalides or polyvinylidene halides containing units corresponding to the following formula:

$$-\left(\begin{matrix} H & & X \\ | & & | \\ C & - & C \\ | & & | \\ Y & & Y \end{matrix}\right)_a - \left(\begin{matrix} H & & X \\ | & & | \\ C & - & C \\ | & & | \\ Z & & Y' \end{matrix}\right)_b -$$

wherein X is selected from among chlorine, bromine or iodine; Y and Y' are independently selected from X or hydrogen; Z is selected from among Y, alkyl of about 1 to 8 carbon atoms, phenyl or alkyl substituted phenyl, the alkyl having about 1 to 8 carbon atoms; and a and b represent the mol percent of each of the components within the acid generating polymer and have an aggregate value equal to 100%.

In Figures 2.1a, 2.1b and 2.1c are shown different forms the image can take upon variation of one or more factors. For example, where there is a predominant amount (generally in excess of about 50% by weight of acid degradable polymer in the imaged regions of the film), degradation of such regions can produce an imaging member wherein such imaged areas are substantially devoid of polymeric material (Figure 2.1c). On the other hand, in the event that only moderate amounts (generally from about 25 to 40% by weight) of acid degradable polymer are present in the regions of film subjected to activating electromagnetic radiation, the degradation of the polymer will only result in plasticizing of these exposed regions (Figure 2.1a).

Where only very small amounts (generally less than about 10% by weight) of acid degradable polymer are present in the film, its degradation upon irradiation may go undetected until the imaged areas of the film are thermally developed. Heating such films will result in an expansion of the gaseous products of polymer degradation, thus forming a vesicular image of the type shown in Figure 2.1b.

FIGURE 2.1: VARIATION IN IMAGE FORMS

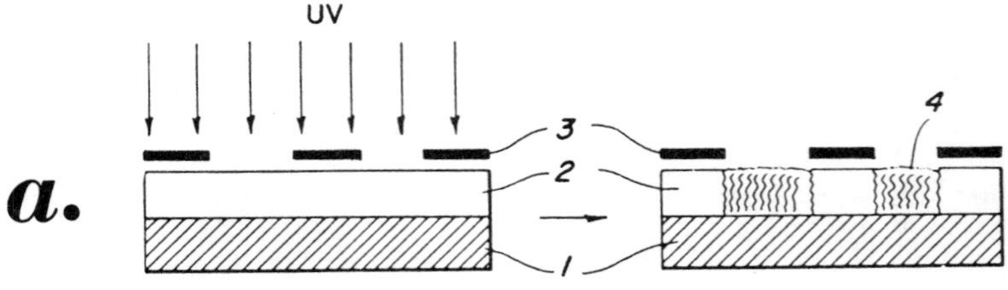

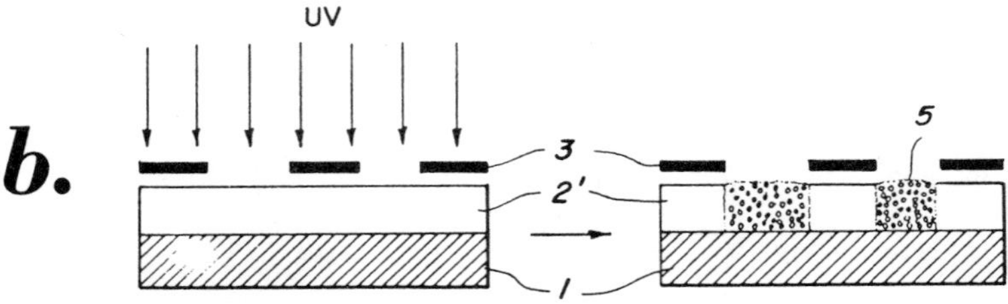

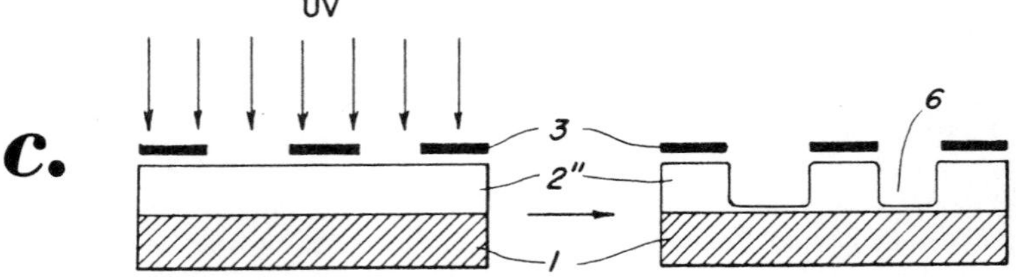

Source: U.S. Patent 3,915,704

Prolonged delay between imaging and thermal development should be avoided since the gaseous degradation products can diffuse from the composition and, thus, subsequent thermal treatment may not produce the desired vesicular image.

In the examples a series of thin films are prepared by dissolving the degradable polymer, the acid generating polymer and the electron acceptor in the proper proportions in tetrahydrofuran (THF) and then casting the resulting solution on a suitable substrate (e.g., glass, aluminum, Mylar etc.). The dry film thickness of such layers is allowed to vary within the range of 0.1 to 300 microns. Each of the films thus prepared is dried prior to use.

Where the degradable polymer is inherently unstable at elevated temperature drying is carried out at room temperature and under a vacuum. Each of the films is selectively irradiated with ultraviolet light (UV SL-25, 10 watt output, Ultraviolet Products) through a quartz glass transparency from a distance of one foot for a period of 60 seconds. The text which follows provides the recipes of such a series of films and the effects of irradiation on the exposed areas of such films.

Example 1: The acid degradable polymeric composition contains 90 parts by weight poly(acetaldehyde) (acid degradable polymer); 10 parts by weight poly(vinylchloride) (acid generating polymer); and 0.1 part by weight tetracyanoethylene (organic electron acceptor). The film thickness is 40 microns. The effects of imagewise irradiation with ultraviolet light are as follows. After selective exposure of film **2''** prepared from the above composition, depressions **6** occur within those regions which are subjected to such irradiation (as shown in Figure 2.1c).

Example 2: The procedures of Example 1 are repeated except for a shift in the relative concentration of acid degradable polymer to acid generating polymer from 90:10 to 50:50, and the concentration of Lewis acid adjusted accordingly. The effects of such a shift are manifest in the type of image produced. Whereas, in Example 1 the exposure of the film prepared therein results in the formation of depressions in the irradiated areas, similar exposure of film **2** prepared from the composition of this example only results in plasticizing of these same regions **4** thus making them soft and tacky (as shown in Figure 2.1a).

Example 3: The procedures of Example 1 are repeated except for a shift in the relative concentration of acid degradable polymer to acid generating polymer from 90:10 to 10:90, and the concentration of Lewis acid adjusted accordingly. The effects of such a shift are manifest in the type of image produced.

Whereas, in Example 1 the exposure of the film prepared therein results in the formation of depressions in the irradiated areas, similar exposure of film **2'** prepared from the composition of this example does not result in any appreciable change in the film's physical properties. Shortly after imaging, the imaging layer of this example is gently heated with a hot air gun until bubbles begin to appear within irradiated regions **5** of the film thus forming a vesicular image (as shown in Figure 2.1b).

The acid degradable polymer of the formula shown in the preceding patent, where R and n are as described there, are used in the process of *W.W. Limburg and D.G. Marsh; U.S. Patent 3,917,483; November 4, 1975; assigned to Xerox Corporation.*

In the preferred embodiments, a thin film of a composition comprising at least one degradable polymer of the formula and a latent acid are subjected to selective illumination with ultraviolet light for a brief interval whereupon a visible image is formed within the thin film.

The preferred acid degradable polymer is poly(acetaldehyde); and the preferred latent acid is beta-naphthol. It is also preferable that irradiation of such films be performed with ultraviolet light at a wavelength of less than about 3000 A.

Good results are obtained wherein the composition contains anywhere from about 0.1 to 5 parts by weight latent acid per 100 parts by weight acid degradable polymer. Latent acids which are especially preferred for use in this composition are capable of undergoing a rapid and efficient transition from the ground to the excited state in response to electromagnetic radiation, the duration of exposure and intensity of such radiation being insufficient by itself to cause any change in the degradable polymer or other polymeric materials present in the composition. Latent acids which are representative of materials having the above properties include hydroxyl functional naphthalene compounds (e.g., beta-naphthol); phenols; and the halogen substituted naphthols and phenols (e.g., p-chlorophenol).

A series of thin films are prepared as described in U.S. Patent 3,915,704 except that the electron receptor of that process is replaced with a latent acid as described above.

Example 1: Following these procedures, imaging layer **2''**, prepared from the following composition, is formed on aluminum substrate **1**. The composition of the imaging layer is 100 parts by weight poly(acetaldehyde) (acid degradable polymer) and 2 parts by weight beta-naphthol (latent acid). The film thickness of the imaging layer is 15 microns.

Subsequent to curing of imaging layer **2''**, a quartz glass transparency **3**, is placed above the imaging member and the imaging layer selectively irradiated in conformity with the information on the transparency. The exposed regions of the imaging layer undergo degradation in these irradiated areas, forming depressions **6** within the layer as shown in Figure 2.1c.

Example 2: Following the procedures outlined above, imaging layer **2**, prepared from the following composition, is cast on an aluminum plate **1**. The composition of the imaging layer is 30 parts by weight poly(acetaldehyde) (acid degradable polymer), 70 parts by weight poly(methylmethacrylate) (thermoplastic matrix) and 2 parts by weight beta-naphthol (latent acid). The film thickness of the imaging layer is 15 microns.

Subsequent to curing of imaging layer **2**, a quartz glass transparency is placed above the imaging member and the imaging layer selectively irradiated in conformity with the information on the transparency. The exposed regions of imaging layer **4** become plasticized as a result of degradation of the acid degradable polymer, thus becoming soft and tacky as shown in Figure 2.1a. A portion of the plasticized polymer in the exposed regions of the imaging layer can be selectively removed by simply pressure-contacting this layer with a sheet of paper thereby offsetting plasticized polymer from the imaging layer to the paper. This paper can then be treated with a standard conversion fluid (Offset Electrostatic Conversion Solution No. 44-1050, A.B. Dick Inc.). This conversion fluid renders the background areas (the nonpolymer-bearing areas of the paper) hydrophilic. The polymer deposits on the paper remain oleophilic. This sheet of paper bearing the polymer image is now suitable for use as a lithographic master in either a direct or offset lithographic printing process.

Example 3: Following the procedures outlined above, imaging layer **2**, prepared from the following composition is cast on an aluminum substrate **1**. The composition of the imaging layer is 10 parts by weight poly(acetaldehyde) (acid degradable polymer), 90 parts by weight poly(methylmethacrylate) (thermoplastic matrix) and 2 parts by weight beta-naphthol (latent acid). The film thickness of the imaging layer is 15 microns.

Subsequent to curing of imaging layer **2'**, a quartz glass transparency is placed above the imaging member and the imaging layer selectively irradiated in conformity with the information on the transparency. The exposed regions of the imaging layer **5** become somewhat diffusive indicating the presence of small microbubbles within the bulk of the irradiated regions of the film. This image is intensified by gentle heating of the imaging layer with a hot air gun. The resulting vesicular image thus produced is shown in Figure 2.1b.

In related work *W.W. Limburg and D.G. Marsh; U.S. Patent 3,915,706; Oct. 28, 1975; assigned to Xerox Corporation* provide an imaging system which comprises exposing to activating radiation in an imagewise manner a film comprising a halogenated polymer capable of releasing hydrogen halide, the polymer having dispersed therein: (1) a degradable polymeric composition containing segments characterized by the formula:

$$\left[-\overset{\displaystyle H}{\underset{\displaystyle R}{\overset{|}{\underset{|}{C}}}}-O- \right]$$

where R is H, an alkyl radical of 1 to 6 carbon atoms, a chlorine or fluorine substituted radical of 1 to 6 carbon atoms or a cyano substituted aliphatic hydrocarbon radical of 1 to 5 carbon atoms; and (2) a photoactive reagent which upon activation is capable of abstracting a hydrogen atom from the polymer backbones of the degradable polymeric composition and halogenated polymer.

When the degradable polymer, halogenated polymer and photoactive agent are formed into a thin layer, a cloudy, translucent film results. This is probably due to the mutual incompatibility of the polymers. When the film is exposed to activating radiation, the degradable polymer breaks down with such breakdown resulting in a change in the compatibility of the polymers and a consequent change in optical density of the exposed areas. When lower molecular weight degradable polymers are used, the change in optical density results in the film changing from translucent to clear in the exposed areas to provide a positive working system. With higher molecular weight degradable polymers, the exposed areas become more translucent than the background even to the point of being opaque, thus providing a negative working system.

When homopolymers of the abovedescribed aldehydes are used in the process, a degree of polymerization within the range of from 20 to 20,000 is preferred for use in the process.

In addition to homopolymers of the desired aldehydes, copolymers and block copolymers containing degradable segments characterized by the foregoing formula can be employed in the process. For example, copolymers and block copolymers may be prepared from one or more of the aldehydes previously described and other polymerizable constituents such as styrene, isoprene, α-methylstyrene, methylmethacrylate, phenyl isocyanate and ethyl isocyanate. In addition, the degradable segments may occur as side chains appended from the backbone of another polymer. Suitable halogenated polymers are those which conform to the formula:

$$-\left[\begin{array}{cc} H & X \\ | & | \\ -C- & -C- \\ | & | \\ Y & Y \end{array}\right]_{n\%}-\left[\begin{array}{cc} H & X \\ | & | \\ -C- & -C- \\ | & | \\ Z & Y' \end{array}\right]_{m\%}-$$

In the above formula, X is chlorine or bromine, Y and Y' are X or hydrogen and Z is Y or an alkyl, aryl or alkaryl constituent containing from 1 to 8 carbon atoms. The symbols n and m represent numbers which designate the relative mol percent composition of the individual units in the polymer and can vary from 0 to 100 with the sum of n percent and m percent being 100.

A preferred class of photoreactive reagents is made up of those compositions which, when subjected to activating radiation, assume a $^3(n,\pi^*)$ or $^1(n,\pi^*)$ state. Preferably, the degradable polymer will make up from 1 to 49 weight percent of the composition. A preferred concentration of photoactive agent is from 0.1 to 5 weight percent of the composition.

Example 1: A solution comprising 1 weight percent poly(acetaldehyde), 10 weight percent poly(vinylchloride) and 1 weight percent benzophenone in THF is spread on a glass substrate with a doctor blade having an 8 mil gate to provide a film having a thickness of from 10 to 25 μ. Duplicate experiments are carried out using poly(acetaldehyde) with molecular weights of approximately 85,000 and 103,000 respectively. The films appear translucent upon drying due to the incompatibility of the polymers.

The films are imaged by exposing them to the unfiltered light from a PEK 112 lamp operated at 100 watts for 60 seconds. After exposure, it is observed that the film is clear in the exposed areas thereby providing a positive working imaging system.

Example 2: The procedure of Example 1 is repeated with the exception that the poly(acetaldehyde) has a molecular weight of approximately 250,000. After exposure, it is observed that the exposed areas are more translucent than the nonexposed areas thereby providing a negative working system.

In related work *D.G. Marsh; U.S. Patent 3,964,907; June 22, 1976; assigned to Xerox Corporation* provides a method for the preparation of a relief printing master which is based on irradiation of a film of a composition comprising the degradable polymeric composition of U.S. Patent 3,915,706 above and one of the preferred photosensitizers of that patent.

Example: A printing plate is prepared by applying a solution of 0.5 gram poly(acetaldehyde) having a molecular weight of approximately 447,000 and 0.004 gram of benzophenone in 10 milliliters of benzene to an aluminum substrate. The solvent is allowed to evaporate leaving a film of approximately 25 μ in thickness on the substrate.

The plate is irradiated through an aluminum stencil with a 100 w high-pressure mercury arc emitting in the ultraviolet region at about 3650 A to provide ultraviolet radiation to the exposed areas having an intensity of approximately 1.2×10^{17} photons/cm^2/sec. Visible depressions are observed in the exposed areas after about 20 seconds. After 60 seconds, irradiation is terminated with depressions of approximately 15 to 25 μ being observed in the exposed areas. Upon placing the plate on a revolving drum and contacting it with paper while applying pressure the polymer is removed in the exposed areas resulting in depressions down to the substrate in the exposed areas.

The plate is inked and used as a printing master to provide ink transfer in the nonexposed areas while the areas corresponding to the exposed portions of the plate remain blank.

A relief printing master is also provided in the process of *D.G. Marsh; U.S. Patent 3,923,514; December 2, 1975; assigned to Xerox Corporation.* The substrate on the printing plate in this case comprises (a) a degradable polymer containing segments characterized by the formula:

$$\left[-\overset{R_1}{\underset{R_2}{\overset{|}{\underset{|}{C}}}} - O - \right]$$

wherein R_1 is hydrogen or methyl and R_2 is hydrogen, an alkyl radical of 1 to 6 carbon atoms, a chlorinated or fluorinated aliphatic radical of 1 to 6 carbon atoms or a cyano substituted radical of 1 to 5 carbon atoms provided that when R_1 is methyl, R_2 is also methyl, and (b) a photo-oxidant which upon activation is capable of abstracting one or more electrons from one or more of the oxygen atoms in the polymer.

Suitable photo-oxidants include an aromatic carbonyl compound, a pyrylium salt, anthracene or a derivative thereof, a diazonium salt, a para quinoid compound, an unsaturated anhydride, a tosylate salt, a diaza heterocyclic compound or a mixture thereof.

Example: A thin film consisting of 10 parts poly(acetaldehyde) having a molecular weight of approximately 500,000 and one part of maleic anhydride in a 5% (by weight of solids) benzene solution is cast onto a Pyrex slide. Upon evaporation of the solvent, the film is exposed through a stencil.

The exposure is accomplished through the use of a UVS-12 mineral light emitting light at 254 nm and providing incident light energy of approximately 7.7×10^3 ergs/cm^2/sec. An exposure of 60 seconds results in the formation of a relief printing master in which the polymer in the light struck areas is removed down to the glass substrate. The plate is inked and used as a printing master thereby transferring ink in the configuration of the nonexposed areas.

Tetraallylsulfonamide and Polythiol

The process of *S.V. Dighe and R.W. Bush; U.S. Patent 3,915,825; October 28, 1975; assigned to W.R. Grace & Co.* is directed to: (a) N,N,N'N'-tetraallyl diphenyl ether-4,4'-disulfonamide (TADEDS); (b) a number of related sulfonamides; and (c) the preparation of such sulfonamides from diphenyl ether-4,4'-disulfonyl chloride (DEDSC) or nuclear substituted DEDSC and diallylamine.

The compounds of the process are used to form a curable composition with a liquid polythiol component having molecules containing at least two thiol groups per molecule, the equivalent ratio of $-CH{=}CH_2$ to $-SH$ being 1:0.8 to 1.2. The curable composition also contains a photocuring rate accelerator present in an amount from 0.05 to 25% by weight of the curable composition.

Example 1: Materials: 36.7 grams (0.1 mol) diphenyl ether-4,4'-disulfonyl chloride; 8.0 grams (0.2 mol) sodium hydroxide; 20.0 grams (0.2 mol) diallylamine.

In a 500 milliliter, three-necked round-bottom flask equipped with a reflux condenser, thermometer, addition funnel and a mechanically driven stirrer were placed 8.0 grams of sodium hydroxide dissolved in 100 milliliters of water and 20.0 grams of diallylamine. The resulting mixture was heated to 50°C while stirring. Diphenyl ether disulfonyl chloride (DEDSC) dissolved in 200 milliliters of tetrahydrofuran was added dropwise from the funnel. After a few milliliters of the solution had been added, the temperature rose to 55°C. At this point heating was stopped and the rate of addition so adjusted as to maintain a temperature of 50° to 55°C during addition.

After the addition was complete, the reaction mixture was refluxed for four hours and then allowed to cool to room temperature. The cold solution separated into two layers. It was added to a large excess (about 1 liter) of water with vigorous stirring. A white precipitate separated; said precipitate was collected on a filter and dried in vacuo at 50°C. The thus dried material was recrystallized from tetrahydrofuran-pentane mixture. The white recrystallized product melted sharply at 85° to 86°C and weighed 42.5 grams corresponding to a conversion (1 pass) yield of 88% of theory. The recrystallized product was identified as TADEDS by NMR.

Example 2: A composition was prepared by admixing TADEDS prepared as in Example 1, supra, with pentaerythritol tetrakis(3-mercaptopropionate) in a mol ratio of 1:1 and incorporating into said mixture 2% by weight of benzophenone (based on the weight of the TADEDS).

The resulting mixture was melted, admixed thoroughly and spread over a thin metal sheet to form a thin film (about 15 to 20 mils thick). The resulting film was cooled to room temperature and exposed to the light of a 4,000 watt Ascorlux pulsed xenon arc printing lamp through a photographic negative having an image thereon. Exposure time was 2.6 minutes, the xenon light being about 30 inches from the surface of the film. A cured image was present on the exposed areas of said film when the film was developed by washing with a suitable solvent (a mixture of acetone and water in a weight ratio of about 3 parts acetone per part of water). The unexposed portions of said film were dissolved and washed away by the solvent, while the exposed portions of said film were not dissolved (and not washed away) by said solvent. The resulting printing plate was mounted on a printing press using double-face pressure-sensitive tape and printing was carried out in the same way conventional metal photoengraved plates are employed. The printing results obtained were superior to those with conventional plates.

Aromatic ortho-Nitrocarbinol Ester Groups

H. Barzynski, M. Marx, G. Storck and D. Saenger; U.S. Patent 3,926,636; December 16, 1975; assigned to Badische Anilin- & Soda-Fabrik Aktiengesellschaft, Germany describe a photocurable composition that consists of the following:

(a) at least one substance which contains at least two aromatic or heteroaromatic o-nitrocarbinol ester groupings of the formula

```
       |
       C=O
       |
       O
       |
     X-C-H   NO2
        \    /
        ( Y )
```

where Y is an aromatic or heteroaromatic ring system of 5 to 14 ring members and X is hydrogen or alkyl of 1 to 8 carbon atoms or unsubstituted or substituted aryl or aralkyl of up to 12 carbon atoms;

(b) at least one compound containing at least two aziridine groupings or isocyanate groupings in the molecule; and also optionally

(c) one or more conventional natural or synthetic binders, fillers, reinforcing materials, pigments, sensitizers, plasticizers and/or other auxiliaries conventionally used in the coating agents industry.

The photocurable composition is particulary suitable for the production of coatings and printing plates.

Example: A homogenous coating mixture is prepared from 10 parts of a 60% solution in ethyl gylcol acetate of a polyester having 2-nitro-3,4-dimethoxybenzyl ester groups and 1.2 parts of 4,4'-diethyleneurea-3,3'-dimethyldicyclohexyl-2,2-propane.

The dissolved polyester is built up from isophthalic acid, adipic acid, neopentyl glycol and trimethylolpropane and contains 10.5% by weight of 2-nitro-3,4-dimethoxybenzyl ester groups. The mixture is knife coated onto a metallic substrate and after drying leaves a layer having a thickness of 125 microns. A mask is applied and through this the layer is exposed to light from a mercury high pressure lamp for three minutes and then washed with acetone. The image formed by the light-permeable zones of mask remains as a yellowish brown raised image and can be used as a block for block printing.

Graft Polymer

D.S. Breslow and D.A. Simpson; U.S. Patent 3,926,642; December 16, 1975; assigned to Hercules Incorporated discovered a process for the preparation of printing plates, which process is not inhibited by oxygen. The process depends upon oxygen being present during the exposure step.

The process comprises the steps of providing the surface of a polymer film with a photooxygenation sensitizer, said film being a film of a polymer containing extralinear olefinic unsaturation of the type in which there is no more than one hydrogen atom on each of the double-bond carbons and in which there is at least one allylic hydrogen on at least one of the carbons adjacent to the double-bond carbons, exposing selected areas of the sensitized film to light having a wavelength of from about 2000 to 12000 angstroms in the presence of oxygen and subjecting the exposed film to contact with a reactant capable of forming a graft polymer structure in the exposed areas of the film. This reactant may be either hydrophilic or oleophilic.

The process essentially involves the grafting of a hydrophilic or oleophilic reactant onto the surface of a film of an unsaturated polymer, and this may be accomplished by two related procedures. The initial reaction in both procedures involves the photosensitized oxidation of a suitably substituted unsaturated polymer, resulting in the formation of hydroperoxide groups on or near the surface of the polymer film.

The polymer hydroperoxides formed in the light-struck areas of the film should be thermally stable and are used in one procedure to graft polymerize a vinyl monomer onto the surface of the film. In the other procedure, the photooxidized film is contacted with a polymeric reactant to form a graft of the reactant on the surface of the film.

Example: Atlac 382E (propoxylated bisphenol-A fumarate polyester resin of MW 3,000) was modified with 2,3-dimethyl-1,3-butadiene (DMB) in a Diels-Alder reaction. Twenty-five grams Atlac 382E (0.059 mol unsaturation) and 9.70 grams of DMB (0.118 mol, 100% excess), were dissolved in 25.0 grams of reagent grade toluene in a 200.0 milliliter polymerization bottle. The reaction was run under air.

To prevent crosslinking of the polyester, about 1% hydroquinone was added as an inhibitor. The reaction mixture was heated at 100°C for 24 hours. (Analysis for unreacted DMB by gas-liquid chromatography indicated the reaction was complete after 22.5 hours.)

The polymer was precipitated by pouring the reaction mixture into about 800.0 milliliters of rapidly stirred hexane. The solvent was decanted and the gummy polymer was redissolved in benzene, filtered through glass wool, reprecipitated by pouring into hexane and dried. A study of the product, and of Atlac 382E and hydrogenated Atlac 382E, by nuclear magnetic resonance indicated the polyester was modified 100 ± 4% with DMB.

Films of this polymer were prepared and crosslinked through its terminal groups with a trifunctional isocyanate. The following procedure is representative: 1.80 grams of DMB-Atlac 382E, 0.50 gram of Desmodur N-75 (the reaction product of 3 mols of hexamethylene diisocyanate and 1 mol of water, named as the biuret of hexamethylene diisocyanate) and 0.050 gram zinc octoate (8% Zn) were dissolved under a dry nitrogen atmosphere in 2.70 grams of Cellosolve acetate.

This solution was used to cast several films on 8 x 8 x 0.003 inch sheets of hard aluminum foil using a 10 mil casting knife. The films were cured at 130°C for 1½ hours. Casting and curing operations were carried out under a dry nitrogen atmosphere. Cured film thickness was about 3 to 4 mils.

The cured films then were coated with methylene blue sensitizer from a 50/50 (volume per volume) solution of chloroform/methanol (3.34×10^{-3} mol/l methylene blue). The sensitizer solution was applied to the films using a camel's hair brush. Methylene blue concentration was about 5.6×10^{-8} mol/cm^2. Sensitizer coating and all subsequent operations were carried out in the dark under a safe light.

A dried, methylene blue coated film was attached to a glass plate and covered with a half-tone, positive, photographic transparency. The film was exposed for 60 seconds from a distance of 30 centimeters to a 375 watt Sylvania R32 photoflood lamp. During exposure the film was cooled by an air blower. Immediately following exposure the transparency was removed and the film was wiped with a methanol-soaked nonwoven fabric to remove the sensitizer.

A grafting solution was prepared from 15.0 grams of acrylic acid, 0.150 gram of vanadium oxyacetylacetonate (1.0% based on monomer) and 45.0 grams of anhydrous methanol. The resulting solution containing 25% by weight of acrylic acid, was degassed at −70°C by evacuation-nitrogen flush cycles. The exposed film was placed in a shallow dish, under a nitrogen atmosphere, and covered with the grafting solution. After ten minutes contact, the film was removed and well rinsed with methanol to remove any residual monomer.

At this point, an image with excellent half-tone definition was clearly visible as a result of grafting to the light-struck areas. Amplification of the grafted areas of the film with a suitable cationic polymer gave a surface useful for lithographic printing. Amplification was achieved by wiping the acrylic acid grafted film with a 5% aqueous solution of Dow PEI 1,000 (polyethylenimine of 50,000 to 100,000 MW) containing a small amount of Ultrawet 30-DS.

The film was then covered with a nonwoven fabric soaked with the PEI solution. After 15 minutes the wipe was removed and the film was rinsed well with water. The dried film was tested as a printing plate on a conventional lithographic press. The amplified surface printed sharp images with good half-tones and excellent ink holdout in the light-struck areas.

Another dried, methylene blue coated film was exposed through a Stauffer 21 Step Sensitivity Guide (No. AT 20 x 0.15) and grafted as described above. Acrylic acid grafting was clearly visible through Step No. 13, indicating that an image could be produced with a one-second exposure under the conditions of the experiment.

Composition Developed with Aqueous Alcoholic Alkali

The process of *J.A. Arcesi and F.J. Rauner; U.S. Patent 3,929,489; Dec. 30, 1975; assigned to Eastman Kodak Company* is directed to a radiation-sensitive composition comprising a soluble condensation polymer having first and second dicarboxylic acid derived repeating units. The first dicarboxylic acid derived repeating units contain nonaromatic ethylenic unsaturation capable of providing crosslinking sites for the purpose of insolubilizing the polymer upon exposure of the composition to actinic radiation.

As the improvement of this process, the polymer incorporates second, aromatic dicarboxylic acid derived repeating units containing disulfonamido units containing monovalent cations as amido nitrogen atom substituents, thus rendering the polymer in its unexposed form soluble in the aqueous alcoholic alkaline developer.

Example: Two and two-tenths grams (0.005 mol) of dimethyl 3,3'-[(sodioimino)disulfonyl]dibenzoate, 26.03 grams (0.095 mol) diethyl p-phenylenediacrylate and 35 grams (0.17 mol) 1,4-(β-hydroxyethoxy)cyclohexane were weighed in that order into a 200 milliliter polymerization flask.

The side arm of the flask was fitted with a cork and the flask itself fitted with a glass tube reaching the material in the flask such that nitrogen gas could be bubbled through the reaction mixture during the first stage of heating. The flask was also fitted with Vigreux column for reflux return of high-boiling material during this first heating stage, but such that the generated alcohols were distilled.

The contents were melted by inserting the flask in a silicone oil bath held at 235°C. Two drops (1/20 milliliter) of titanium isopropoxide was added to the melt and the flask and contents were heated under reflux for 4 hours. The Vigreux column, inert gas tube and the cork were removed and the side arm connected to a vacuum system in series with two dry ice-acetone traps. A stainless steel crescent shaped stirrer, fitted with a vacuum-tight ball joint, was inserted into the reaction melt to stir the polymer.

The pressure was gradually lowered to 0.05 mm Hg with stirring, at which pressure the polyester was stirred for 40 minutes, collecting distillate in the two dry-ice traps. The final inherent viscosity was determined by monitoring the final stage of the reaction with a Cole-Parmer Model 4425 Constant Speed and Torque Control Unit operating at 200 rpm and terminating the reaction when the desired inherent viscosity had been reached. A glassy amber polymer was obtained. The isolated polymer had an inherent viscosity of 0.59.

A formulation was prepared from the polymers comprising

Polyester	2.5 g
(2-benzoylmethylene)-1-ethyl-β-napthothiazoline	0.10 g
2,6-di-tert-butyl-p-cresol	0.10 g
Dichloroethane	100 cc

The formulation was filtered through a coarse filter paper. The solution was whirl-coated at 100 rpm on phosphoric anodized aluminum support for 2 minutes plus additional drying for 15 minutes at 50°C. The dried coating was exposed imagewise to a line negative on a xenon source exposure device. The exposed plate was swab-developed by applying the developer having the composition shown below to the plate surface and allowing it to soak 15 seconds, followed by swabbing for 30 seconds. Desensitizer gum was then applied, followed by hand inking.

The aqueous alcohol alkaline developer developed the coatings. The composition of this developer is as follows.

Glycerol	33 cc
2-phenoxyethanol	4 cc
2-ethoxyethanol	10 cc
Distilled water	50 cc
2-diethylaminoethanol	3 cc

Photographic speed was assessed by exposure through a step tablet having 14 steps with a step density increment of 0.3. The number of steps crosslinked to give a full ink image was 10. The number of additional steps yielding some degree of visible ink image in printing was 2. The nonimage areas were clean and the ink receptivity was good.

Lithographic printing plates prepared as described above exhibit no image loss after 10,000 press-run impressions on a 1250 Multilith duplicator press employing either conventional or Dahlgren (alcoholic solution of gum arabic) fountain solution.

Arylglyoxyacrylate Groups

T.M. Muzyczko and D.W. Fieder; U.S. Patent 3,930,868; January 6, 1976; assigned to The Richardson Company describe a light-sensitive composition comprising arylglyoxyalkyl acrylates that exhibits useful light sensitivity. The basic structure of the compositions, which may also themselves be polymerized, is as follows:

$$\mathrm{Ar{-}\overset{\overset{\displaystyle O}{\|}}{C}{-}\overset{\overset{\displaystyle O}{\|}}{C}{-}O{-}R_1{-}O{-}\overset{\overset{\displaystyle O}{\|}}{C}{-}\overset{\overset{\displaystyle R_2}{|}}{C}{-}R_3}$$

Wherein Ar represents an aromatic structure selected from the group consisting of benzene, naphthalene and substituted products of each, R_1 represents an alkyl group having from one to ten carbon atoms, R_2 represents a grouping selected from the group consisting of hydrogen, or a lower alkyl group having from one to five carbon atoms and R_3 represents an alkenyl group having from one to ten carbon atoms and singular unsaturation.

The light-sensitive compositions may themselves be utilized in photochemistry as photopolymers; they may be combined with suitable solvents and additives or polymerized with suitable backbone polymers to provide substances which can be used as light-sensitive coatings. These coatings may be placed on substrates and in one instance as presensitized lithographic plates.

Example 1: Synthesis of Phenylglyoxyethyl Methacrylate – Fifty grams (0.333

mol) of phenylglyoxylic acid, 52.3 grams (0.40 mol) 2-hydroxyethyl methacrylate which contained approximately 85% hydroxyethyl methacrylate according to gas chromatographic measurements, and 700 milliliters of benzene were charged into a one-liter one-neck flask. A Dean-Stark trap and water condenser were fitted to the flask.

The reaction mixture was then refluxed 25 hours and 24 minutes, after which 4.5 milliliters of water were collected. The cooled light-green solution recovered was washed five times with 250 milliliters of a 10% by weight $NaHCO_3$ solution.

The basic material was dried over Na_2SO_4 and stripped on a flash evaporator at less than 40°C to constant weight; 53.2 grams of light-yellow liquid was recovered. Next 45.3 grams of the light-yellow liquid was placed in a separatory funnel to which 500 milliliters of carbon tetrachloride was added and washed ten times with 250 milliliters of tap water.

The washed material was dried over Na_2SO_4. The material was then filtered and stripped on a flash evaporator at less than 45°C. The filtered material was a yellow liquid and weighed 15.8 grams. Upon examination of this material by infrared spectrographic analysis it was determined that the synthesis of phenylglyoxyethyl methacrylate had been performed.

Example 2: Light Sensitivity Testing – A solution of 3 grams of the end product of Example 1, plus 97 grams of a 50/50 mixture of methyl ethyl ketone and acetone was prepared. The solution was coated on an aluminum brush grained plate two times. The plate was exposed 10 minutes to pulsed xenon light through a step wedge. The exposed plate was then developed with a sodium silicate solution. Small print on the step wedge was clear.

3-Sorboyloxy-2-Hydroxypropyl Groups

G.E. Green, U.S. Patent 3,936,366; February 3, 1976; assigned to Ciba-Geigy Corporation describes compounds having at least three 3-sorboyloxy-2-hydroxypropyl groups directly attached to ether oxygen atoms which are polymerized by exposure to actinic radiation, preferably in the presence of a sensitizer such as Michler's ketone or benzoin.

The compounds may be obtained by the reaction either of sorbic acid with a substance having at least three glycidyl ether groups or of glycidyl sorbate with a substance having at least three phenolic or alcoholic hydroxyl groups. If desired, not all of the glycidyl groups may be consumed, so that, after actinically-induced polymerization the epoxide-containing polymer may be crosslinked by reaction with a curing agent for epoxide resins. The compounds are useful in making printed circuits or printing plates for offset printing.

Example 1: A mixture of 85 g of an epoxy novolak resin (having an epoxide content of 5.48 equiv/kg and being a polyglycidyl ether made from a phenol-formaldehyde novolak of number average molecular weight 420), sorbic acid (56 g), triethylamine (1.4 g), hydroquinone (0.14 g), and toluene (400 g) was heated under reflux for five hours, by which time the epoxide content of the mixture had fallen to a negligibly low value. Toluene (562 g) and acetone (321 g) were added to give a clear (10%) solution of the polysorbate. To this solution was added Michler's ketone as sensitizer (6.75 g, that is, 5% of the weight of the polysorbate) and the composition was used to prepare a printed circuit in the following manner.

A copper-clad laminate was coated with the composition and the solvent was allowed to evaporate, leaving a film about 10 μm thick. This film was irradiated for 30 seconds through a negative using a 125-watt medium pressure mercury lamp at a distance of 230 mm. After irradiation the image was developed in a mixture of acetone and toluene (1:3), washing away the unexposed areas to leave a good relief image on the copper. The uncoated copper areas were then etched using an aqueous solution of ferric chloride (60% w/v $FeCl_3$) containing concentrated hydrochloric acid (10% v/v).

Example 2: A mixture of 50 g of the tetraglycidyl ether of 1,1,2,2-tetra(p-hydroxyphenyl)ethane (having an epoxide content of 5.2 equiv/kg), sorbic acid (28.8 g), triethylamine (0.8 g), hydroquinone (0.1 g) and toluene (200 g) was heated under reflux for five hours, by which time the epoxide content was negligible. A solution (10%) was made up as in Example 1 and tested. A good image was obtained after three minutes' irradiation and was developed with ethanol.

Polyester-Polyether Block Copolymer

J. Ibata, H. Kobayashi, K. Toyomoto, K. Suzuoki, Y. Hayashi and M. Kurihara; U.S. Patent 3,960,572; June 1, 1976; assigned to Asahi Kasei Kogyo Kabushiki Kaisha, Japan provide a photopolymerizable composition comprising a polyester-polyether block polymer which is useful for preparing relief images and printing plates.

Example: Two hundred grams of polyethylene adipate diol having a number average molecular weight of about 2,000 were reacted with 34.8 grams of a mixture of 2,4-tolylene diisocyanate and 2,6-tolylene diisocyanate in a weight ratio of 80 to 20 in the presence of 0.5 gram of dibutyltin dilaurate at 70°C for two hours under a nitrogen atmosphere with stirring to give an isocyanate-terminated polyethylene adipate.

Then the resulting polyethylene adipate was further reacted with 100 grams of polypropylene glycol diol having a number average molecular weight of about 2,000 at 70°C for two hours under a nitrogen atmosphere with stirring to give an isocyanate-terminated polyester-polyether block polymer having a number average molecular weight of about 6,500. Then 300 grams of the resulting polyester-polyether block polymer were reacted with 25 grams of 2-hydroxyethyl methacrylate in the presence of 0.1 gram of hydroquinone at 70°C for two hours to give a prepolymer.

To 200 grams of the prepolymer, there were added 40 grams of 2-hydroxethyl methacrylate, 30 grams of n-butyl acrylate, 5 grams of acrylamide and 4.0 grams of benzoin ethyl ether and the mixture was thoroughly mixed to give a uniform photosensitive composition.

On a transparent glass sheet, 10 millimeters in thickness, there was placed a negative film and the negative film was covered with a polyester film, 10 microns in thickness, and a spacer 3 millimeters in thickness was placed thereon. Then, the resulting photosensitive composition was charged thereto and a transparent glass sheet, 10 millimeters in thickness, was placed on the spacer. Subsequently the resulting assembly was exposed to a 270 watt super high pressure mercury lamp set at a distance of 30 centimeters from each of the glass sheets at room temperature, first the upper side was exposed for 18 seconds and then the under side was exposed.

Then the glass sheets, the negative film and the film covering the negative film were removed from the assembly and the unexposed portions of the photosensitive layer were washed out with a 2% sodium hydroxide solution and the resulting plate was dried in hot air. The relief images of the flexographic printing plate thus obtained had a height of 2.4 millimeters and a Shore hardness A of 60 and the resolution of the plate was 0.1 millimeter or more and the inking of the plate was also good.

The physical properties of the flexographic printing plate were as follows:

Tensile strength	135 kg/cm^2
100% modulus	54 kg/cm^2
Elongation	480 %
Elongation set	4 %
Impact resilience	52 %

Using the printing plate obtained by backing the resulting plate with a rubber sheet, 3 millimeters in thickness, a printing test was carried out with corrugated cardboards to give at least 500,000 clear and precise prints.

Complete descriptions of the polyesters and polyethers usable in the process and many additional examples are included in the patent.

Alkoxyaromaticglyoxy Groups

T.M. Muzyczko and T.H. Jones; U.S. Patent 3,969,119; July 13, 1976; assigned to The Richardson Company provide a photoreactive composition containing an effective amount of a compound having at least two alkoxyaromaticglyoxy substituents per molecule. The substituents have the following general formula:

$$-[C(R)_2]_n-O-Ar-\overset{\overset{\displaystyle O}{\|}}{C}-\overset{\overset{\displaystyle O}{\|}}{C}-OM$$

wherein R is selected from the class consisting of H, aryl, alkyl, halo and aralkyl having up to 10 carbon atoms, n is an integer from 1 to 18, Ar is an aromatic substituent, and M is selected from the class consisting of H, alkali metal, ammonium and substituted ammonium. These compositions are useful in a wide variety of photochemical and photomechanical processes and are particularly suited for use as photopolymers, photoinitiators and photosensitizers in light-sensitive coatings of presensitized lithographic plates.

Example 1: Preparation of Glycerol-1,3-Diphenyl Ether – Phenyl glycidyl ether (151 grams or 1 mol), phenol (97.5 grams or 1.03 mols), and powdered potassium hydroxide (0.6 gram or about 0.2% of the total charge), were combined in a flask equipped with a condenser, thermometer, stirrer and nitrogen inlet. A slow flow of nitrogen gas was maintained over the reaction mixture throughout the reaction. The mixture was heated slowly to 110° to 130°C and held there for two hours. It was further heated to 150°C for one and one-half hours to complete the reaction. The reaction mixture was cooled to about 60°C and poured into a large tray whereupon it crystallized rapidly. The product was ground up and slurried into water and filtered, washed with dilute sodium hydroxide to remove excess phenol and washed with several liter portions of water to give the crude product. The crude product was recrystallized from a liter of 80% alcohol and water to give 208 grams, or 85% yield, of purified crystals melting at 81° to 83.5°C.

Preparation of Poly(Phenyl Glycidyl Ether) – In a resin kettle equipped with condenser, stirrer, and nitrogen inlet is placed 200 milliliters of xylene, glycerol-1,3-diphenyl ether (14.7 grams or 0.06 mol, prepared as above), phenyl glycidyl ether (270.4 grams or 1.8 mols), and powdered potassium hydroxide (0.65 gram or about 0.2% of the total charge).

A slight flow of nitrogen is maintained during the entire reaction. The mixture is heated rapidly to about 110° to 120°C and held there for several hours and then heated up to the reflux temperature, or about 145°C. Then the condenser is reversed and the xylene is distilled off of the reaction mixture. The remaining reaction mixture is heated slowly to 160°C with continued nitrogen flow and vacuum is applied to remove any excess phenyl glycidyl ether remaining behind in the polymer. The reaction mixture is cooled and poured into a jar.

The yield of polymer is nearly quantitative. The polymer is used without further purification. This polymer has a theoretical degree of polymerization of about 30 since an excess of 30 mols of phenyl glycidyl ether was used relative to the glycerol-1,2-diphenyl ether.

Preparation of Poly(p-Glyoxy Phenyl Glycidyl Ether) – In a resin kettle equipped with a motor driven stirrer, nitrogen inlet, condenser and gas outlet, 250 milliliter addition funnel and thermometer is placed 300 milliliters of dry nitrobenzene and 26.6 grams or 0.18 equivalent of poly(phenyl glycidyl ether).

The apparatus is flushed with nitrogen gas and the reaction vessel is cooled with ice and 25 grams (or 0.18 mol) of ethyl oxalyl chloride is added. In the addition funnel is placed a solution of 39 grams (or 0.29 mol) of anhydrous aluminum chloride dissolved in about 120 milliliters of nitrobenzene. The flow of dry nitrogen gas is maintained throughout the reaction.

The reaction kettle is then cooled down to 0° to 5°C and the addition of the aluminum chloride solution is started. The solution is added over a period of about 20 minutes, with continued stirring. After completion of the addition of the aluminum chloride solution, the reaction mixture is allowed to warm slowly to room temperature. Shortly after the addition is complete, the mixture thickens up to a soft gel, but stirring is maintained for the remainder of the day and is then discontinued overnight.

The following day the gel-like reaction mixture is dropped in small portions into a mixture of ice and dilute hydrochloric acid with vigorous stirring. The resulting solution is then steam distilled to remove all of the nitrobenzene. Ethyl acetate is then added to the warm solution remaining after the steam distillation to dissolve the suspended polymer. When the polymer is completely dissolved, the solution is then placed in a separatory funnel and the two layers separated.

The aqueous layer is then extracted with two more portions of ethyl acetate. The combined ethyl acetate extracts were dried over anhydrous sodium sulfate; the aqueous layer was then discarded. The ethyl acetate is filtered to remove the sodium sulfate drying agent and then a solution of 80% saturated sodium bicarbonate is added and these are stirred together vigorously. The solution is poured into a separatory funnel and the aqueous layer separated from the ethyl acetate layer.

The ethyl acetate layer is then washed with two more portions of sodium bicarbonate solution. The combined sodium bicarbonate solutions are then vacuum evaporated to remove all the ethyl acetate. The aqueous solution is then cooled with ice and hydrochloric acid is added in sufficient quantity to bring the pH to one thus precipitating the polymer from the aqueous solution. The polymer is filtered off, washed repeatedly with deionized water and dried. The yield of polymer was 27.4 grams, having a neutralization equivalent of 294. From this it can be calculated that the percent of glyoxylation was 67% and the yield of polymer from the reaction was 78% of the theoretical.

Example 2: One gram of polymer prepared in Example 1 was dissolved in 25 grams of acetone and 24 grams of methyl ethyl ketone to make a coating solution. This solution was flow coated on a brushed grain aluminum plate previously treated to render the surface hydrophilic and was allowed to drain and dry at room temperature, then for several minutes in a warm oven.

The resulting plate was then exposed for eight minutes on a Nuarc FT 40 Flip-Top Platemaker with a pulsed xenon light source having 4,000 watts input power to the lamp through a negative. The exposed plate was then developed merely by swabbing it with a standard lithographic desensitizer and then gumming it with a standard gum asphaltum etch.

The plate was then mounted on a printing press. Plate exposure was such that seven solid steps printed on a lithographic sensitivity guide. Printing was continued until the first evidence of image wear, which occurred on a 300-line 20% screen after 38,000 impressions.

RESIST COMPOSITIONS

Gold Compositions

The composition of *J.J. Felten; U.S. Patent 3,877,950; April 15, 1975; assigned to E.I. du Pont de Nemours and Company* is a printable paste compatible with thick-film techniques used for fabricating electrically conductive patterns and layers on substances for electronic circuits.

The conductor composition comprises finely divided gold and an inorganic binder therefor dispersed in a photosensitive vehicle. The vehicle comprises by weight

(a) 15 to 45 parts of a polymer having an inherent viscosity, in chloroform at 25°C, in the range of 0.15 to 0.95, said polymer being selected from the class consisting of polymethyl methacrylate, polyethyl methacrylate, and mixtures thereof,

(b) 55 to 85 parts of a solvent for the above polymer, the solvent being selected from the class consisting of dihydroterpineol, benzyl alcohol, tetralin, and mixtures thereof,

(c) a monomer selected from the class consisting of di-, tri-, and tetraethylene glycol diacrylate and mixtures thereof, the weight of monomer being 3 to 20% of the total weight of polymer plus solvent,

(d) a photoinitiator for the monomer in an amount effective to initiate polymerization thereof.

When gold compositions are referred to herein, compositions which are predominately gold, but may comprise other noble metals such as platinum or palladium (as mixtures or alloys with gold), are intended.

Where relatively thick films are to be printed the photosensitive vehicle in said gold composition additionally comprises

(e) an organometallic compound of a metal selected from the class consisting of Hf, Zr, Ti, P and mixtures thereof.

In the gold compositions the amount of organometallic compound is preferably such that the weight ratio of metal in the organometallic compound to photoinitiator is in the range of 1:10 to 1:2.5. The optimum ratio is in the range of 1:8.3 to 1:4.

Where the solvent is dihydroterpineol, benzyl alcohol or a mixture thereof, said vehicle may additionally comprise 1 to 12 parts of polymeric rheology modifier selected from the class consisting of ethylcellulose and polybutyl methacrylate.

Preferred gold compositions are those wherein the vehicle comprises 20 to 35 parts polymer, 65 to 80 parts solvent and an amount of monomer equal to 4 to 16% of the total weight of polymer plus solvent. It is preferred that the polymer have an inherent viscosity, in chloroform at 25°C, in the range of 0.15 to 0.5 and that the weight of photoinitiator be about 5 to 50% of the weight of the monomer.

In the gold composition it is preferred that the weight of gold plus inorganic binder comprise 75 to 85% and the vehicle 15 to 25%. It is further preferred that the inorganic binder be 0.5 to 10% of the weight of the gold (or gold plus any other noble metals, if any).

In the examples all parts, percentages and ratios are by weight, unless otherwise stated. The same mixture of monomers is used throughout the examples, a 3:2 mixture of diethylene glycol diacrylate and tetraethylene glycol diacrylate. The same binder was used in all, a 45:55 mixture of a glass (59.6% CdO, 7.3% Na_2O, 16.5% B_2O_3, 2.3% Al_2O_3, 14.3% SiO_2) and bismuth trioxide.

Example 1: A gold powder with spherical particles ranging in size from 0.4 to 4 microns was dispersed in a mixture of solvent, inorganic and polymeric binders, monomer and sensitizer by passing over a three-roll mill until dispersed. Composition A is an example of this invention; Composition B is a comparative example not of the process.

The photoinitiator was benzoin methyl ether (BME), for Composition A as a 50% weight solution in dihydroterpineol (DHT) and for Composition B as a 33% solution in tetralin (Tet). The polymer in Composition A was a 40% solution in Tet of low molecular weight polyethyl methacrylate with inherent viscosity of 0.2 in chloroform at 25°C; the polymer in Composition B was a 20% solution in DHT of very high molecular weight polyisobutyl methacrylate (inherent viscosity of 1.1 in chloroform at 25°C). Compositions are as set forth in the table below.

	Composition A	Composition B
Gold	75.0	74
Inorganic binder	3.0	3.1
Polymer solution	20.0	20.7
Polymer	(8.0)	(4.1)
Solvent	(12.0)	(16.7)
Monomers	1.4	1.5
Initiator solution	0.6	0.5
Initiator	(0.2)	(0.25)
Solvent	(0.4)	(0.25)
Total solvent	12.4	16.9
Silane dispersing agent	-	0.1

The compositions were screen printed to cover the area in which the pattern was to be generated, using a 230 mesh nylon screen. Coatings were dried at 50° to 60°C for 15 to 20 minutes. The weight of the dried coatings was 40 to 50 milligrams per square inch. Substrates then were exposed for 30 to 45 seconds through a photographic mask bearing an image of the conductor pattern to be generated, exposing with a collimated light beam from a mercury lamp. During exposure, the coated substrate was kept under a nitrogen blanket to exclude oxygen, which inhibits the reaction occurring during exposure.

Sample A was developed (the unexposed portion was removed) by a perchloroethylene spray from a hand spray gun driven by 100 psi air pressure. Distance of gun from substrate was six inches; development time was 4 to 6 seconds. The developed pattern was dried with a jet of compressed air. Substrates were fired in a three-zone moving belt furnace, spending 5 minutes residence in each zone (450°C, 650°C and 850°C) with 5 more minutes for cooldown.

In the samples produced using Composition A of this process, the fired thickness was typically 0.15 to 0.20 mils. Sheet resistivity was 10 to 15 milliohms per square at this thickness. Fired gold films were readily bonded to gold wire by thermal compression. Resolution of 1-mil wide lines on 4-mil centers (the centers of each of two parallel lines were 4 mils apart) was readily and cleanly achieved.

In the case of comparative Composition B, development with high pressure spray (100 psi) completely washed the exposed pattern off the substrate. When a low pressure (less than 10 psi) perchloroethylene spray propelled by gas from an aerosol can was used, 1-mil lines on 4-mil centers could not be cleanly resolved without breaks occurring in the lines, resulting in useless structures. Two-mil lines on 5-mil centers were not reproduced as well with Composition B as with Composition A. More undeveloped material remained between the conductor lines when Composition B was used than when Composition A was used. Occasional breaks were seen even with 2-mil lines of Composition B.

Example 2: This illustrates the effect of addition of an organometallic titanium compound on adhesion of developed patterns during the firing process, where the fired patterns are about twice as thick as those of Example 1.

Two typical compositions of this process were formulated, one with no titanium (A), the other with about 1% Engelhard titanium resinate (B), based on the total composition weight. The ratio of the weight of the Ti to weight of initiator was 1:5. The compositions were prepared as in Example 1. The polymer solution was 25.5 parts low molecular weight polyethyl methacrylate (I.V. 0.2 in chloroform at 25°C) plus 3 parts very high molecular weight polyisobutyl methacrylate (PBMA, I.V. 1.1 in chloroform at 25°C) in 71 parts DHT. Compositions were as follows, where TBAQ is tert-butylanthraquinone:

	Composition A	Composition B
Gold	78.1	77.3
Inorganic binder	4.6	4.6
Polymer solution	15.6	15.5
Polymer	(3.9)	(3.9)
Solvent	(11.2)	(11.2)
PBMA	(0.5)	(0.5)
Monomers	1.3	1.3
Initiator (TBAQ)	0.37	0.37
Titanium resinate	0	0.98

Substrates were printed through an 83-mesh nylon screen and dried 45 minutes at 55°C. Coated substrates were exposed through a mask to an ultraviolet lamp as in Example 1, but for 60 seconds, under nitrogen and developed with a hand spray gun (100 psi) using tetrachloroethylene solvent. Substrates thus developed were then fired as in Example 1.

Substrates with Composition A showed extensive catastrophic delamination of sintered gold at pattern edges, while substrates coated with Composition B showed no delamination. Fired thickness was about 0.4 mil; fired resistivity was 3.0 to 3.5 milliohms per square at this thickness.

Arylated Polysulfones

M.J.S. Bowden and E.A. Chandross; U.S. Patent 3,884,696; May 20, 1975; assigned to Bell Telephone Laboratories, Inc. describe a method of forming a patterned resist on the surface of a substrate which comprises exposing a portion of an arylated polysulfone film on said substrate to ultraviolet radiation having a wavelength within the range of 1700 to 4000 Angstroms for a time period sufficient to lower the molecular weight of the polysulfone and selectively removing the exposed portion from said substrate. The polysulfone is selected from the group consisting of

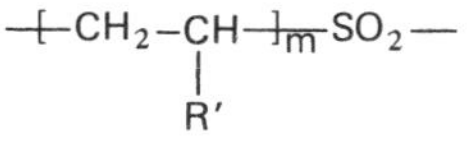

$$-\!\!\left[-CH_2-\underset{\underset{R'}{|}}{CH}-\right]_m\!-SO_2-$$

and

$$-\!\!\left[-\underbrace{CH-CH}_{R'}-\right]_m\!-SO_2-$$

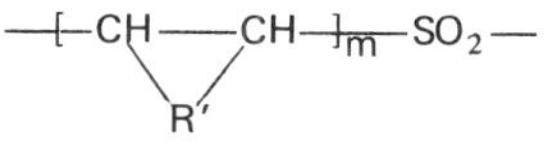

wherein m is an integer having a value of at least 1, R is an aromatic hydrocarbon radical selected from the group consisting of phenyl, biphenyl, naphthyl, phenanthryl, fluoroanthryl and anthracyl radicals and R' is a 1,8 naphthylidene radical.

Example: A solution of acenaphthylene (8.75 grams, 0.057 mol) in freshly distilled styrene (30 grams, 30.29 mols), containing azobisisobutyronitrile (0.1 gram) and dimethylformamide (5 milliliters) was outgassed by several freeze-thaw cycles. Then approximately 0.8 mol (50 grams) of sulfur dioxide was added to the solution and its container sealed, immersed in a water bath and maintained at 50°C for a time period of 72 hours. The resulting poly(styrene-acenaphthylene-sulfone) (3:1:1) was precipitated into methanol. Then, it was redissolved twice in dioxane, reprecipitated and finally dried under vacuum at room temperature.

The resultant polymer was dissolved in methoxyethyl acetate and spun upon a substrate comprising silicon dioxide on silicon using a conventional photoresist spinner to a thickness of approximately 5000 Angstroms. The film was prebaked at 180°C for 30 minutes and then irradiated through a mask with a collimated light from a 200 watt high pressure mercury arc source. The light was filtered through a Corning 7-56 filter which transmits the 220 to 400 nm range of the optical spectrum, the intensity of the resist surface being 20 milliwatts per square centimeter as measured with a thermopile.

Development of the exposed film was then effected by spraying the irradiated areas of the film with a solution comprising 60% dioxane and 40% isopropyl alcohol for 30 seconds.

Resolution was determined by exposing the resist through a contact mask and developing the image as described. Finally, the resist was postbaked at 180°C for 30 minutes and etched in buffered hydrofluoric acid. The resist absorbed strongly at 294 nm and continued to absorb out to 330 nm, the sensitivity being approximately 500 mJ/cm^2.

Preformed, Alkali-Soluble Binder

M.N. Gilano, R.E. Beaupre and M.A. Lipson; U.S. Patents 3,887,450; June 3, 1975 and 3,953,309; April 27, 1976; both assigned to Dynachem Corporation developed a light-sensitive composition which can be readily developed by means of an alkaline aqueous solution to yield a product which is useful for printing plates and photoresists. These compositions are particularly useful for the manufacture of printed circuits because the resists formed therefrom are impervious to conventional plating solutions.

The advantages of the process are obtained by selecting a preformed compatible macromolecular polymeric binding agent which is a copolymer of a styrene-type or a vinyl monomer and an unsaturated carboxyl-containing monomer. The use of this composition completely eliminates the need for organic solvents and provides a highly solvent-resistant resist.

The photopolymerizable compositions are composed of from 10 to 60 parts by weight of a conventional addition photopolymerizable nongaseous ethylenically unsaturated compound; 40 to 90 parts by weight of the aforesaid binding agent; from 0.001 to 10 parts by weight of a conventional free radical initiator; and from

0.001 to 5 parts by weight of a conventional thermal addition polymerization inhibitor. Additionally, the compositions may contain suitable dyes and pigments and other additives, such as plasticizers and adhesion promoters.

The vinyl monomer constituent of the polymeric binder is nonacidic and has the general formula

$$\begin{array}{c} CH_2{=}C{-}Y \\ \quad | \\ \quad X \end{array}$$

wherein, when X is hydrogen, Y is $OOCR_1$, OR_1, OCR_1, $COOR_1$, CN, $CH{=}CH_2$, $CONR_3R_4$ or Cl; when X is methyl, Y is $COOR_1$, CN, $CH{=}CH_2$, or $CONR_3R_4$; and when X is chlorine, Y is Cl; and wherein R_1 is an alkyl group having from 1 to 12 carbon atoms, a phenyl or a benzyl group and R_3 and R_4 are hydrogen, an alkyl group having 1 to 12 carbon atoms or a benzyl group.

The second comonomer may be one or more unsaturated carboxyl containing monomers having from 3 to 15 carbon atoms, preferably from 3 to 6. Most preferred compounds are acrylic acid and methacrylic acid.

The ratio of the styrene or vinyl component to the acidic comonomer is selected so that the copolymer is soluble in the aqueous alkali medium. As a convenient criterion, the binder copolymer should be such that a 40% solution in ketones or alcohols will have a viscosity of from 100 to 50,000 centipoises.

Representative comonomer ratios are 70:30 to 85:15 for styrene-acrylic acid or methacrylic acid; 35:65 to 70:30 for styrene-monobutyl maleate and 70:30 to 95:5 for vinyl acetate-crotonic acid. The degree of polymerization of the binder copolymer is such that binder forms a nontacky continuous film after exposure and development.

Broadly, the molecular weight is from 1,000 to 500,000. The ranges for the copolymer ratios and the degree of polymerization for the particular binders can be readily ascertained by testing the solubility in the dilute alkali solution of representative polymers. This represents a molecular weight of from about 1,000 to 500,000.

Example: The following solution was coated onto a one-mil thick polyester film and dried in air. The dry thickness of the sensitized layer was about 0.001 inch. The dried layer was covered with a one-mil thick polyethylene film.

	Grams
Copolymer*	67.0
Trimethylopropane triacrylate	22.0
Tetraethylene glycol diacrylate	11.0
Benzophenone	2.3
4,4'-bis(dimethylamino)benzophenone	0.3
2,2'-methylenebis(4-ethyl-6-tert-butyl phenol)	0.1
Methyl violet 2B base	0.07
Benzotriazole	0.20
Methyl ethyl ketone	140.0

* 37% styrene and 63% monobutyl maleate, average MW 20,000, viscosity of 10% aqueous solution of ammonium salt = 150 cp

A piece of copper clad, epoxy-fiber glass board was cleaned by scouring with an abrasive cleaner, swabbing and thoroughly rinsing in water. It was then given a 20-second dip in a dilute hydrochloric acid solution (2 volumes water plus 1 volume concentrated hydrochloric acid), a second rinse with water and then dried with air jets.

The polyethylene cover film was removed from a section of the sandwiched photopolymerizable element. The bared resist coating with its polyester support was laminated to the clean copper with the surface of the photopolymerizable layer in contact with the copper surface. The lamination was carried out with the aid of rubber covered rollers operating at 250°F with a pressure of 3 pounds per lineal inch at the nip at a rate of 2 feet per minute.

The resulting sensitized copper clad board protected as it is by the polyester film could be held for later use if need be. Actually it was exposed to light through a high-contrast transparency image in which the conducting pattern appeared as transparent areas on an opaque background. The exposure was carried out by placing the sensitized copper clad board (with its polyester film still intact) and the transparency into a photographic printing frame.

The exposure was for a period of 45 seconds to a 400-watt, 50-ampere mercury vapor lamp at a distance of 12 inches. The polyethylene terephthalate support film was peeled off and the exposed resist layer developed by agitating the board in a tray containing 2% sodium carbonate in water for 3½ minutes followed by a water rinse. The resulting board contained a dyed resist pattern of the clear areas of the exposed transparency.

The board was now etched with a 45° Baumé solution of ferric chloride, then rinsed and dried. The resist was removed from the remaining copper by dipping for 2 minutes in a 3% solution of sodium hydroxide in water at 70°C. The result was a high quality printed circuit board.

Photoelectropolymerization Using Zinc and Alkali Metal Sulfite

The process of *S. Levinos, U.S. Patent 3,909,255; September 30, 1975; assigned to Keuffel & Esser Company* provides a system of photoelectropolymerization which can utilize the desirable zinc oxide photoconductors in their most effective cathodic roles, yet can achieve polymerization at a support surface by providing an anodic polymerization reaction.

In accordance with the process, a conductive zinc or zinc-containing support is coated with a layer of a composition comprising a polymerizable vinyl monomer and an alkali metal sulfite. Combining the resulting sheet material in the usual manner with a zinc oxide photoconductor layer on a substantially transparent conductive support, and a source of electric current, yields a system of potential electrolysis which can be activated by exposure of the photoconductor to light, usually in the form of the image to be reproduced.

As the result of illumination, and thus the completion of the electrical circuit in the system, electrolysis of the imaging sheet composition causes the generation of zinc ions at the carrier anode with formation of zinc sulfite in the immediate vicinity of the anode and a resulting initiation of polymerization in that region of the composition.

The initial lack of zinc ions throughout the composition mass renders an acidic environment in the polymerizable material unnecessary, yet the noted reaction at the acid electrode, that is, the anode, provides the desirable acid conditions during the period of actual polymer formation. Thus, the polymerizable material may be maintained in a more storage-stable neutral pH condition. The material is particularly useful in negative-working imagery and in the preparation of patterned resist layers.

Example 1: The following formulation was prepared:

Deionized water	100 ml
Gelatin (inert, high Bloom)	4.5 g
Sodium dodecylbenzenesulfonate (20% solution)	5 ml
Polyvinyl pyrrolidone	0.8 g
Sodium sulfite (anhydrous)	1.0 g
Acrylamide	2.1 g
N,N'-methylenebisacrylamide	0.4 g
Glycerin	2.5 g

Each of the ingredients was added to the deionized water in the order given, with stirring to effect complete dissolution of all the solid components. The pH of this mixture was 11.5 and it was adjusted with dilute sulfuric acid to a value of 7.0.

The composition was coated at a wet thickness of about 150 μm on a clean sheet of zinc metal. The coating was then allowed to air dry. A metal linotype printing slug was placed on the coating and was connected to the negative terminal of a DC power supply. The zinc sheet was made the anode of the system by connecting it to the positive terminal of the power supply. A potential of 20 volts was applied between the two electrodes for a period of 15 seconds.

Nonelectrolyzed areas of the coating were then removed by washing with water at a temperature of about 30°C. Polymerized areas were then rendered visible by staining with a 1% aqueous solution of methylene blue. A short water rinse was used to remove remnant dye solution from the surface of the plate and provide a vivid image upon the zinc sheet.

Example 2: The coated side of a sheet prepared in the manner disclosed above was placed in intimate contact with the carbon coating of a dye-sensitized zinc oxide photoconductive plate such as earlier described. With the zinc carrier of the imaging sheet as anode and the conductive surface of the glass panel as cathode, these two elements were arranged in electrical circuit with a 75-volt DC potential. The glass-plate panel of this assembly was then exposed to a projected 15X negative image (500-watt tungsten lamp source) for a period of about 5 seconds.

The coated zinc sheet was then removed from the assembly and the coating was washed in clear warm water for about 1 minute, during which time portions of the coating corresponding to the unexposed areas of the photoconductor layer were removed from the carrier sheet. There thus remained, upon the carrier, a good quality, enlarged polymeric positive image of the original negative.

Specific Glass Transition Temperature

The process of *R.J. Faust and K.W. Klüpfel; U.S. Patent 3,930,865; Jan. 6, 1976; assigned to Hoechst Aktiengesellschaft, Germany* relates to a photopolymerizable copying composition comprising at least one polymerizable compound, at least one photoinitiator, and at least one copolymer of (a) an unsaturated carboxylic acid, (b) an alkyl methacrylate with at least 4 carbon atoms in the alkyl group and (c) at least one additional monomer which is capable of copolymerization with monomers (a) and (b), the homopolymer of the additional monomer having a glass transition temperature of at least 80°C.

According to a preferred embodiment, the copying composition comprises a terpolymer in which the component (c) is styrene, p-chlorostyrene, vinyl toluene, vinyl cyclohexane, acrylamide, methacrylamide, N-alkylacrylamide, phenyl methacrylate, acrylonitrile, methacrylonitrile, or benzyl methacrylate, styrene being preferred.

Monomer component (a) is used in concentrations between 10 and 40% by weight, component (b) is preferably used in concentrations ranging from 35 to 83% by weight, and component (c) is used in concentrations between about 1 and 35% by weight.

The copying layers produced with the copying compositions according to the process are distinguished in that they possess an excellent adhesion to metallic supports, especially to copper, after exposure, are very flexible, and have an excellent adsorption capacity for liquid monomers. Further, the copolymers have the important advantage that the flexibility of the resist layer can be adjusted as desired, that is, the desired consistency of the layer may be selected within a wide range, by selecting adequate ratios between the concentrations of the monomers.

Example: A solution of 5.6 parts by weight of a copolymer of 50 parts by weight of methacrylic acid, 100 parts by weight of 2-ethylhexyl methacrylate and 15 parts by weight of acrylonitrile, 5.6 parts by weight of a monomer produced by reacting 1 mol of 2,2,4-trimethylhexamethylene diisocyanate with 2 mols of hydroxyethyl methacrylate, 0.2 part by weight of 9-phenylacridine, 0.15 part by weight of triethylene glycol dimethacrylate, 0.015 part by weight of Michler's ketone and 0.06 part by weight of 2,4-dinitro-6-chloro-2'-acetamido-5'-methoxy-4'-(β-hydroxyethyl-β'-cyanoethyl)aminoazobenzene in 13.0 parts by weight of methyl ethyl ketone and 40.0 parts by weight of ethylene glycol monoethyl ether is whirler-coated onto a 25-micron thick biaxially stretched polyethylene terephthalate film and dried for two minutes at 100°C. The resulting layer is 18 microns thick. The layer thus applied is flexible and its surface is not tacky at room temperature.

The material thus produced may be used in this form as a dry resist film. For this purpose, it is laminated at a temperature between 115° and 130°C onto a phenoplast laminate to which a 35-micron thick copper foil had been applied. The plate is then exposed for 10 to 30 seconds in a xenon copying apparatus (type Bikop, model Z, 8 kilowatts), the distance between the lamp and the printing frame being 80 centimeters. A screen test plate having lines of a width from 4 millimeters down to 5 microns is used as the original. After development, the polyester layer is pulled off and the copying layer is developed with an aqueous alkaline developer having the following composition: 1,000 parts by weight of water, 15 parts by weight of sodium metasilicate nonahydrate, 3 parts by weight of polyglycol 6000, 0.6 part by weight of levulinic acid and 0.3 part by weight of strontium hydroxide octahydrate.

The pH value of this solution is 11.3. Development is effected by wiping for 60 to 100 seconds or by spraying. Finally, the plate is rinsed with water. Very sharp edged resist lines of the same dimensions as in the original are thus produced, lines down to a width of 50 microns being accurately reproduced.

The above described polymer binder may be replaced by the same quantity of a terpolymer of methacrylic acid, decyl methacrylate and 1-vinyl naphthalene (50:90:15 parts by weight), or of a terpolymer of methacrylic acid, decyl methacrylate and p-chlorostyrene (45:80:25 parts by weight), or of a terpolymer of methacrylic acid, n-hexyl methacrylate, and 1-vinyl naphthalene (40:75:15 parts by weight), or of a terpolymer of methacrylic acid, n-hexyl methacrylate and acrylonitrile (95:175:50 parts by weight). If the above described procedure is followed, in each case very sharp edged etch or electroresist masks of excellent adhesion are obtained after exposure and development.

Anhydride-Containing Group

F.W. Steele; U.S. Patent 3,933,746; January 20, 1976; assigned to Ball Corporation provides compositions which can be readily coated on supporting surfaces useful in making printing resists via photopolymerization. The process relates to polymer compositions comprising as an essential recurring unit an anhydride-containing group represented by the structure:

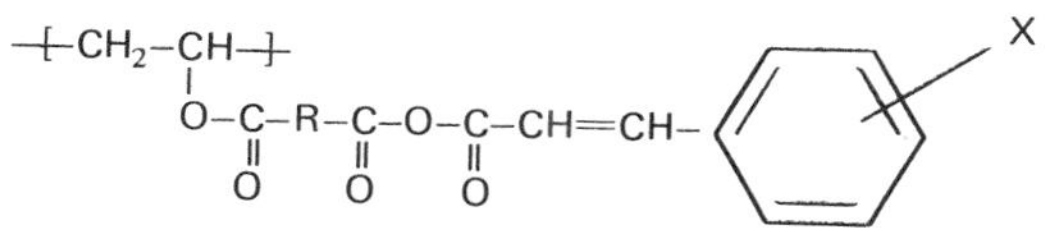

$$-\!\!\left[CH_2-\underset{\displaystyle O-\underset{\underset{\displaystyle O}{\|}}{C}-R-\underset{\underset{\displaystyle O}{\|}}{C}-O-\underset{\underset{\displaystyle O}{\|}}{C}-CH=CH-C_6H_4X}{\overset{|}{CH}}\right]\!\!-$$

wherein R is a residue of a dibasic acid selected from the group consisting of aliphatic dicarboxylic acids and aromatic dicarboxylic acids and X is selected from the group consisting of hydrogen, halogen, lower alkoxy, lower alkyl and nitro.

The polymers comprise at least 10 mol percent of the recurring unit of said polymer. Generally and more preferrably the polymer may comprise from 10 to 85 mol percent of said unit. The remaining recurring structural units are non-anhydride units and include any polymeric units derived from polymerization of vinyl alcohols with other polymerizable materials. The polymers herein may comprise from about 10 mol percent to about 50 mol percent of the recurring unit having the above structure.

Example 1: In a glass vessel 1 mol of polyvinyl alcohol having an average molecular weight of about 10,000 was slowly stirred for about 12 hours in 750 milliliters anhydrous pyridine at 80°C. One mol of Δ4-tetrahydrophthalic acid anhydride was dissolved in 850 milliliters anhydrous pyridine and added to the vessel containing the polyvinyl alcohol. The reaction mixture was heated at about 80°C, under constant stirring, for about 24 hours. Thereafter, the reaction mixture was cooled to 0°C and about 50 milliliters cinnamoyl chloride was added dropwise. After complete solution the reaction mixture was stirred at room temperature for 8 hours, filtered through cotton and the polymer thus formed precipitated into approximately 10 liters of cold distilled water.

The collected polymer was thoroughly washed several times with distilled water and vacuum-dried at room temperature. The polymer was thereafter dissolved in 1% aqueous ammonium hydroxide solution to obtain a weight-volume concentration of 5%. Thereafter, about 0.5 gram 2-benzoylmethylene-1-β-naphthothiazoline was added and thoroughly mixed therein.

Example 2: The polymer solution obtained from Example 1 was coated onto four different metal plates having a fine grain and one plastic substrate, namely, aluminum, zinc, magnesium, copper and a polyester film (Mylar). The coatings were allowed to thoroughly dry and then they were exposed to an ultraviolet light source under a graduated density step tablet to insolubilize the polymer coating in the areas of exposure. Thereafter, the unexposed polymer of the coating was removed from each plate by treatment with a dilute 0.5% ammonium hydroxide. The substrate thus treated contained an excellent visible image of the original step tablet.

Depolymerizable Aromatic 1,2-Dialdehyde

A radiation sensitive element is provided by *R.W. Nelson; U.S. Patent Application (Published) B 462,893; February 24, 1976; assigned to Eastman Kodak Co.* which comprises a depolymerizable polymer, such as a polymer or copolymer of an aromatic 1,2-dialdehyde and, optionally, a radiation-sensitive substance or combination of substances which, upon absorption of radiation, is capable of initiating the depolymerization reaction. The radiation-sensitive element may also contain a binder and/or a substance or combination of substances which is capable of forming a colored or fluorescent reaction product with the monomer produced in the depolymerization reaction.

This process also comprises a process of exposing to radiation a composition comprising a depolymerizable polymer, preferably o-phthalaldehyde and, optionally, a photosensitive substance or combination of substances which, upon absorption of radiation, gives a product capable of initiating the depolymerization reaction (for example, an acidic product).

After exposure, the composition is heated to a temperature of up to about 150°C, forming a monomer such as an aromatic 1,2-dialdehyde. A visible image can be formed during heating by the reaction of the released monomer with a color-forming substance or substances included in the element, by heating in contact with a separate element containing the color former, or by heat treatment in the presence of a gaseous color-forming reagent, such as ammonia. Alternatively, a fluorescent image may be produced by an appropriate choice of a substance or combination of substances which reacts with the monomer.

A positive-working radiation-sensitive process comprises exposing a radiation-sensitive element to radiation following which the element is heated or contacted with a solvent which removes the exposed areas but permits the unexposed areas to remain as a positive resist.

It is an advantage of this process that the depolymerization of polyphthalaldehyde, once it has been initiated, is irreversible above –43°C.

Example 1: A solution was prepared from

Polyphthalaldehyde	75.0 mg
Poly[4,4'-isopropylidene(bis-phenyleneoxyethylene)-co-ethylene terephthalate] 50/50	150.0 mg
Tetrabromomethane	20.0 mg
Triphenylmethane	20.0 mg
Diphenylamine	20.0 mg
Dichloromethane	1.6 ml
Chlorobenzene	0.6 ml

It was coated on a polyester film support with a doctor blade set at 5 mils and dried in air at 43°C. The resultant coating was then cured at 95°C for three minutes.

A sample of this coating was exposed to UV light for 10 seconds in a 3M Filmsort "Uniprinter 086" Copier through a step wedge test object. Following exposure, the sample was heated at about 90°C for 5 seconds and then was placed for 30 seconds in a chamber containing ammonia vapor at about 125°C. The resulting near-neutral negative image had a maximum visible density of 2.00, a minimum visible density of 0.06 and an average contrast of 1.41.

Example 2: A solution similar to that of Example 1 but containing α,α-dibromo-p-nitrotoluene in place of the combination of tetrabromomethane, triphenylmethane and diphenylamine was coated on a support as in Example 1. A sample of this coating was exposed for 10 seconds in a copier through a negative-appearing microphotograph.

The sample was developed by heating at about 90°C for 5 seconds, followed by 30 seconds of treatment with ammonia vapor at about 125°C. There resulted in the sample a sharp, grainless, positive-appearing copy of the original microphotograph, with minute details faithfully reproduced.

Cyclization Product of Butadiene plus Organic Solvent

M. Ichikawa, Y. Takeuchi, T. Miura, Y. Harita, M. Tashiro and T. Tsunoda; U.S. Patent 3,948,667; April 6, 1976; assigned to Japan Synthetic Rubber Co., Ltd., Japan provide solvent-developed type photosensitive compositions containing as an essential ingredient at least one selected from the group of the cyclization product of butadiene polymer or copolymer and organic solvent, with or without at least one member selected from the group consisting of photosensitizer and photosensitive crosslinking agent, both soluble in an organic solvent.

The solvent-developed type photosensitive composition of this process containing cyclization product of butadiene polymer or copolymer as an essential ingredient is so photosensitive, in comparison with conventional photosensitive materials containing "cyclized rubber" or cyclization product of polyisoprene as base polymer, that it may be applied effectively to chemical milling, electroplating, printed circuit and other metal-working treatment.

Example: One gram of a cyclization product (degree of cyclization 66%) of 1,4-cis-polybutadiene was dissolved in 100 milliliters of xylene and, after removal of impurities by filtration, predetermined amounts of 2,6-di(4'-azidobenzal)-4-methylcyclohexanone (1) and 1,9-benzanthrone (2) were added to the solution.

Each of the photosensitive solutions prepared in this way was applied dropwise over an aluminum plate to form a photosensitive film by rotary coating method. The film was dried and then was exposed to UV for 30 seconds through Photographic Step Tablet No. 2 by contact printing method. Then the surface was irradiated with ultraviolet rays for 30 seconds. The unexposed parts were removed by a toluene developer and specific sensitivity (number of resist steps) was determined. The results were as tabulated below:

Example No.	Photosensitive Crosslinking Agent (1), grams	Photosensitizer (2), grams	. . Specific Sensitivity . . . (1)	(2)	(1) + (2)
1	0.03	0	13.73	0	-
2	0.02	0.01	9.64	0	13.73
3	0.01	0.02	6.46	1.00	9.64
4	0.005	0.025	6.46	1.00	9.64
5	0	0.03	0	1.41	-

Keto-Olefin-SO_2 Copolymers

R.J. Himics, S.O. Graham and D.L. Ross; U.S. Patent 3,964,909; June 22, 1976; assigned to RCA Corporation found that films of copolymers of SO_2 and certain keto-olefins are excellent positive photoresists having useful sensitivity, adhesion and etch resistance characteristics. The photoresists form nontacky films from solution that do not require either prebaking or postbaking steps, with concomitant savings in processing.

The copolymers of the process comprise a copolymer of SO_2 and a keto-olefin of the formula

$$R_1-\overset{\overset{O}{\|}}{C}-\left[\underset{\underset{R_2}{|}}{CH}\right]_x-\overset{\overset{R_3}{|}}{C}=C\begin{matrix}R_4\\ \\ R_5\end{matrix}$$

wherein R_1 is alkyl, aryl or substituted aryl; R_2 is hydrogen or methyl; x is an integer of 1 or 2; R_3 is hydrogen or is part of a carbocyclic ring with R_4 or R_5 with the proviso that when x is 2, R_3 must be hydrogen; R_4 is hydrogen, alkyl or aryl and R_5 is hydrogen or (when R_4 is hydrogen) is alkyl or aryl, or R_4 and R_5 together are part of a carbocyclic ring.

Example: Preparation of Copolymer – One-tenth mol (9.81 grams) of 5-hexane-2-one was charged to a polymerization tube containing 15 milliliters of liquified SO_2. The tube was sealed and maintained at -78°C for sixteen hours. The resultant polymer softened at 100°C. Elemental analysis was as follows–Calculated for $C_6H_{10}SO_3$: C, 44.44%; H, 6.17%; S, 19.75%. Found: C, 44.37%; H, 6.51%; S, 19.36%.

The white polymer was dissolved in DMF and poured into ice water. It was soluble in 1,1,2,2,-tetrachloroethane and cyclopentanone and partly soluble in dioxane and 1,2-dichloroethane.

Preparation and Exposure of Photoresist – One solution of the copolymer prepared as above was made by dissolving 0.7 gram of the copolymer in 10 milliliters of cyclopentanone. A second solution was prepared containing 0.35 gram of the copolymer and 0.1 gram of benzophenone sensitizer in 5 milliliters of cyclopentanone.

Films of the resist about 1 micron thick were spun onto glass slides coated with a chromium-nickel film at 2,000 rpm with both solutions. The films were exposed through a typical metal television aperture mask using a mercury-xenon lamp having a wavelength range of about 2600 to 4800 Angstroms. The films were developed by immersing the exposed slides in a 2½% by volume mixture of toluene/1,2-dichloroethane for ten minutes.

The resist prepared from the copolymer alone gave a pattern which was not completely defined down to the substrate. The resist prepared from the sensitized copolymer gave a well-formed 12 to 13 mil (0.030 to 0.033 centimeter) hole pattern.

Polymerizable Siloxanes

E.D. Roberts and J.F. Steggerda; U.S. Patent 3,969,543; July 13, 1976; assigned to U.S. Philips Corporation describe a method of providing a patterned silicon-containing oxide layer on a substrate. The method comprises applying a coating of a polymerizable cyclosiloxane containing methyl groups bonded to silicon and an ultraviolet sensitive sensitizer capable of being converted to a basic reacting form capable of promoting polymerization of the polymerizable cyclosiloxane to an insoluble state in a solvent for the polymerizable cyclosiloxane, exposing desired areas of said coated substrate to ultraviolet radiation to thereby cause polymerization of the polymerizable cyclosiloxane to an insoluble state, treating said surface with a solvent capable substantially only of dissolving unexposed polymerizable cyclosiloxane and then converting the polymerized cyclosiloxane to a silicon-containing oxide.

Example: A slightly prepolymerized polyvinyl-cyclosiloxane mixture is prepared by hydrolysis of vinyltrichlorosilane followed by a treatment with a little ammonia, using the method described in United Kingdom Patent 1,316,711.

The polycyclosiloxane formed consists mainly of chain-units of siloxane rings with attached vinyl groups. They also contain an –H and –OH group in end-positions, so that the chain has an –OH group at each end. The average degree of polymerization lies between 2 and 6 chain units per molecule. The polycyclosiloxane solution has the form of a colorless viscous liquid. The viscosity is reduced by means of a suitable solvent using three parts by weight of methyl isobutyl ketone per part by weight of polycyclosiloxane.

To the polycyclosiloxane solution is added a sensitizer which, according to one embodiment, mainly consists of 2,6-di(4′-azidobenzal)cyclohexanone, or according to the other embodiment, consists mainly of 2,6-di(4'-azidobenzal)-4-methylcyclohexanone.

Ten milligrams of the sensitizer are added per milliliter of the 25% polyvinyl-cyclosiloxane solution. In order to obtain a silicon oxide pattern on a substrate (for example, on a silicon body as a step in the manufacture of semiconductor devices) a film of the polyvinyl-cyclosiloxane solution with sensitizer is provided on the substrate and the solvent is evaporated from said film. The film is then exposed to ultraviolet radiation in known manner through a suitable mask which has a pattern with radiation-impervious parts.

The use of exposure times of a few seconds with ultraviolet radiation in irradiation apparatus conventionally used for manufacturing photoresist patterns is sufficient. By a treatment with acetone the unexposed siloxane film parts are dissolved and the exposed film parts remain. By heating in an oxygen-containing atmosphere in a furnace, the remaining siloxane film parts are converted into silicon dioxide. The temperature and the time used are such that the uncovered parts of a silicon substrate are oxidized at most to a small extent. Any very thin oxide layer formed in that case can be removed rapidly, if desired, by means of a dip-etching treatment in a buffered HF solution while maintaining the silicon pattern obtained by conversion of the siloxane substantially throughout its layer thickness.

In the above-described manner, for example, patterns of the polyvinyl-cyclosiloxane material on silicon are obtained in layer thicknesses of about 6000 A or more. By conversion into silicon dioxide, patterns of silicon dioxide which have a masking effect during diffusion of phosphorus and boron are obtained on the silicon substrate.

ADDITIVES

Diacetone Acrylamide to Improve Adhesion

Photoresists with improved adhesion to a substrate are provided by *Y. Oba and T. Tsunoda; U.S. Patent 3,884,703; May 20, 1975; assigned to Hitachi, Ltd., Japan* by addition of diacetone acrylamide to a mixture of a water-soluble organic resin and a bisazide compound as a photochemical reaction initiator.

Examples of the known water-soluble organic resins are as follows: Polyvinylpyrrolidone; polyacrylamide; gelatin; water-soluble polyacrylates such as sodium salt; methylcellulose; poly-L-glutamate water-soluble salts such as sodium salts, ammonium salts, etc.; copolymers of vinyl alcohol and maleic acid; copolymers of vinyl alcohol and acrylamide.

The blending ratio of these known water-soluble organic resins and bisazide varies depending on the kind of raw materials. Any known photoresists in which a bisazide is used in place of bichromate as a photochemical reaction initiator can be used in this process.

Example: A photoresist solution was prepared by dissolving 1.5 parts by weight of polyacrylamide (Olefloc NP-1) and 0.03 part by weight of sodium 4,4'-diazidestilbene-2,2'-disulfonate in 100 parts by weight of water. The resultant photoresist solution was divided into four parts. To each of the four solutions was added 0, 0.15, 0.30 and 1.5 parts by weight of diacetone acrylamide, respectively, to obtain four kinds of photoresist solutions.

Each of them was uniformly coated on glass substrates (50 x 50 millimeters) and dried to form films of about 1 micron in thickness. Thereafter, a mask which has holes of 0.35 millimeter in diameter two-dimensionally disposed at a pitch of 0.6 millimeter was put on each film. These films were exposed through the holes of the mask at 800 lx for 4 minutes with an ultra-high-pressure mercury lamp and developed with water spray (water pressure 0.5 kg/cm^2) for 30 seconds and with flowing water for 3 minutes.

For comparison of adhesion force of the hardened photoresists, from a nozzle of 1 millimeter in diameter water having a pressure of 40 mm Hg was vertically thrown on the surface of each film (which is 160 mm away from the nozzle) for 2 minutes and peeling of dots of the hardened photoresist was examined. The results are shown in the table below. It is clear from these results that diacetone acrylamide has an effect of improving the adhesion.

Amount of Diacetone Acrylamide Added (parts by weight)	Number of the Removed Dots	Proportion of Number of the Removed Dots (percent)
0	26	100
10	21	81
20	10	38
100	0	0

Maleic Anhydride and Dye to Intensify Image

M. Stolka; U.S. Patent 3,888,670; June 10, 1975; assigned to Xerox Corporation describes an imaging method based upon the intensification of an image comprising a polyene containing at least two conjugated double bonds per molecule dispersed in an imagewise manner upon a suitable substrate. The polyene is contacted with a solution of maleic anhydride to form a reaction product between the polyene and the maleic anhydride.

The so-formed reaction product is then contacted with a base to convert the carboxylic acid anhydride group of the maleic anhydride to its corresponding salt and the salt is then contacted with a solution of a dye containing a colored cation capable of replacing the cation of the salt.

Example: A sandblasted aluminum foil substrate is coated with a film of polyvinyl chloride containing approximately 1% by weight of tetracyanoethylene (TCNE) by applying a solution of the polymer and the TCNE in tetrahydrofuran to the substrate and evaporating the solvent to form a photographic plate.

The film is exposed through a simple negative to ultraviolet light from a Mineral Light for 10 minutes and heat developed at about 70°C for 10 minutes. The plate, which bears a brown image in the exposed areas, is then dipped into a warm solution of 5% by weight maleic anhydride in benzene for several seconds. Next the plate is washed with acetone, exposed to aqueous ammonia vapors and immersed in a 2% by weight solution of methylene blue in methanol. The image on the plate is intensified by the formation of a deep-blue color in the image-bearing areas.

Latent Catalyst Precursor for Epoxy Resist

In accordance with the process of *J. Roteman; U.S. Patent 3,895,954; July 22, 1975; assigned to American Can Company*, a mixture is formed of an epoxide material in admixture with an organohalogen compound as a latent catalyst precursor and an organometallic compound as an enhancer therefor. The composition, at a convenient time, subsequently is exposed to electromagnetic radiation or other energy sources to release an active catalyst or catalysts which effects the polymerization of the epoxide material.

The precursor is an organohalogen compound, wherein the organo radicals are alkyl, aryl, alkaryl, aralkyl, alkoxy and aroxy in combination with an organometallic compound having the formula R_3M wherein M is P, Sb, As or Bi and R is a hydrocarbyl radical or hydrogen with the proviso that at least one R is a hydrocarbyl radical.

It has been found that satisfactory results are obtained by providing an organohalogen compound in amounts by weight of from 5% to about 50% relative to the weight of the polymerizable epoxide material. The organometallic compounds are employed in amounts of 1 to 3 mols organometallic compound per mol of organohalogen.

For an imaging system, the mixture, which may contain a suitable solvent in substantial proportions, is coated on a metal plate, dried if necessary to remove solvent present and the plate is exposed to ultraviolet light through a mask or negative. The light initiates polymerization which propagates rapidly in the exposed image areas. The resulting polymer in the exposed areas is resistant to many, or most, solvents and chemicals while the unexposed areas can be washed with suitable solvents to leave a reversal image of the polymer in this embodiment.

Example: Two formulations were prepared to contain (a) 3.62 grams allyl glycidyl ether-glycidyl methacrylate copolymer, 2.7 grams toluene, 8.2 grams methyl ethyl ketone and 1.63 grams iodoform; (b) same as formulation (a) to which 0.55 gram triphenyl bismuthine is added.

Separate coatings were made on Redicote aluminum using a No. 12 Mayer rod. Sample strips were exposed through a No. 2 step tablet for 30 seconds and then developed in methyl ethyl ketone. The coating with iodoform alone reproduced 15 steps while the coating with triphenyl bismuthine also present reproduced 17 steps. Both coatings could be dyed with a trichloroethylene solution of Orasol black. When similarly exposed to an xenon lamp for 15 seconds and developed, the two coatings yielded 5 and 8 steps respectively.

Divinyl Urethane plus Carboxylic Polymer as Development Aid

The process of *T. Yonezawa and N. Kita; U.S. Patent 3,907,574; September 23, 1975; assigned to Fuji Photo Film Co., Ltd., Japan* relates to a photopolymerizable composition for a printing relief or a photoresist.

The composition comprises: (a) 80 to 50% by weight of a divinyl urethane compound represented by the following general formula,

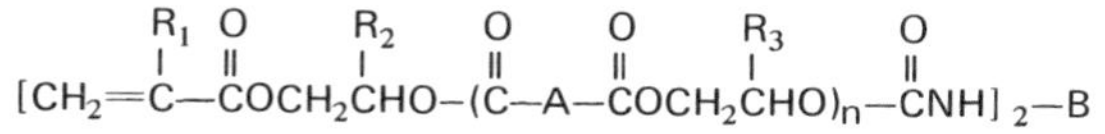

$$[CH_2{=}\overset{R_1}{\overset{|}{C}}{-}\overset{O}{\overset{\|}{C}}OCH_2\overset{R_2}{\overset{|}{C}}HO{-}(\overset{O}{\overset{\|}{C}}{-}A{-}\overset{O}{\overset{\|}{C}}OCH_2\overset{R_3}{\overset{|}{C}}HO)_n{-}\overset{O}{\overset{\|}{C}}NH]_2{-}B$$

in which R_1, R_2 and R_3 each represents a hydrogen atom or a methyl group, n represents 1, 2 or 3, A represents a divalent residue of a cyclic carboxylic acid anhydride having 4 to 10 carbons, B represents $-R_4-$ or $-R_4NHCOOR_5OOCNHR_4-$, in which R_4 represents an alkylene group with 4 to 13 carbons, cycloalkylene group or arylene group and R_5 represents an alkylene chain with 2 to 8 carbons or an oxyalkylene chain with 2 to 4 carbons; (b) 20 to 50% by weight of an organic high molecular weight polymer miscible with the divinyl urethane compound and having carboxyl groups on the side chains and an acid number of 30 or more; and (c) 0.01 to 10% by weight of a light sensitizer capable of initiating the photopolymerization of an ethylenically unsaturated compound upon exposure to active radiation.

The composition is suitable for the production of a printing plate for a relief image having a relief height of 0.2 millimeter or more. In addition, this composition is suitable for use as a light-sensitive layer for a photoresist having a relief height of 1 to 10 microns.

In these cases, an organic high molecular weight polymer having carboxyl groups with an acid number of 30 or more on the side chains is used in order to provide a composition whereby the nonexposed areas can be dissolved in an aqueous weak alkali solution or aqueous weak alkali solution containing an organic solvent soluble in water.

Such an organic high molecular weight polymer is not used as a binder, in contrast to its generally employed use. Thus the organic high molecular weight polymer is used in a smaller quantity than the divinyl urethane compound. The divinyl urethane compound as a hardener can be used in a relatively large quantity, since it is a viscous liquid or a solid at room temperature.

Example: Three hundred ninety-eight grams (1.0 mol) of hydroxyethylphthalyl methacrylate represented by the structure:

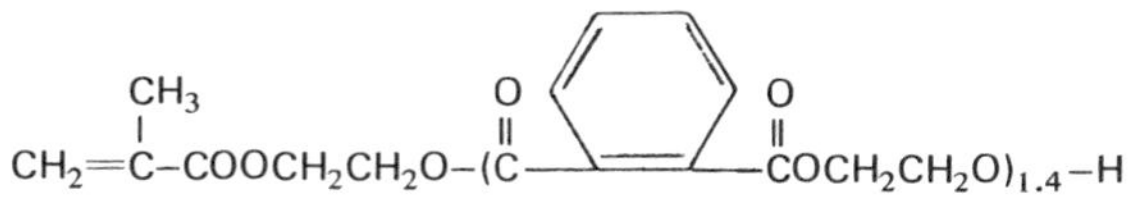

$$CH_2{=}\overset{CH_3}{\overset{|}{C}}{-}COOCH_2CH_2O{-}(\overset{O}{\overset{\|}{C}}{-}C_6H_4{-}\overset{O}{\overset{\|}{C}}OCH_2CH_2O)_{1.4}{-}H$$

(OH number: 141; average molecular weight: 398), 94 grams (0.5 mol) of xylylene diisocyanate, 1.5 grams of benzoic acid as a catalyst and 0.25 gram of 2,6-di-t-butylcresol as a polymerization inhibitor were mixed and reacted at 80°C for seven hours while bubbling air through the mixture to obtain a divinyl urethane compound having a residual isocyanate number of 5.2.

To 60 parts of this divinyl urethane compound were added 30 parts of hydroxypropylmethylcellulose hydrogen phthalate (HP-55, acid number 127), 10 parts of a styrene-maleic anhydride copolymer partially esterified with a lower alcohol (Styrite CM-2L), 1 part of benzoin ethyl ether as a sensitizer, 0.05 part of 4,4'-thiobis(3-methyl-6-t-butylphenol) as a thermal polymerization inhibitor and 40 parts of acetone and 17 parts of methanol as a solvent.

The solution was allowed to stand at 40°C for 24 hours to defoam the mixture. The thus defoamed light-sensitive liquid was coated onto a polyester film of a thickness of 100 microns so as to give a film thickness of 0.6 millimeter on a dry basis using a knife coater and then dried at room temperature for one day and in hot air at 50°C for one day. The dried sheet was laminated at 80°C on a grained aluminum plate having a thickness of 0.3 millimeter.

After the polyester film of 100 microns was stripped from the laminated light-sensitive plate, a photographic negative film having alphabetic designations was contacted with the light-sensitive plate and exposed for 10 minutes using a vacuum printer in which chemical fluorescent lamps of 20 watts were placed at a distance of 6 centimeters. After the exposure, the nonexposed area was dissolved off with weak aqueous alkali solution containing 0.5% isopropyl alcohol and 0.2% sodium hydroxide to obtain a sharp relief image.

Nitroso Dimer plus Cr as Inhibitor System

D.H. Scheiber and R.V. Weaver; U.S. Patent 3,914,128; October 21, 1975; assigned to E.I. du Pont de Nemours & Company provide a film-forming photohardenable paste composition comprising from 20 to 75 weight percent of finely divided inorganic material capable on firing of being fused into an element of an electronic circuit, from 2 to 35 weight percent of a polymeric binder, not more than about 10 weight percent of a photopolymerizable monomer, from about 0.02 to about 1 weight percent of an organic sensitizer and from 20 to 75 weight percent of a solvent.

As an inhibitor system there is used from about 0.01 to about 1.0 weight percent of a nitroso dimer having a dissociation constant in solution at 25°C of from about 10^{-2} to about 10^{-10} with a rate of dissociation in solution with a half life comparable to the exposure time of the paste composition, and from about 0.05 to about 7.0 weight percent chromium present as either the free metal or an oxide thereof. Examples of the nitroso dimers include

(1) [S ring]–N(→O)=N(→O)–[S ring]

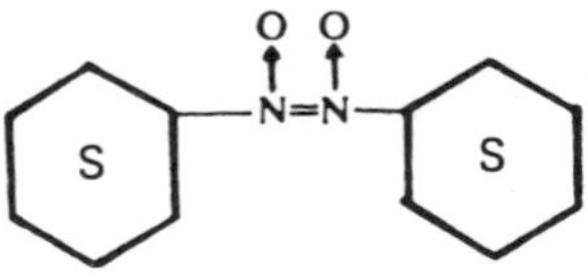

(2) $(CH_3)_3CCH_2CH[C(CH_3)_2ONO_2]–N(\rightarrow O)=N(\rightarrow O)–CH[C(CH_3)_2ONO_2]CH_2C(CH_3)_3$

(3) [cyclobutane ring, CH_3, CN]–N(→O)=N(→O)–[cyclobutane ring, CH_3, CN]

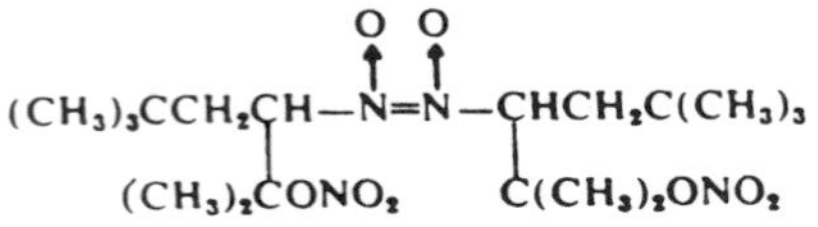

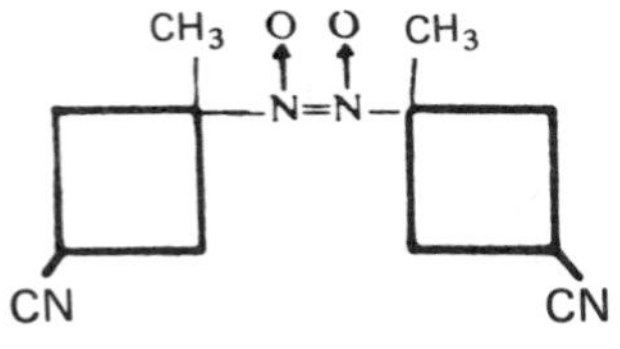

(4)

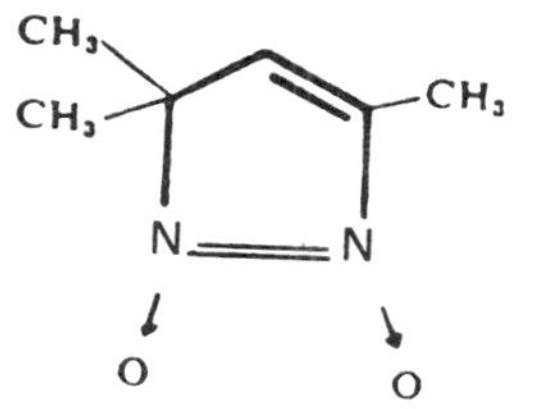

CH_3 CH_3 CH_3

N═══N

↙ ↘

O O

(5) $$HO(CH_2)_6\overset{\overset{O}{\uparrow}}{N}{=}\overset{\overset{O}{\uparrow}}{N}(CH_2)_6OH$$

(6) $$[(CH_3)_3C]_2CH\overset{\overset{O}{\uparrow}}{N}{=}\overset{\overset{O}{\uparrow}}{N}CH[C(CH_3)_3]_2$$

In the inhibitor system the inhibitor species is produced relatively slowly and in small amounts by dissociation of a noninhibiting species. Thus, there is created a threshold value of illumination for photohardening. Below the threshold value inhibitor species are produced as fast as they are consumed by reaction with the active initiator radical. Thus, photohardening will be retarded for a prolonged period until all of the source of inhibiting species has been decomposed. Above the threshold value of illumination the concentration of inhibiting species is maintained at an extremely low level and photohardening can proceed. By proper balancing of the initiator, the inhibitor source, opacity of the paste, exposure intensity and time, improved resolution can be attained.

For the following examples, a hydrogenated terpene vehicle was prepared by the hydrogenation of a mixture of alpha- and beta-terpineol. The paste compositions were prepared for the following examples by first dispersing a monomer, polymeric binder and sensitizer in the vehicle.

The inorganic solid constituent was pretreated by dispersion in the vehicle; then dried and redispersed with the monomer/binder/sensitizer dispersion in the vehicle; milled to form a paste on a three-roll mill; and then screened to remove undispersed particulate material.

Examples 1 through 16: A paste composition was prepared according to the above procedure containing a 20% by weight solution of polyisobutyl methacrylate dissolved in dihydroterpineol, 19.5 parts by weight; glass powder having a particle size of 1 to 12 microns, 21.60 parts by weight; diethylene glycol diacrylate/tetraethylene glycol diacrylate monomers 60/40 (w/w), 1.27 parts by weight; benzoin methyl ether sensitizer (50% by weight solution in dihydroterpineol), 0.36 part by weight.

The mixture was roll milled until a smooth paste containing well dispersed glass particles were obtained. The particulate chromium compounds and the nitroso dimers were added to the paste composition and mulled until thoroughly mixed. Coatings of the compositions below were applied to 96% alumina substrates by screen-printing the resultant paste compositions using a 63-mesh Nytex screen and dried for 15 minutes in a stream of air at 60°C to form a film. The print and dry sequence was repeated to provide a total coating thickness of 2.0 to 2.2 mils on the substrates.

The coated substrates were then masked with a clear photographic negative containing a pattern of opaque squares 10, 5, 3, 2 and 1 mil on a side, placed in a vacuum frame under a vacuum of 1 to 2 torrs and the coating was imagewise exposed using a 250-watt medium-pressure mercury vapor lamp at a distance of 10 inches.

The coatings exposed to the image were developed in a spray of perchloroethylene at 75 psi and a distance of 2 to 3 inches for 5 to 7 seconds. The developed coatings were dried in a stream of air and fired at 850° to 925°C to produce fired films on the substrate. The fired films substantially retained the resolution of the developed unfired films listed in the table. All percentages are by weight based on the total weight of the composition.

Example*	Paste (g)	Cr_2O_3	Dimer	Exposure	Resolution (mils)
Control	4.27	–	–	1/50–10 sec	–
A	4.27	0.038 g	--	1/50–60 sec	10
B	4.27	–	0.004 g**	1-30 sec	–
1	4.27	0.038 g	0.004 g**	5-35 sec	5
2-7	4.27	0.038 g	0.004 g	10-25 sec	5
8	4.31	0.023 %	0.02 %	5 sec	10
9	4.31	0.023 %	0.58 %	60 sec	10
10	4.31	1.15 %	0.02 %	5 sec	10
11	4.31	1.16 %	0.12 %	60 sec	5
12	4.31	4.43 %	–	30 sec	5
13	4.31	4.43 %	0.02 %	45 sec	3
14	4.31	4.43 %	0.11 %	30 sec	5
15	4.31	4.43 %	0.55 %	5 min	5
16	42.70	0.24 g	0.04 g	30 sec	5

*The nitrosocyclohexane dimer of Formula 1 was used in Examples B, 1, 2 and 8 through 16. The nitroso dimers of Formulas 2 through 6 were respectively used in Examples 3 through 7. In Examples 10 and 16 CrO_2 was used in place of Cr_2O_3.
**Dissolved in 2 cc of dihydroterpineol.

In the control example exposure times were varied from 1/50 to 10 seconds. For exposure times of less than one second, polymerization was incomplete in the exposed areas and the coating was partially washed away during the development. For exposure times greater than one second the exposed areas remained intact but the unexposed areas could not be washed out.

In Example A exposure times up to 15 seconds gave incomplete polymerization in the exposed areas with subsequent destruction of the coating. For exposure times in excess of 15 seconds, only the unpolymerized areas 10 mils on a side could be washed free of glass, even under intense spraying which caused disruption of the coating in the polymerized areas.

In Example B, exposure times of one second resulted in the exposed areas of the coating having holes after development, indicating incomplete polymerization, yet the unexposed areas could not be washed free of glass. After exposure times greater than one second, the unexposed areas could not be washed out without disturbing the exposed areas.

In Example 1 exposure times of 5 seconds produced 10, 5 and 3 mil areas which could be washed clean of glass but small holes in the exposed areas of the coatings indicated incomplete polymerization. Exposure times of 10 to 35 seconds produced 10 to 5 mil unexposed areas cleanly resolved and washed free of glass.

The exposed areas were polymerized and undisturbed by the development of the unexposed areas. In Examples 2 through 7, using the six different dimers described above and exposure times of 10 to 25 seconds, cleanly resolved 5-mil vias in the unexposed areas (which were readily washed free of glass without disturbing the exposed polymerized areas) were obtained.

Examples 8 through 16 resulted in cleanly resolved vias in the exposed areas at the exposure times given. The unexposed polymerized areas were undisturbed by development.

Increasing Polysulfone Sensitivity

According to the process of *E. Gipstein, W.M. Moreau and O.U. Need III; U.S. Patent 3,916,036; October 28, 1975; assigned to International Business Machines Corporation,* the polysulfone resists of the prior art are sensitized by the addition of either an energy transfer agent or a free radical source or, most preferably, both types of sensitizers. It has been found that the addition of such materials increases the sensitivity of the polysulfone resist to radiation with electron beams and also serves to make the resist useful as a photoresist for lithographic production of semiconductor circuits by contact or projection printing.

Example: 3% solutions of polysulfone and free radical or triplet sensitizers in nitromethane were prepared and spin-coated on SiO_2/Si wafers at 2,000 rpm. The wafers were placed in a vacuum oven at 65°C for 12 hours to dry and to remove the solvent. The wafers were then exposed with a 150-watt ultraviolet lamp for 1 hour through a mask using a Pyrex filter to cut off all light below 3300 Angstroms. The lamp distance from the wafer was 5 inches.

Polysulfone controls (without additives) were also prepared and exposed 3 hours. The images were developed with 1,4-dichlorobutane solvent. The time of development varied from sample to sample. The image was considered developed when the subsurface of the silicon wafer could be observed. The following results were obtained.

Additive	Type of Image	Quality of Image
CBr_4	+	Excellent
Azulene	+	Good
2,4,7-Trinitrofluorenone	+	Good
Fluorene	+	Good
Diphenylamine	+	Good
p-Nitroaniline	+	Good
Azobenzene	+	Good
Control (polycyclopentene sulfone)	No image developed after exposure for three hours	

Cyclic Amide as Gelation Inhibitor in Epoxide

S.I. Schlesinger; U.S. Patent 3,951,769; April 20, 1976; assigned to American Can Company provides improved stabilized polymerizable compositions comprising epoxides and mixtures of epoxides with monomers selected from the group consisting of lactones and vinyl-containing compounds containing radiation sensitive catalyst precursors and a class of gelation inhibitors which, upon admixture with the polymerizable materials, inhibit gelation of the reaction composition prior to irradiation. This is accomplished by the inclusion of a small quantity of a cyclic amide, unsubstituted in the 1-position, for example, caprolactam, as gelation inhibitor. Such compositions have greatly extended storage or pot-life, premature reaction in the dark or at minimal levels of radiation being inhibited so that the mixtures may be retained for extended periods of time before application.

Thus, in accordance with the process, a mixture first is formed of the polymerizable materials, a Lewis acid catalyst precursor, and the cyclic amide inhibitor. The resulting mixture, at a convenient time, is subsequently exposed to an energy source such as actinic or electron beam irradiation to release the Lewis acid catalyst in sufficient amounts to initiate the desired polymerization reaction.

Notable among the compounds found to be especially effective as the premature-gelation inhibitor in the process are heterocyclic amides which may be represented by the generalized formula

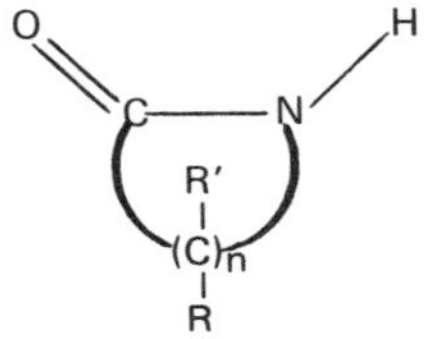

O H
C—N
R'
|
$(C)_n$
|
R

where each of R' and R is hydrogen or an unreactive group, commonly alkyl, and n is a low integer, preferably from 3 to 11, the bond as shown to an R group alternatively may be a double bond to an adjacent carbon atom.

The polymers produced by the polymerizing process of the method are useful in a wide variety of applications in the field of graphic arts, due to their superior adhesion to metal surfaces, excellent resistance to most solvents and chemicals, and capability of forming high resolution images.

Example: Three epoxide blends, A, B and C were formulated to contain equal amounts of p-chlorobenzene diazonium hexafluorophosphate and varying amounts of caprolactam. A fourth formulation, D, served as a control with no caprolactam added.

	Formulation (grams)			
	A	B	C	D
1,4-Butanediol diglycidyl ether	25.0	25.0	25.0	25.0
3,4-Epoxycyclohexylmethyl-3,4-epoxycyclohexane carboxylate	25.0	25.0	25.0	25.0
p-Chlorobenzenediazonium hexafluorophosphate	0.500	0.500	0.500	0.500
Caprolactam	0.0164	0.0323	0.0643	-

Within two hours an exothermic reaction began in formulation D, with gelling commencing. This formulation finally set to a hard solid within four hours. Formulation A gelled to a solid within two days, while formulations B and C gelled within five days. Coatings were cast with formulations A to C, after aging for one day, on paperboard samples using a No. 9 wire wound rod.

Upon exposure to a 360-watt mercury arc ultraviolet lamp for 15 seconds at 18 centimeters distance, coatings derived from formulations B and C cured to dry, hard, glossy coatings. Coatings derived from formulation A also cured to a dry, hard, glossy coating but with a minimum of five seconds of exposure.

PROCESSES

Two-Exposure Process Using Nitroso Dimer

The process of *G.R. Nacci; U.S. Patent 3,885,964; May 27, 1975; assigned to E.I. du Pont de Nemours & Company* relates to producing an image on a substrate by,

(a) coating the substrate with a photopolymerizable composition which comprises:

  a nongaseous, ethylenically unsaturated compound capable of forming a high polymer by free-radical initiated chain addition propagation,

  3 to 95% by weight, based on the total composition of an organic polymeric binder having a molecular weight of at least 4,000,

  1 to 10% by weight, based on total composition, of an organic, radiation-sensitive, free-radical generating system and

  0.1 to 2% by weight, based on the total composition, of a thermally dissociable nitroso dimer having a dissociation constant of about 10^{-2} to 10^{-10} and a dissociation half-life of at least about 30 seconds in solution at 25°C, the weight ratio of nitroso dimer to free-radical generating system being less than 2 to 1,

(b) exposing the photopolymerizable coating to radiation having wavelengths essentially limited to about 3400 to 8000 A through an image-bearing transparency at a temperature of about 50° to 70°C;

(c) allowing the coating to cool to a temperature below about 45°C to reduce the concentration of nitroso monomer;

(d) reexposing a greater portion of the coating, including the portion struck by radiation during the first exposure, to radiation having wavelengths essentially limited to about 3400 to 8000 A at a temperature below about 45°C; and

(e) developing the resulting image.

Nitroso dimers are not free-radical polymerization inhibitors, but dissociate to active inhibiting mononitroso species.

The process involves two exposures. During the first imagewise exposure, photopolymerization does not occur. The equilibrium concentration of nitroso monomer at the elevated temperature is sufficient to prevent the chain propagation required for polymerization. During this exposure the free radicals formed from the initiator by absorption of radiation are consumed by the nitroso monomer. Between the first and second exposures, the temperature is reduced to below about 45°C. Since nitroso dimer is in thermal equilibrium with monomer,

$$(RNO)_2 \overset{\Delta}{\rightleftharpoons} 2RNO$$

a decrease in temperature of the photopolymerization system shifts the equilibrium thereby decreasing the relative concentration of nitroso monomer molecules present. Hence, during the second exposure at a lower temperature, the concentration of nitroso monomer is insufficient to prevent radical chain propagation of monomer molecules and imaging occurs in the areas radiation-struck in the second exposure but not struck during the first exposure.

In the example, the coating solutions were prepared by dissolving the reactants in methylene chloride at 25°C. The solutions were coated with a doctor knife onto a one-ounce copper-clad circuit board, 100 mils thick. The copper surfaces of the boards were cleaned with pumice powder and water just before coating with the photopolymer solutions. The coatings were dried at 25°C and those coatings so identified were coated with 1% by weight polyvinyl alcohol solution (Elvanol 51-05) in water using a cotton ball dampened with the polymer solution. Coating thicknesses (dried) of these topcoats were 0.05 mil or less.

Samples were exposed in a glass vacuum frame at 1 millimeter pressure or under nitrogen at atmospheric pressure to a medium pressure mercury resonance lamp (100 w AH_4) held four inches away from the sample. The system was evacuated for two minutes prior to exposure and during the exposure. Silver image film transparencies of a 1951 Air Force test pattern were used with the emulsion side of the patterns in contact with the photopolymerizable coatings.

After the exposures, the samples were washed with cold water to remove the polyvinyl alcohol coatings and then spray-developed using methylchloroform in a spray gun held two inches from the samples. The developed samples were examined optically.

Example: A stock solution of a mixture of 2.90 grams of trimethylolpropane triacrylate containing 245 parts per million hydroquinone inhibitor, 0.88 gram of conventional plasticizers, 0.44 gram of triethylene glycol diacetate, 5.24 grams of polymethyl methacrylate resin, 0.40 gram of 2-o-chlorophenyl-4,5-diphenylimidazole dimer, 0.03 gram of tris(4-diethylamino-2-methylphenyl)methane, 0.02 gram of an adhesion promoter, and 0.01 gram of Michler's ketone dissolved in 40 milliliters of methylene chloride was prepared. To one-eighth of this solution was added 0.015 gram of nitrosocyclohexane dimer and the resulting solution coated onto a copper-clad circuit board.

The solvent was evaporated at 25°C to leave a coating 2 mils thick. The plate was exposed through a line negative under a nitrogen atmosphere, as described, at 50°C for 0.5 minute. The exposed plate was cooled to 25°C and reexposed, without the negative, for 0.5 minute. After development as described, a positive image was obtained. A similar experiment in which the exposure temperatures were 60° and 25°C gave essentially similar results. In another experiment, the nitrosocyclohexane dimer was replaced with di-t-butylnitrosomethane dimer and the first exposure was carried out through a line negative at 50°C for four minutes, followed by reexposure without the negative at 0°C for two minutes. After development as described, a positive image was obtained which was developed down to the base of the copper plate.

Two-Exposure Process Yielding Reverse Image

S.-Y.L. Lee; U.S. Patent 3,888,672; June 10, 1975; assigned to E.I. du Pont de Nemours & Company provides compositions capable of yielding reverse photopolymer images since relatively intense radiation prevents polymerization, while less intense radiation yields photopolymerization. The first (intense) exposure is imagewise and the second (less intense) exposure is nonimagewise. The compositions are useful in the graphic arts where a positive-working system is required, for example, for relief or planographic printing plates, direct positive copying films, and the like.

The photosensitive composition consists essentially of:

(a) a hydrogen- or electron-donor compound;

(b) a hexaarylbiimidazole;

(c) an ethylenically unsaturated compound capable of forming a high polymer by free-radical initiated, chain propagating addition polymerization;

(d) an organic polymeric binder.

The concentrations (which are critical) of the components are, by weight, less than 0.4% (a), at least 1% (b), 30 to 70% (c) and 68 to 28% (d), with the ratio of (b) to (a) being 10 to 1 or greater, said composition being capable of yielding reverse images by photopolymerization.

High intensity exposure of the above-described composition destroys polymerizability without polymerization taking place, while medium or low intensity exposure yields polymerization. The method, therefore, includes the process of imagewise exposing the composition to radiation which destroys polymerizability of the composition in the exposed areas and subsequently exposing the entire composition to radiation which induces photopolymerization in the previously unexposed areas.

Example: A coating composition was prepared containing the following ingredients:

Trichloroethylene	10.8 g
Acetone	1 ml
Polyethylene glycol dimethacrylate (average molecular weight 736)	1.2 g
Polymethyl methacrylate*	1.2 g
2,2′-bis(o-chlorophenyl)-4,4′,5,5′-tetrakis(m-methoxyphenyl)biimidazole	0.1 g
5-(p-dimethylaminobenzylidene)rhodanine	0.00134 g

*Inherent viscosity = 1.0 for a solution of 0.25 g polymer in 50 ml chloroform as measured at 20°C with a No. 50 Cannon-Fenske Viscometer.

After thorough mixing, a portion of the photopolymerizable composition thus formed was coated by means of a doctor knife set at a clearance of 0.002 inch on a 0.002-inch thick polyethylene terephthalate film. The coating was air-dried for 30 minutes to permit evaporation of solvent and a sheet of 0.0005-inch thick polyethylene terephthalate film was then applied by hand over the tacky coating surface. The dried, tacky photosensitive coating was 0.0004 inch (approximately) in thickness. The photosensitive layer prepared in this way was exposed both by transmission and reflex exposures.

Transmission Exposures – The photosensitive layer was exposed to a 1,000-watt Colortran tungsten-iodine light source, at a distance of 54 inches, through a high contrast transparency which was in contact with the cover sheet. Two exposures of quite different intensities were made. After exposure, the cover sheet was removed and the layer dusted with a green pigment toner which adhered to the tacky, unpolymerized portion, but not to the photopolymerized regions.

When the exposure was of short duration, about 30 seconds, and the source was modulated by a Kodak 1A Wratten filter (opaque to radiation below 310 nm, transparent above 380 nm, reduced intensity 310 to 380 nm) a positive image was obtained on toning. However, exposure of about 3 minutes, in the absence of any filter, gave a negative image on toning.

The above results, exemplary of this process, are interpreted as follows: The shorter exposure through the filter was obviously less intense (the filter decreasing intensities in the region where hexaarylbiimidazoles absorb), and merely resulted in photopolymerization as known in the art. That is, the transparent portions of the process transparency transmits radiation sufficient to induce photopolymerization in the photosensitive layer, while the opaque portions screen the layer from the radiation; thus, the screened portions remain tacky and (after removal of the cover sheet) will accept a toner. The result is a positive image, since the toned (dark) areas correspond to the opaque areas of the process transparency.

The longer, unfiltered exposure was obviously more intense and gave a negative image after toning. Thus, the high-intensity radiation transmitted through the transparent portion of the transparency did not induce photopolymerization. Apparently, the hydrogen-donor component (the rhodanine derivative) was deactivated. The polymerization found under the opaque portions of the transparency probably resulted from scattered radiation of lesser intensity, but of intensity sufficient to induce photopolymerization.

Reflex Exposures – The photosensitive layer was placed between the light source (Colortran, at 54 inches, as above) and an opaque original (that is, printed paper) with the cover sheet in contact with the opaque original. As above, subsequent to the exposures described below, the cover sheet was removed and the coating dusted with a green pigment toner.

As with the transmission exposures, a lower intensity exposure (about 15 seconds with a Kodak 1A Wratten filter inserted between the photosensitive layer and light source) gave a positive image. On the other hand, a high-intensity exposure (about 3 minutes without any filter) gave a negative image.

Variable Depth Contour Images

J.F. Pazos; U.S. Patent 3,901,705; August 26, 1975; assigned to E.I. du Pont de Nemours & Company found that variable depth, positive, contour images are produced on substrates by coating the substrate with a photopolymerizable coating composition containing: a nongaseous, ethylenically unsaturated compound capable of polymerization by free-radical initiated chain propagation; an organic light-sensitive free-radical generating system; and a photodissociable nitroso dimer having a dissociation constant of about 10^{-2} to 10^{-10} and a dissociation half-life of at least about 30 seconds in solution at 25°C.

These images are produced by exposing the photopolymerizable coating to light having wavelengths both above and below 3400 A through a transparent film of variable optical density to light having a wavelength of less than about 3400 A, and developing a positive polymeric image by removing the nonpolymerized portion of the polymer coating.

The process is based on the fact that nitroso dimers are not free-radical polymerization inhibitors but are dissociated to active inhibiting mononitroso species by ultraviolet light of wavelength about 2800 to 3400 A. The dimers are relatively unaffected by light of longer wavelength.

On the other hand, light-sensitive free-radical initiators absorb light of longer wavelength to provide sufficient radicals for polymerization of the monomer in the absence of an appreciable concentration of mononitroso species. The nitroso monomer formed by irradiation of nitroso dimer with short wavelength light interferes with the normal free-radical induced polymerization process by reaction with radicals or with photoactivated nitroso monomers to form stable nitroxide radicals which do not propagate a radical chain process. Hence these nitroxide radicals serve as efficient chain terminators.

Since the extinction coefficients of the initiator systems are relatively low at the longer wavelengths, generation of initiating radicals is not as depth dependent as is generation of mononitroso inhibitors. Hence, selection of a transparent film of variable optical density toward short wavelength ultraviolet light permits selective dissociation of nitroso dimer as a function of depth of the photopolymerizable layer resulting in a graded polymerization.

The transparency passes light of longer wavelength which effects polymerization. The amount of polymer formed on the base is inversely proportional to the optical density of that part of the transparency through which the light passes and a contour image is produced.

Example: A stock solution was prepared by mixing 9.65 grams of trimethylolpropane triacrylate, 0.25 gram of benzoin methyl ether and 0.1 gram of nitrosocyclohexane dimer. The solution was used to prepare films on four microscope slides which contained a 10-mil Mylar polyester film rim. The coated microscope slides were covered with quartz plates and ultraviolet step wedge negatives.

The films were irradiated with a 275-watt RS sunlamp for 5 minutes from a distance of 12 inches. The negatives were prepared by staggering six layers of cellulose acetate butyrate film, each layer absorbing 50% of short wavelength light as shown below. The numbers represent percent of short wavelength light transmitted.

Step	1	2	3	4	5	6	7
% Transmission	100	50	25	12.5	6	3	1.5

Unpolymerized monomer was removed with an acetone washout. The films showed seven visible steps with the most complete polymerization occurring where the least amount of short wavelength light was transmitted. Each step differed on the average by 0.5-mil thickness.

	Steps (thickness in mils)						
Run	(1)	(2)	(3)	(4)	(5)	(6)	(7)
1	6.0			8.0			9.5
2	6.0			7.9			9.1
3	6.1			8.0			9.0
4	5.3			10.0			11.0

Bicyclic Amidine

H. Fukutani, K. Miura, C. Eguchi, Y. Takahashi and K. Torige; U.S. Patent 3,923,703; December 2, 1975; assigned to Mitsubishi Chemical Industries Ltd., Japan provide a preparation for a photosensitive polymer which comprises reacting a polymer having an active halogen with an α,β-unsaturated carboxylic acid having the formula:

$$Ar{-}(YC{=}CX)_n{-}\overset{\displaystyle O}{\overset{\|}{C}}{-}OH$$

wherein X and Y represent a hydrogen atom, halogen atom, cyano group or nitro group, Ar represents an aryl group, n is 1 or 2; in an aprotic solvent in the presence of a bicyclic amidine.

Photosensitive compositions can be prepared by dissolving the photosensitive polymer of the process in a suitable solvent such as chlorobenzene, xylene, methyl Cellosolve acetate, ethyl Cellosolve acetate, Pentoxone or the like and adding a conventional sensitizer. The photosensitive compositions can be used as photoresists for the preparation of printed circuits, for chemical milling, for printing plates for relief printing and intaglio processes and for photoprinting plates. They can also be used as a photocurable paint or as photoadhesives.

Example: Ten grams of polyepichlorohydrin having a reduced specific viscosity of 0.97 (0.2 g/dl benzene at 30°C) was dissolved in 342 ml of dimethylformamide and 20.8 g of cinnamic acid and 21.4 g of 1,8-diazabicyclo[5.4.0]undecene-7 were added. A homogenous solution was formed. The reaction was conducted at 85°C for 10 hours with stirring in an argon atmosphere. The reaction mixture always remained homogenous.

A part of the reaction mixture was removed and the amount of chlorine ion formed by the reaction was measured with silver nitrate by a potentiometric titration method. It was found that 58% of the chlorine atoms of the polyepichlorohydrin was replaced with cinnamic acid radicals. The reaction mixture was then poured into about 2 liters of methanol to precipitate the polymer and the cake polymer was cut and washed with methanol and then dried.

Support for Liquid Photosensitive Resin

Y. Kitanishi and H. Yamada; U.S. Patent 3,948,666; April 6, 1976; assigned to Teijin Limited, Japan provide a support for use in preparing a printing plate using a liquid photosensitive resin. The support includes a flexible self-supporting base plate and an adhesive layer thereon for applying a layer of a photosensitive resin. The adhesive layer is a layer of a crosslinked polyester-polyurethane resin formed by reacting a linear polyester-polyurethane resin with a polyfunctional isocyanate on the surface of the base plate.

Example: A polyester-polyurethane resin having a molecular weight of about 40,000 obtained by chain-extending polyethylene adipate having a hydroxyl group at both ends and having an average molecular weight of 1,300 was dissolved in a mixed solvent consisting of ethyl acetate and methyl ethyl ketone to form a 35% solution. This polyester-polyurethane resin had an end-blocking agent added to its termini and had a hydroxyl value of not more than two. Ten parts of a 75% ethyl acetate solution of an adduct of trimethylol propane and TDI (Coronate L-75, isocyanate group content 13.4%) was added to 100 parts of the resulting 35% solution.

Then, an oil-soluble yellow dye (Color Index No. 11390) was added to the mixture in an amount of 2% based on the resin content. A solvent was further added to form a solution having a solids content of 15%. The solution was thoroughly stirred in a ball mill until its viscosity became 20 poises. The coating solution was then coated by a bar coater uniformly on the surface of a 0.15 millimeter thick aluminum plate degreased with trichloroethylene.

The coated base plate was immediately placed in a hot-air dryer held at 100°C and dried for about 2 minutes to drive off the solvent. The coated layer had an average thickness of about 8 microns. At this time the surface of the coated layer was almost free from tackiness and could come into contact with other materials without blocking.

When the resulting product was allowed to stand for 5 days in a dry room held at 40°C, the coated layer was sufficiently crosslinked and was hardly swollen with ethyl acetate.

For comparison, 30 parts of Coronate L-75 (the molar ratio of hydroxyl to isocyanate being about 1) was added to 100 parts of a 40% ethyl acetate solution of the same polyethylene adipate as used above (hydroxyl value 87) and 0.2%, based on the resin content, of a tertiary amine catalyst was added to form a coating solution. The solution was coated on the same base plate and dried under the same conditions. The coating was very tacky and its handling was extremely inconvenient unless a releasing paper was laminated thereon. When the coating was heated for 40 hours at 80°C, the surface became nontacky and crosslinking proceeded efficiently.

A semitransparent liquid photosensitive composition consisting of 35 parts of methylcellulose, 65 parts of triethylene glycol diacrylate, 3 parts of anthraquinone, and 0.03 part of hydroquinone was uniformly coated on each of the two plates obtained above to a thickness of 0.7 millimeter. The plate was immediately exposed to ultraviolet rays through a negative film adhered intimately to the photosensitive resin layer via a 12-micron thick polypropylene sheet. At this time, the temperature of the surface of the film was 35°C. Those portions of the resin layer which were not polymerized by exposure were washed out with water to prepare a printing plate. Furthermore, the printing plate was sufficiently exposed to solidify the reliefs completely. The solidified portions were in the form of a square having each side measuring 10 millimeters and a height of 0.7 millimeter; and a rectangle having a size of 10 x 20 millimeters.

The test pieces obtained were each wound around a cylinder having a radius of curvature of 45, 20, 16, 12, 9 and 6.5 millimeters respectively at 20°C and the state of separation of the relief portions was observed. The minimum radius with which the separation occurred is shown in the table for each of the test pieces.

	Relief (10 x 10 mm)	Relief (10 x 20 mm)
Example	9 mm	16 mm
Comparative Example	16 mm	45 mm

In the comparison, separation occurred between the adhesive layer and the resin layer and the printing plate could not be used.

PLASTICS

PLASTIC COMPOSITIONS

Polyene plus Polythiol

The process of *J.L. Guthrie and S.V. Dighe; U.S. Patent 3,883,598; May 13, 1975; assigned to W.R. Grace & Co.* is directed to thiol-containing compounds having no hydrolytically sensitive groupings which are useful as polymerization transfer agents in curable polymer systems containing polyenes and a free radical initiator.

The thiol-containing compounds are formed by reacting a polyol with an allylic halide to form a poly(unsaturated ether) whose double bonds are converted to thioesters by free radical catalyzed addition of SH groups from a thiolcarboxylic acid, e.g., thioacetic acid and the like. Hydrolysis of the thioester yields the desired thiol or polythiol derivative. Alternatively the poly(unsaturated ether) is reacted with H_2S to give the desired thiol or polythiol directly in a single step.

Example 1: Triallyl Ether of Pentaerythritol – Into a 5 liter three-necked flask fitted with condenser and addition funnel was placed a solution of 650 g (16.25 equivalents) of sodium hydroxide in 650 ml of water. To this was added 272 g (two mols) of pentaerythritol. This mixture was stirred by means of a magnetic bar and heated to 70°C. Then 1,936 g (1,385 ml, 16 mols) of allyl bromide was added over an 8-hour period at such a rate that the temperature stayed between 70° and 80°C.

Following this, heating was resumed, keeping the temperature at 80° to 82°C for an additional four hours. Volatile materials were removed by distillation at atmospheric pressure until the temperature of the condensing vapor reached 98°C.

One liter of water was added to the hot residue (to prevent crystallization of the salts). The product was cooled to room temperature and the layers were separated. The water layer was extracted twice with 300 ml portions of diethyl ether. The combined organic layers were dried over anhydrous magnesium sulfate and then distilled at atmospheric pressure to remove the diethyl ether. The triallyl ether of pentaerythritol product weighed 451 g (88% conversion). The infrared and NMR spectra were those expected for the triallyl ether of pentaerythritol. The triallyl ether of pentaerythritol product had a boiling range of 120° to 121°C at 1 mm, n_D^{24} 1.4625.

Example 2: Conversion of Triallyl Ether to Trithiol — Ten drops of tert-butyl hydroperoxide was added to 85.2 g (1 equivalent of unsaturation) of pentaerythritol triallyl ether from Example 1 in a 500-ml flask equipped with a condenser and magnetic stirrer. This mixture was heated to 40°C and 76 g (1 mol) of thiolacetic acid was added during 1 hour at such a rate that the temperature did not exceed 90°C. After the addition was complete, the product was kept at 80°C for an hour and then allowed to cool to room temperature overnight.

To the product was added a solution of 100 g (2.5 equiv) of sodium hydroxide in 200 ml of water. This mixture was heated under reflux for three hours and then cooled to room temperature and diluted with 300 ml of ether to facilitate separation of the layers. After separation, the water layer was acidified to pH 2 to 5 with HCl and then extracted twice with 400-ml portions of ether. Ether was evaporated from the combined organic layers, and the residue was diluted with an equal volume of toluene.

The resulting solution was washed with 5% aqueous sodium bicarbonate, then with 5% aqueous hydrochloric acid, and with water. The toluene and other volatile contaminants were removed by distillation at 0.1 mm Hg until the temperature of the residue reached 225°C. The product [pentaerythritol tris(γ-mercaptopropyl) ether] weighed 119 g (100% conversion) and had a mercaptan content of 7.09 meq/g. This is 84.6% of the theoretical amount.

Part of the trithiol was distilled at 0.1 mm Hg. The distillate had a boiling range of 243° to 245°C and a mercaptan content of 7.88 meq/g (94% of theoretical value).

Example 3: A polyene was prepared in the following manner. A round bottom flask is fitted with a stirrer, thermometer, dropping funnel, nitrogen inlet and outlet. The flask can be placed in a heating mantle or immersed in a water bath as required.

Two mols (428 g) of trimethylol-propane diallyl ether were mixed with 0.2 cc of dibutyltin dilaurate under nitrogen. Then one mol of toluene diisocyanate was added to the mixture, using the rate of addition and cooling water to keep the temperature under 70°C. The mantle was used to keep the temperature at 70°C for another hour. Isocyanate analysis showed the reaction to be essentially complete at this time resulting in the following viscous product

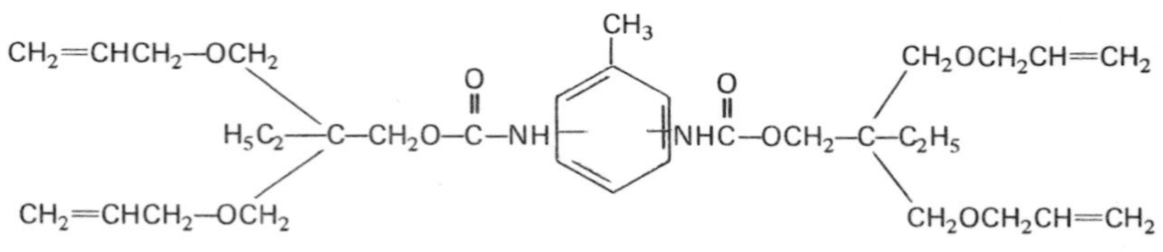

```
                                        CH3
CH2=CHCH2-OCH2                           |                    CH2OCH2CH=CH2
              \            O                       O         /
               \           ||                      ||       /
         H5C2---C-CH2O-C-NH-[benzene ring]-NHC-OCH2-C-C2H5
               /                                            \
              /                                              \
CH2=CHCH2-OCH2                                                CH2OCH2CH=CH2
```

which will be referred to hereinafter as Prepolymer A. 15.0 g (0.1 equiv of unsaturation) of Prepolymer A was admixed with 11.9 g (0.1 equiv of thiol) of the polythiol from Example 2 along with 0.3 g of benzophenone (a photocuring rate accelerator).

A 40 mil thick film of the admixture was placed under a 275 watt sunlamp delivering 4,000 microwatts/cm^2 at the film surface at a distance of 10 inches and irradiated with UV light for 5 minutes. The resulting elastomeric film had a tensile strength of 588 psi, elongation at failure of 54% and a tensile modulus at 1% elongation of 3,220 psi.

Example 4: An improved polythiol crosslinking agent was prepared by mixing 71.6 g (0.2 mol) of the polythiol from Example 2 with 17.4 g (0.1 mol) of toluene diisocyanate and thereafter allowing the mixture to stand for 1 day at room temperature, 25°C. The resulting viscous hexathiol was used to make a cured polymer film as in the previous example using 15.0 g (0.1 equiv) of Prepolymer A and 14.8 g (0.1 equiv thiol) of hexathiol.

After irradiation as in the previous example, the resulting glassy film had a tensile strength of 5,990 psi, an elongation at failure of 6% and a tensile modulus at 1% elongation of 152,000 psi.

Fatty Acid Modified Vinylated Polyester

T. Watanabe, K. Murata and K. Tsubouchi; U.S. Patent 3,882,007; May 6, 1975; assigned to Kansai Paint Co., Ltd, Japan provide a photohardenable fatty acid-modified vinylated polyester resin composition comprising

(A) from 10 to 60% by weight of a vinyl monomer, vinylidene monomer or a mixture thereof,

(B) from 40 to 90% by weight of a fatty acid-modified vinylated polyester resin prepared by reacting 1.0 mol of diisocyanate compound with 0.7 to 1.3 mols of vinyl or vinylidene monomer having a hydroxyl group to provide an isocyanate-terminated adduct, and then reacting 0.4 to 1.2 mols of the isocyanate adduct per 1.0 mol of a pendant hydroxyl of a fatty acid-modified polyester resin to obtain the fatty acid-modified vinylated polyester resin having from 0.2 to 3.0 α,β-olefinic unsaturation units per 1,000 units molecular weight of the resin, in which the fatty acid-modified polyester resin has a number average molecular weight of 800 to 5,000, a fatty acid content of about 5 to 35% by weight based on a total weight of the fatty acid and unmodified polyester resin, a number of hydroxyl groups of 0.5 to 3.0 units per 1,000 units molecular weight and an acid value of 5 to 30,

(C) metal salt dryer selected from cobalt, manganese, calcium, zinc, iron, zirconium, lead or copper salt in amount of 0.1 to 2.0% by weight based on the total amount of the fatty acid-modified vinylated polyester resin and vinyl monomer, and

(D) a photosensitizer in an amount of 0.05 to 10% by weight based on the total amount of the fatty acid-modified vinylated polyester resin and vinyl monomer.

In the examples parts are by weight unless otherwise indicated.

Example 1: 199 g of isophthalic acid and 88 g of adipic acid as the saturated dibasic acids and 84 g of neopentyl glycol and 164 g of trimethylol propane as the polyhydric alcohols were placed in a flask equipped with a thermometer, a water separator provided with a fractionator, a nitrogen introducing tube and a stirrer and heated to 140°C in an atmosphere of nitrogen gas.

Successively the temperature of the reaction system was gradually elevated from 160° to 220°C in 2 hours. The system became transparent after heating for one hour at 220°C, then 112 g of linseed oil fatty acid and hydroquinone as the polymerization inhibitor in an amount of 500 ppm based on the amount of linseed oil fatty acid were added, and the reaction was continued for 2 hours at 200°C. Then the reaction was switched from fusion process to solvent process by adding 25 g of toluene to the system, and the reaction was continued under dehydration until the acid value reached 15 and the viscosity reached F on a Gardner bubble viscometer (measurement of viscosity was carried out after dilution with Cellosolve acetate so as to obtain a resin content of 50 weight percent at 25°C; all viscosity herein was determined in the same manner).

The fatty acid-modified saturated polyester resin was solid at room temperature, and was changed to solution with a resin content of 50 weight percent by adding methyl methacrylate in order to facilitate handling.

Example 2: 174 g of tolylene diisocyanate was placed in a flask equipped with a reflux condenser, a thermometer and a dropping funnel and heated to 60°C. 143 g of 2-hydroxyethyl methacrylate was added dropwise for 2.5 hours. The speed of addition and agitation requires precise control since the reaction is exothermic.

After the addition of 2-hydroxyethyl methacrylate is completed, the reaction was continued until the isocyanate value reached 128 which was half of the theoretical value (the reaction required one hour in this example). The reaction product was solid at room temperature, and was therefore changed to a solution by adding methyl methacrylate so as to obtain a content of isocyanate-terminated adduct of 80 weight percent.

Example 3: 925 g of solution of fatty acid-modified saturated polyester resin prepared in Example 1, 208 g of isocyanate-terminated adduct obtained in Example 2 and 2,6-di-tert-butylhydroxytoluene as the polymerization inhibitor in an amount of 300 ppm based on the total amount of the fatty acid-modified saturated polyester resin and the isocyanate-terminated adduct were mixed and then reacted for 2 hours at 80°C.

The reaction was stopped when the viscosity of the reaction mixture reached D to E on a Gardner bubble viscometer at 25°C. Successively, 143 g of 2-hydroxyethyl methacrylate was added and reacted for 1 hour at 80° to 85°C to remove unreacted free isocyanate groups. There was obtained a solution of the fatty acid-modified vinylated polyester hereinafter referred to as Varnish A.

A solution of 100 parts of Varnish A, 0.2 part of benzoin ethyl ether and 0.5 part of a metal salt dryer consisting of 0.05 part cobalt naphthenate and 0.45 part lead naphthenate was prepared.

A test piece was prepared by applying a sample on a polished soft-steel plate of 0.5 mm thickness so as to obtain a coating thickness of 40 to 50 microns and hardening the coating by means of irradiation for 5 minutes with a high-pressure mercury lamp of 2 kw at a 15 cm distance from the coating.

The coating had a hardness of 3H and an adhesion rating of 100, an impact resistance of 40 cm (Parlin, ½", 500 g), an Erichsen depth of 5.0 mm. The compositions according to the process can be applied in various industrial fields such as molding, injection, laminating, finishing boards, adhesives and paints.

Polyester-Polythiol Compositions

In the process of *D.W. Larsen; U.S. Patent 3,959,103; May 25, 1976; assigned to W.R. Grace & Co.* a polar unsaturated polyester is photocured by irradiating (ultraviolet radiation) an admixture of a polythiol, an unsaturated polyester, and a photocuring rate accelerator. The unsaturated polyester is prepared by reacting mono- or polycarboxylic acids with mono- or polyhydric alcohols.

The double bonds are furnished by any of these components. At least one of the double bonds is situated internally to the main backbone chain of the polyester. The photocure is achieved by the addition of the sulfhydryl groups of the polythiol to the reactive double bonds of the polyester.

The polythiol is present in the composition in an amount between 5 and 70% by weight; the unsaturated polyester is present in an amount between 30 and 95% by weight; and the photocuring rate accelerator is present in an amount between 0.0005 to 10%, preferably 0.1 to 5.0% by weight of the polyester, and polythiol. Nearly stoichiometrical amounts of polythiol and unsaturated polyester, are usually used.

Whereas the products obtained by prior art vinyl polymerization techniques are brittle, the corresponding products obtained by the thioether forming reaction of this process are relatively less stiff, more flexible and more impact- or abuse-resistant.

Useful ultraviolet (UV) radiation has a wavelength in the range of about 2000 to 4000 A units. When ultraviolet radiation is used for the curing reaction a dose of 0.0004 to 6.0 watts/cm^2 is usually employed.

Example 1: 20 grams of dipropylene glycol maleate, 10 grams of Mercaptate Q-43 Ester, and 0.15 gram of benzophenone were thoroughly admixed in a glass container. Mercaptate Q-43 is pentaerythritol tetrakis(β-mercaptopropionate). The admixture was spread in a film about 40 mils thick and irradiated with UV light from a Westinghouse sunlamp (No. RS 275 watt) delivering 4000 μw/cm^2 at the film surface at a distance of 10 inches for 25 minutes. The admixture was thoroughly photocured to a tough and flexible product.

Example 2: 20 grams of a copolymer based on diethylene glycol, propylene glycol, isophthalic acid and maleic acid (anhydride), 5 g of Mercaptate Q-43 Ester and 0.15 gram of benzophenone were thoroughly admixed. The admixture was irradiated (for 25 minutes) as in Example 1. The admixture was thoroughly photocured to a tough product.

Example 3: Example 2 was repeated, except that half of the pentaerythritol tetrakis(β-mercaptopropionate) was replaced with 2.5 grams of ethylene glycol bis(β-mercaptopropionate). The admixture was thoroughly photocured to a tough and nontacky product.

Solid Polythioethers

The process of *D.W. Larsen; U.S. Patent 3,976,553; August 24, 1976; assigned to W.R. Grace & Co.* relates to curable liquid polyene-polythiol compounds, the means for preparing the compounds, and to methods for forming solid polythioether products using the compounds.

The curable liquid polyene-polythiol compounds provided by the process may be represented by the following general formula: $(HS)_p-(G)-(X)_q$ wherein X is $-(CR_2)_n-CR{=}CR_2$ or $-(CR_2)_n-C{\equiv}C-R$, n being a numeral from 0 to 9; p and q are numerals of at least one, and preferably at least 2, the sum of p plus q being always at least 3 and preferably at least 4; and G is a polyvalent organic moiety having a valence of at least 3, free of reactive carbon-to-carbon unsaturation, free of highly water-sensitive members, and necessarily containing at least one and preferably at least 2 polar groups selected from the group consisting of urethane, thiourethane, urea, carbonate ester, ester, thioester, amide, substituted amide, and tertiary amine.

The substituent on the substituted amide may be selected from the radical X as defined above and the members of the group from which the radical R is selected as set forth below. Where ester, thioester, amide and substituted amide are included as the contained group they are included in branches of the member G with the members –SH and –X being connected to the branches. Additional polar functionality such as, for example, carboxylic acid (–COOH), ether (–O–), thioether (–S–), nitrile (–C≡N), quaternary amine salt, chloride, bromide, etc., may be included.

The member G may include atoms selected from carbon, oxygen, nitrogen, sulfur, hydrogen, silicon, phosphorus, chlorine, bromine, etc. Typically, G consists of atoms selected from carbon, oxygen, nitrogen, sulfur and hydrogen.

R is a radical selected from the group consisting of hydrogen, chlorine, fluorine, furyl, thienyl, pyridyl, phenyl, substituted phenyl, benzyl, substituted benzyl, alkyl, substituted alkyl, alkoxy, substituted alkoxy, cycloalkyl, and substituted cycloalkyl. The substituents on the substituted members are selected from the group consisting of nitro, chloro, fluoro, acetoxy, acetamide, phenyl, benzyl, alkyl, alkoxy and cycloalkyl. Alkyl and alkoxy have from 1 to 9 carbon atoms and cycloalkyl has from 3 to 8 carbon atoms.

Example: A reaction mixture was formed by adding to a steam-jacketed stainless steel reaction kettle 103 grams (1 mol) of diethylenetriamine, 285 grams (2.5 mols) of allyl glycidyl ether, 150 grams (2.5 mols) of ethylene sulfide and 0.5 gram Ionol (to prevent premature –SH addition to the carbon-to-carbon double bond). A nitrogen blanket was provided on the kettle and a steel cover was secured thereon so as to seal the kettle. Steam was introduced into the jacket resulting in increasing the temperature of the reaction mixture to 95° to 100°C and increasing the kettle pressure to about 35 psig. Reaction was continued with heating and stirring until the reaction pressure was decreased to less than 5 psig.

Thereafter the reactor was vented to atmospheric pressure and excess volatile material was carried off using nitrogen as the carrier. The thus formed liquid polyene-polythiol product included an average of 2.5 equivalents per mol of product of $-\!\!(NCH_2CH_2SH)$ and 2.5 equivalents per mol of product of

$$-\!\!(NCH_2-\underset{\displaystyle OH}{\underset{|}{CH}}-CH_2-O-CH_2-CH{=}CH_2)$$

After dissolving 1 gram benzophenone (UV sensitizer) in the product, the sensitized liquid composition was spread into a thin film, and cured under a sunlamp to a solid elastomeric polythioether polymer.

The polyene-polythiol compounds, upon exposure to a free-radical generator, e.g., actinic radiation and preferably ultraviolet light, form cured polythioether products having many and varied uses. Examples of some uses include but are not limited to adhesives; caulks; elastomeric sealants; coatings; encapsulating or potting compounds; liquid castable elastomers; thermoset resins; impregnants for fabric, cloth, fibrous webs and other porous substrates; laminating adhesives and coatings; mastics; glazing compounds; fiberglass reinforced composites; sizing or surface finishing agents; filleting compounds; cure in place gasketing compounds; rocket fuel binders; foamable thermosetting resins or elastomers; molded articles such as gaskets, diaphragms, balloons, automobile tires, photoresists, photocurable printing plates, etc.

The increased polarity of the polyene-polythiols aids in forming cured polythioethers having greatly improved performance characteristics such as higher tensile strength, greater elongation, improved adhesiveness to substrates, and the like.

SPECIAL APPLICATIONS

Orthopedic Cast

J.A. Corvi and D.C. Garwood; U.S. Patent 3,881,473; May 6, 1975; assigned to Merck & Co., Inc. provide an orthopedic cast which is formed by a single exposure in a short period of time to ultraviolet light.

The cast comprises a highly light and air permeable nonirritating textile of interknit or interwoven yarns and an ultraviolet light curable resin system carried by the yarns, the textile being of a mesh having essentially regularly spaced generally rectilinear openings with side dimensions between about 0.045 and 0.150 inch and wales or warp yarns separating adjacent openings to provide an average thickness to the textile of between about 0.035 and 0.060 inch.

The yarns are preferably composed of a plurality of glass filaments having an individual diameter less than about 0.00021 inch. This diameter has been found to be less irritating to the skin of the body member. Preferably, the fabric is of a Raschel knit which affords a lateral stretchability without lengthwise stretchability in the tape.

The preferred resin system includes a polyester containing carbon-to-carbon unsaturation, an ethylenically unsaturated monomer, an inhibitor, and benzoin ethers and esters. The unsaturated monomer is normally employed in an amount equal to 5 to 50% by weight, based on the weight of the polyester. The photoinitiator is used in an amount from 0.1 to 5% by weight of the polyester.

The preferred ultraviolet band for the hardening of the resins is at a narrow band having a maximum energy and peak wavelength from about 3650 to 3720 A. This preferred band provides at least 50% of the total energy being concentrated in the band and a curing time to achieve the desired strength in as little as 2 or 3 minutes.

Nail-Coating Preparation

I. Rosenberg; U.S. Patent 3,896,014; July 22, 1975; assigned to Clairol Inc. describes a photocurable liquid nail lacquer composition comprising in photocurably effective amounts as separate components distinct from each other from 1 to 98.5% by weight of a mixture of a polyene component and 0.5 to 1.0% by weight of a polythiol component, 1 to 10% by weight of a photocuring rate accelerator component and a surfactant which is soluble in the liquid composition.

The surfactant is selected from the group consisting of sorbitan sesquioleate, sorbitan dioleate, sorbitan trioleate, pentaerythritol dioleate, pentaerythritol trioleate, glyceryl monooleate, glyceryl dioleate, glyceryl trioleate, polyglycerol ester of oleic acid, alkenyldimethylethylammonium bromide, dicocodimethylammonium chloride, and quaternary imidazolinium salt (from stearic acid).

The nail lacquer compositions of the process comprise generally a photocurable liquid system containing as essential ingredients at least one polyene, polythiol, photocuring rate accelerator, and one or more of a class of special surfactants described in more detail below. This liquid system is adapted to form a cohesive film when applied to human nails and exposed to actinic light and particularly ultraviolet light.

In related work *I. Rosenberg; U.S. Patent 3,928,113; Dec. 23, 1975; assigned to Clairol Inc.* provides a two-part nail coating system comprising as a first part (a) a basecoat composition of a water-soluble or water-swellable polymer in a solvent and as a second part (b), a photocurable nail lacquer composition adapted to be cured to a hard nail coating on its exposure to a light source as described above.

When the aforesaid basecoat composition is applied to human nails and allowed to dry and this is then followed with an overcoating of a photocurable nail lacquer which is cured by exposing it to an appropriate light source that both the basecoat and the nail lacquer composition are readily removable by the simple expedient of soaking the nails in water for a short period of time. This causes the basecoat to be released from the nail whereby the whole coating may be easily peeled off.

Filament Wound Article

J. Feltzin; U.S. Patent 3,922,426; November 25, 1975; assigned to ICI America Inc. provides an improved process for the preparation of filament wound articles. The process comprises coating or impregnating the filament with a resinous matrix comprising an ethylenically unsaturated polyester, ethylenically unsaturated monomer, organic peroxide, and at least one sensitizer characterized by the formula

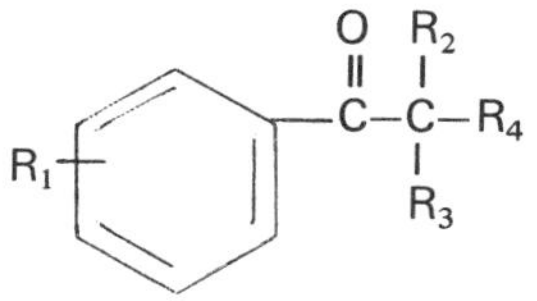

where R_1 is $-H$, $-Br$, or $-CH_3$, R_2 is $-H$ or $-Br$, R_3 is $-H$ or $-CH_3$, and R_4 is $-H$ or $-CH_3$, and wherein at least one of R_1 and R_2 is $-Br$, and subjecting the resin matrix to ultraviolet, electromagnetic radiation to copolymerize the unsaturated polyester and unsaturated monomer.

The resin matrix employed in the process has an almost indefinite pot life as the resin will cure only when activated by subjecting it to ultraviolet radiation. Upon exposure of the resin matrix to ultraviolet radiation, gelation of the resin matrix occurs rapidly, usually in less than 45 seconds. This provides a resin-rich inner layer because the rapid gelation eliminates the squeezing out of the polyester by the pressure of overlapping filaments being wound and the evaporation of the ethylenically unsaturated monomer.

The process also eliminates the need to wait for the resin catalyst to cure after the filament wound structure is fully fabricated. The filament wound structure is curing during the fabrication process itself and thereby results in a nearly fully cured structure at the end of the fabrication process. Furthermore, the process is safer and not as expensive as processes requiring the use of hard radiation. The process requires a relatively inexpensive ultraviolet light source and protection can be achieved with aluminum foil for shielding. In the examples, all parts and percentages are by weight, unless otherwise specified.

Examples 1 through 8: The ethylenically unsaturated polyester which is employed in the following illustrative examples is prepared according to the following procedure: A 3-liter, glass, round-bottom flask is fitted with mechanical stirrer, carbon dioxide inlet tube, temperature indicator, and distillation head. The flask is charged with 1,566 grams of polyoxypropylene (2.2) 2,2-di(4-hydroxyphenyl)propane. While the glycol is warmed and stirred, 512 grams of fumaric acid are added along with 1.04 grams of hydroquinone.

When addition to these ingredients is complete, carbon dioxide is bubbled into the mass, stirring rate is set at 130 rpm, and the reactants are heated to 210°C. These conditions are maintained for 6 hours, at which time the reaction product has an acid number of 31. The product is labeled "Polyester A." "Polyester B" is prepared by the foregoing procedure except that the glycol used is polyoxypropylene (16) 2,2-di(4-hydroxyphenyl)propane. Four parts of Polyester A and one part of Polyester B are dissolved in five parts of styrene and the resulting solution is employed in Examples 1 through 8.

Example No.	Sensitizer	Peroxide	UV Curing Time (minutes)	Light Distance (inches)	Numbered Passes	Barcol* Hardness	Glass (percent)	Split D Tensile Strength (psi)	Average Interlaminar Shear (psi)	Compressive Modular 10% Deflection (psi)**	Inside Diameter (inches)
1	1% BMPP***	1% TBPP†	38	5	6	20-35	68.9	-	-	-	3.5
2	1% BMPP	2.5%††	37	5	6	30-50	63.5	-	-	-	3.5
3	1% BMPP	2.5%††	40	5	6	47-53	65.1	-	-	-	3.5
4	1% BMPP	2.5%††	40	3.5	6	28-22	61.0	92,400	3,875	58	6
5	1% BMPP	1%††	50	3.5	10	48-51	57.8	92,300	4,041	88	6
6	1% BMPP	1% TBPP	50	3.5	10	43-48	59.4	94,100	4,190	85	6
7	1% BMPP	1.25%††	50	3.5	10	38-42	55.0	80,500	4,395	111	6
8	1% BMPP	1.25%††	40	2	10	35-41	49.9	77,000	3,592	128	6

*Inner surface.

**ASTM D-2412.

***BMPP = 2-bromo-2-methylpropiophenone.

†TBPP = 75% solution of tert-butyl peroxy pivalate.

††Fine dispersion of peroxides (UV-50).

A glass filament is passed through a system of pulleys which provide a tension of about 2 pounds and then passed through a small hole which squeezes off excess resin. The filament saturated with resin is then passed through a ring which is attached to a traverse which guides the filament horizontally along a rotating mandrel. The traverse is set so that it takes 5 minutes to travel from one end of the mandrel to the other and the mandrel speed is adjusted so that even winding results with no overlapping and no gaps between each revolution of winding.

As the filament reaches the opposite end of the mandrel, the traverse automatically reverses and winds on top of the previous layer. The number of passes depends on the desired thickness of the filament wound structure. The light source used is a Hanovia high pressure quartz mercury vapor lamp, Model 819A. The ultraviolet light source is placed above the rotating mandrel.

After the last layer is wound, the mandrel and traverse are stopped and the ultraviolet light is left on for 10 minutes to provide curing of the final layer. The ultraviolet light is then turned off and the system is allowed to cool. The resin bath comprises the indicated sensitizer and the indicated organic peroxide dissolved in a resin solution prepared by dissolving one part of the above described polyester solution in one part of styrene.

The table on the preceding page shows various combinations of sensitizer, organic peroxide, distance light is above mandrel, curing time, number of passes or layers, inside diameter of pipe, percent glass, and properties of the resulting filament wound article.

Steel Strip Composite for Transformer Core Laminate

S.H. Schroeter, J. Van Winkle and C.B. French; U.S. Patent 3,924,022; Dec. 2, 1975; assigned to General Electric Company provide a continuous, substantially pollution-free method for uniformly improving the surface resistivity of oriented steel strip having an inorganic material as a surface insulating coating to produce an oriented steel strip-inorganic coating-organic resin composite capable of providing an average Franklin test reading of from 0.1 to 0 amp at a pressure of up to 300 psi.

The process comprises: (1) treating the oriented steel strip with an irradiation curable solventless organic resin to a thickness of up to 0.2 mil, and (2) passing the treated oriented steel strip through an irradiation curing zone at a rate of from 100 to 600 fpm, using an irradiation flux sufficient to effect the cure of the solventless organic resin to an organic resin with a hardness capable of being tested by the Franklin test.

The resin is characterized by having a viscosity of up to 3,000 cp at 25°C, and is a mixture consisting essentially of from 99 to 1% by weight of organic monomer and correspondingly from 1 to 99% by weight of organic polymer, where the irradiation curable solventless organic resin is capable of providing with 10C hydrocarbon oil, an interfacial tension reading of from 30 to 40 dynes/cm in accordance with ASTM D971-50 (1970).

Preferred resins include UV curable wax containing polyesters having a viscosity of from 500 to 3,000 cp containing the following essential ingredients by weight

(A) 20 to 40% of a vinyl aromatic material selected from styrene, vinyl toluene, tert-butylstyrene and mixtures thereof,

(B) 80 to 60% of an unsaturated polyester reaction product of (1) a glycol and (2) an aliphatically unsaturated organic dicarboxylic acid, where:

(1) is a glycol selected from the class consisting of
  (a) a mixture of 20 to 60 mol percent of neopentyl glycol and 40 to 80 mol percent of a member selected from propylene glycol, ethylene glycol, and mixtures thereof, and
  (b) a mixture of (a) and 1 to 40% by weight thereof of trimethylolpropane monoallyl ether, and
(2) is an aliphatically unsaturated organic dicarboxylic acid consisting essentially of a mixture of 50 to 65 mol percent of fumaric acid and 35 to 50 mol percent of a member selected from tetrahydrophthalic anhydride, endomethylene tetrahydrophthalic anhydride and mixtures thereof

where there is utilized in making the unsaturated polyester reaction product of (B), up to at least about 10 mol percent excess of the glycol of (1) over the aliphatically unsaturated organic dicarboxylic acid of (2),

(C) 1 to 5% based on the weight of (A) and (B) of a UV sensitizer, and

(D) 0.05 to 1% based on the weight of (A), (B) and (C) of paraffin wax.

Cure of the organic resin is preferably effected by using UV irradiation which can have a wavelength of from 1849 to 4000 A. The composite can be used to make power transformer core laminate.

Low-Pilling Polyester Fiber Products

K.H. Magosch, R. Feinauer and J. Ruter; U.S. Patent 3,960,686; June 1, 1976; assigned to Chemische Werke Huls AG, Germany describe an improved process for the manufacture of low-pilling effect polyester fibers.

This object is achieved by irradiating threads, fibers or fabrics made of polyesters containing from about 0.05 to 5.0 mol percent of a substituted cyclobutanedimethanol, preferably 2,4-diphenyl-cyclobutanedimethanol-1,3, by means of high energy light. Aside from its content in substituted cyclobutanedimethanol, the polyester suited for the process consists wholly or predominantly of polyethylene terephthalate.

Ultraviolet light, particularly that with wavelengths from about 250 to 400 millimicrons, is suitable as the high energy light.

The polyesters are extrusion spun in conventional manner. Appropriately, the fibers or threads are first stretched and then subjected to UV irradiation. However, subsequent treatment of the fabrics so made is also possible. Irradiation takes place so that the threads, the fibers or the fabric are at rest or moving, i.e., carried underneath or above the irradiation source. The length of irradiation is determined by the desired pilling effect and depends on the distance from and the intensity of the radiation source. Generally this length is from one second to thirty minutes. The distance between the irradiated sample and the radiation source is varied from five to one hundred centimeters. This distance depends on irradiation time and on the intensity of the source, which may be varied from 25 to 1,000 watts. More intense radiation is employed where appropriate.

Example: The following components are introduced into an esterification reactor provided with a stirrer, a double heating jacket and a fractionating column: 1,940 g of dimethyl terephthalate (DMT), 13.4 g of 2,4-diphenyl-cyclobutane-dimethanol-1,3 (corresponding to 0.5 mol percent referred to DMT), 1,240 g of ethylene glycol, and 0.6 g of zinc acetate.

The mixture is heated and at about 150°C, methanol begins distilling off. After methanol dissociation at 200°C has ceased, this generally will be the case in approximately 3 hours, the esterification product is fed into the polycondensation reactor.

Then, 1.2 g of triphenyl phosphate and 0.194 g of antimony oxide are added. Then the temperature is continuously increased to 280°C, while gradually decreasing the pressure to about 0.1 torr. Polycondensation is stopped when power absorption while stirring the reactor contents corresponds to a melt viscosity of about 900 poises at 285°C.

The RSV value of the colorless polyester (measured in phenol/tetrachloroethane 60/40 at 25°C) is 0.60 dl/g and the melting point as determined from differential thermal analysis is 259°C.

The material is spun through an 8 hole spinneret, the holes being 0.25 mm in diameter, at 315°C, and the yarn is wound with a take-off speed of 640 meters per minute. Thereafter the yarn is stretched at a ratio of 4:1. Yarn properties are as follows: titer, 36/8; tear resistance, 2.98 p/dtex; and elongation, 35.7%. The yarns subsequently are irradiated from a high pressure mercury lamp of 125 watts at a distance of 10 cm away for a time of 0 to 10 minutes.

The DKZ [wirekink rupture coefficient, described in *Chemiefasern* 12, 853 (1962), a measure of pilling effect] is determined at single capillaries after various times. The average values from 48 individual measurements are listed below as a function of time of irradiation.

Time of irradiation (min)	0	1	2	5	10
DKZ (lifts)	2,200	2,105	1,130	560	260

For purposes of comparison, polyester fibers with the following properties—titer, 36/8; tear resistance, 3.12 p/dtex; and elongation, 35.7%—containing no 2,4-diphenyl-cyclobutanedimethanol-1,3 are irradiated under the same conditions. As shown by the values listed below, the DKZ decreases much more slowly than for the yarns containing 2,4-diphenyl-cyclobutanedimethanol-1,3.

Time of irradiation (min)	0	1	3	5	10
DKZ (lifts)	2,230	2,100	1,470	1,370	1,150

When the ester interchange catalyst that was used is replaced by calcium or manganese acetate, or when the polycondensation catalyst that was used is replaced by germanium dioxide or gallium lactate, similar results are obtained.

When p-methoxy disubstituted 2,4-diphenyl-cyclobutanedimethanol-1,3 is used instead of 2,4-diphenyl-cyclobutanedimethanol-1,3, a similar result is obtained.

If besides the dimethyl terephthalate 5, 10 or 15 mol percent of isophthalic acid dimethyl ester, naphthalenedicarboxylic acid-2,6-dimethyl ester, respectively, besides the ethylene glycol 5, 10 or 15 mol percent of 1,4-dimethylolcyclohexane, butanediol-1,4 or neopentylglycol are used, then a comparable DKZ decay is obtained.

In related work by *R. Feinauer, K.H. Magosch and J. Rüter; U.S. Patent 3,953,405; April 27, 1976; assigned to Chemische Werke Huls AG, Germany,* improvement comprises the addition of a substituted cyclobutanedicarboxylic acid, preferably α-truxillic acid, to the melt-polymerization at a concentration of about 0.05 to 5 mol percent of the acid component. Melt spinning of the modified polyester is carried out and the fibers are irradiated with high energy light. Similar results are obtained.

Contact Lens with Improved Wettability

The process of *P.C. Hoell; U.S. Patent 3,978,341; August 31, 1976; assigned to E.I. Du Pont de Nemours and Company* provides apparatus for treating a shaped article prepared from a synthetic organic polymer. The apparatus comprises a means for supporting the shaped article in a radiation zone and electrical circuitry capable of producing ultraviolet radiations having a wavelength in the range of about 830 to about 1335 A in an atmosphere of air.

The electrical circuitry consists of first and second electrodes spaced apart for providing a discharge arc in the radiation zone. An electrical circuit is connected to the electrodes for providing energy pulses thereto. The circuit includes a capacitor and an energy source for repetitively energizing the capacitor. An exciting electrode adjacent the first and second electrodes for initiating the discharge may also be provided together with a means for limiting current between the exciting electrode and the capacitor to prevent the capacitor from discharging through the exciting electrode.

Preferably the energy source is a transformer capable of developing a voltage across the first and second electrodes of at least 1 kv per mm of spacing. A switching and timing means are connected to the energy source to energize the transformer primary.

This process also involves a method for improving the surface properties of a shaped article prepared from a synthetic organic polymer comprising the following steps. The article is positioned in the presence of an atmosphere of air in a radiation zone and the surface of the article is repetitively subjected to ultraviolet radiation in the range of about 830 to 1335 A for a predetermined period.

The radiation is generated by a discharge arc between the first and second electrodes which are energized by a voltage of at least 1 kv per mm of spacing between the electrodes.

Preferably the radiation is generated in repetitive bursts of 0.1 to about 1 second, with each burst being followed by a cooling period of about 1 to 6 seconds. The shaped article is preferably fabricated from a copolymer of 20% methyl methacrylate and 80% perfluoroalkylethyl methacrylates and may be used for preparation of a contact lens having improved wettability. Drawings of the apparatus are included in the patent.

Example: A copolymer sample of 20% methyl methacrylate and 80% perfluoroalkylethyl methacrylates was irradiated with the electrical discharge circuit described hereinbefore with the sample placed 3.5 ± 0.5 mm from the plane of the arc discharge. The electrodes were spaced 7 mm apart and the exciting electrode was adjusted to discharge the capacitor at 8 to 12 kv (variable due to electrode conditions). The number of operating cycles, "on" time and "off" time are listed below.

Measurement of the contact angle was made with a horizontal, low magnification microscope equipped with a protractored eye piece. The following results were obtained.

"on" (sec)	"off" (sec)	Number of Operating Cycles	Sample No.	θ_n°	θ_i°
.25	.75	40	1	99-109.5	22.5-29
			2	98.5-106.5	19-24.5
			3	91-93	21-30.5
.5	1.5	20	1	84.5-93.5	19-26
			2	97-98.5	22-25
			3	96-101	16.5-21
1.0	3.0	10	1	96-103.5	22-31
			2	99-101	22-25
			3	99.5-102	16.5-21

θ_n = advancing contact angle of water for non-irradiated sample.
θ_i = advancing contact angle of water for irradiated sample.

Lamp Base Using UV Curable Adhesive

M.G. Shinn and B.G. Fransen; U.S. Patent 3,982,185; September 21, 1976; assigned to General Electric Company provide a lamp-basing process capable of continuous operation.

In the process, at successive stations on a closed loop system, wire lamps are received, an ultraviolet curable adhesive is applied and the wire lamps are joined to the bases, UV radiation is applied to bond the two together, the wire leads are soldered or spot-welded to the base, and the lamps are removed. Alternatively, the adhesive can be applied to the base.

The curing depends upon the intensity, duration, and wavelength of the incident radiation. Also, as is generally true of chemical reactions, the curing is somewhat dependent upon the temperature of the adhesive. The table on the following page summarizes process parameters which must be observed in carrying out this process.

	Range	Preferred
Seal temperature at adhesive application	<300°F	250°F
Base temperature	ambient - 180°F	150°F
UV curing intensity	8-75 μw/cm^2	25 μw/cm^2
UV wavelength	3200-4000 A	3650 A
Cure time	<20 sec	5 sec

ADDITIVES

Quaternary Ammonium Compound

For many industrial purposes, for example the production of putty or for use in the lining of plywood, chipboard or metal panels, it is particularly important to be able to photochemically cure compositions comprising inert fillers incorporated into the unsaturated polyester resins.

S. Vargiu and B. Passalenti; U.S. Patent 3,917,522; November 4, 1975; assigned to Societa Italiana Resine SIR SpA, Italy found it possible to obtain compositions of photochemically curable unsaturated polyester resins which are stable when stored for long periods of time.

The essential aspect of this process is the presence of the quaternary ammonium compounds in the compositions. In order to guarantee that the compositions will have the best characteristics of stability in the course of time, the quaternary ammonium compounds should be present in quantities from 0.005 to 1 part by weight with respect to 100 parts by weight of unsaturated polyester resin.

As quaternary ammonium compounds, compounds belonging to the class which can be represented by the following structural formula can be used:

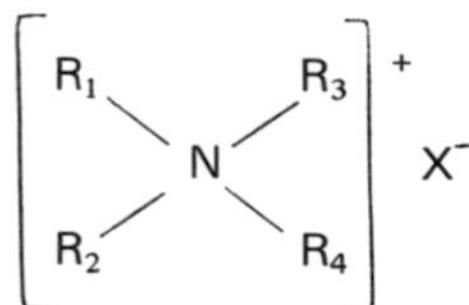

in which R_1, R_2, R_3 and R_4 represent alkyl, aryl (including phenol), or aralkyl radicals, while X represents a halogen.

Trimethylbenzylammonium chloride, triethylbenzylammonium chloride, tetramethylammonium chloride and tetraethylammonium chloride have been found particularly suitable for the purpose.

Example 1: Phthalic anhydride, maleic anhydride and propylene glycol in a molar ratio 1:1.25:2.42 were introduced into a glass vessel fitted with an agitator, thermometer and an inert gas injector system. The temperature was gradually raised to approximately 190°C and these conditions were maintained until a product was obtained which had a Gardner viscosity of U to V (measured in 66% styrene solution) and an acid number of 45 to 50.

This product was then brought to 100°C and diluted with styrene containing 20 ppm hydroquinone, 30 ppm of para-tertiary-butylcatechol and 20 ppm copper naphthenate.

Cooling to ambient temperature brought the viscosity to approximately 1,000 cp, measured by means of a Hubbelhode viscosimeter. To 100 parts by weight of the resin produced in this way, was added 1 part by weight of photosensitive substance, consisting of benzoin isopropyl ether and 0.1 part by weight trimethylbenzylammonium chloride, previously dissolved in propylene glycol in a quantity equal to 20% by weight.

The gelling time of the resin was regulated, by the addition of para-tertiary-butylcatechol, to approximately 6 to 9 minutes at 25°C. The gelling time was determined in the following manner. To 100 g unsaturated polyester resin were added 0.2 ml of cobalt octoate solution with a 6% cobalt metal content.

After blending in a thermostatically controlled bath kept at 25°C, 1.5 ml of 50% methyl ethyl ketone peroxide were added. Rapid homogenization was performed and time measurement commenced.

The progress of viscosity was followed by means of a glass rod every 30 to 60 seconds. The gelling time can be defined as the period of time which elapses from homogenization of the mixture with the peroxide to the moment at which resin raised with the glass rod clings to it.

The product obtained as above, constituted essentially of the unsaturated polyester resin, the photosensitive substance and the quaternary ammonium compound, was then charged to a vessel fitted with an agitator where, while stirring, 50 parts by weight talcum, 25 parts by weight barium sulfate, 70 parts by weight calcium carbonate, 30 parts by weight Asbestine and 1 part by weight silica were added.

Then a further 50 parts by weight styrene were added and the mixture was maintained under agitation until thoroughly homogenized. The resultant composition was particularly suitable for putties.

The accelerated stability was then determined as follows. 50 g of the product were placed in a glass phial of approximately 50 ml capacity. After the phial had been closed, it was placed in a thermostatically controlled oven set at 65°C. The appearance and consistency of the product were checked every 15 to 20 hours, once the sample had been reduced to room temperature. When it was found that the product in the phial had hardened, the test was regarded as completed.

The time elapsing between introduction of the phial into the oven and the appearance of the hardened product was regarded as the accelerated stability time. It is presumed that a stability of 1 day at 65°C corresponds to the stability of 1 month at room temperature. The accelerated stability at 65°C, determined in this way, proved to be 7 days.

Example 2: The same procedure was adopted as is described in Example 1—0.2 part by weight trimethylbenzylammonium chloride being however used to 100 parts by weight of resin. A product was obtained with an accelerated stability of 12 days at 55°C. The hardness was then determined by Albert Koenig's method. For this purpose, part of the composition obtained was spread on a glass slide measuring 8 x 17 cm in the form of a film 200 microns thick.

After the slide has been left exposed to the air for 10 minutes, it was placed in a tunnel and exposed to irradiation from an Osram L40 W73 40 watt lamp for 5 minutes. In particular, a distance of 10 cm was maintained between the source of radiation and the film.

After irradiation, the slide was allowed to cool to room temperature for 10 minutes and then the hardness was determined by means of an Albert Koenig pendulum. For this purpose, the slide was placed on the appropriate bed and the pendulum released.

Once the oscillatory movement had dropped to position 3, counting of the seconds commenced. The hardness, using the Albert Koenig method of evaluation was 51 seconds.

Stabilizer for Unsaturated Polyester

The process of *L. Roskott and A.A.M. Groenendaal; U.S. Patent 3,969,206; July 13, 1976; assigned to Akzona Incorporated* relates to stabilization of unsaturated polyester resins, preferably those containing filler, which are curable by ultraviolet light and which contain a benzoin compound as a photosensitizer and a quinone compound as a stabilizer.

If the unsaturated resins are to be applied as putties or applied on solid substances, such as wood-chip-plates and flaxfiber-plates, fillers must be incorporated in the polyester resins. However, these fillers have an unfavorable influence upon the stability of the polyester resin. It has been found that by incorporating a compound having the general formula:

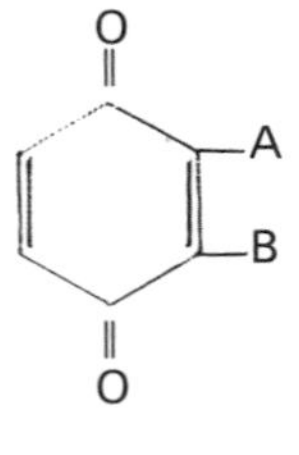

wherein A represents a methyl group and B is a hydrogen atom, or A and B together represent a substituted or nonsubstituted benzo group, into ultraviolet curable unsaturated polyester resins, especially filler-containing polyester resins with a content of benzoin compounds as photosensitizers, a composition may be obtained which has not only good stability, but moreover exhibits good curing under irradiation by ultraviolet light.

Examples of compounds to be used as stabilizers according to the process are methyl-p-benzoquinone and naphthoquinone-1,4. The stabilizers to be used according to the process are incorporated in the unsaturated polyester resin in quantities ranging from 0.5 to 2% by weight, preferably in a quantity of 1.0% by weight, calculated on the photosensitizer.

The stabilizing effect of the compounds to be used according to the process may be still further improved by the additional incorporation of quaternary ammonium compounds into the unsaturated polyester resin, preferably in combination with iron or copper in the form of a compound soluble in the resin, such as copper naphthenate or iron chloride, in quantities preferably from 1 to 4 ppm of iron or copper and from 0.02 to 0.05% by weight of a quaternary ammonium compound calculated on the polyester resin.

Example: 2 parts by weight of n-butyl ether of benzoin, 0.02 part by weight of methyl p-benzoquinone and 0.05 part by weight of alkyldimethylbenzylammonium chloride were dissolved in 100 parts by weight of an unsaturated polyester resin. Subsequently, 100 parts by weight of calcium magnesium silicate were admixed. The stability in the dark at 100°C and the residual styrene-content of the composition thus obtained were determined after illumination for 30 seconds.

Comparable tests were made with a composition containing solely the n-butyl ether of benzoin and the calcium magnesium silicate, and with a composition containing in addition the alkyldimethylbenzylammonium chloride and/or the methyl p-benzoquinone. The compositions used and the results obtained are set forth in the table below.

2% of benzoin n-butyl ether	100% of calcium magnesium silicate	0.05% of alkyldimethyl benzyl ammonium chloride	0.02% of methyl p-benzoquinone	stability in min	residual styrene-content
+	+	−	−	21	1.8
+	+	+	−	28	1.8
+	+	−	+	133	1.9
+	+	+	+	150	2.0

PHOTOINITIATORS

FOR COATING AND MOLDING COMPOSITIONS

Butyric Acid Derivatives

According to *H. Rudolph, H.-G. Heine, K. Fuhr and H. Schnell; U.S. Patents 3,891,524; June 24, 1975; 3,914,166; October 21, 1975; and 3,915,823; October 28, 1975; all assigned to Bayer AG, Germany* γ-hydroxy-γ-phenyl-γ-benzoylbutyric acids, their esters and amides are useful photosensitizers for the photopolymerization of polymerizable compounds or compound mixtures including mixtures of unsaturated polyesters and copolymerizable monomeric compounds. The photosensitizers are compounds of the formula

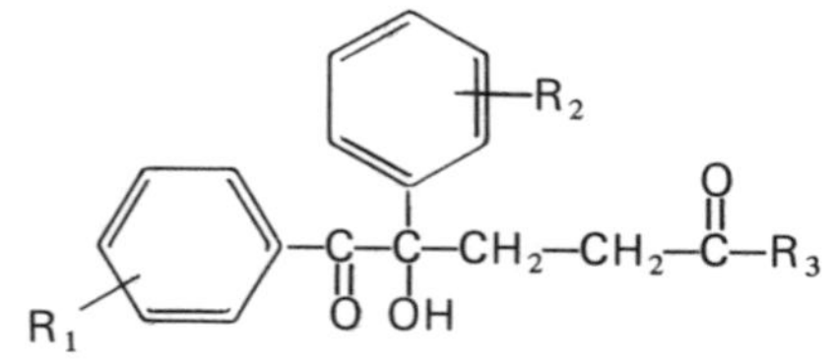

in which R_1 and R_2 are identical or different and represent hydrogen, lower alkyl having 1 to 4 carbons, methoxy or halogen, R_3 represents —OH, —OR_4 or optionally substituted —NH_2, and R_4 represents lower alkyl having 1 to 4 carbons; and the lactones of formula

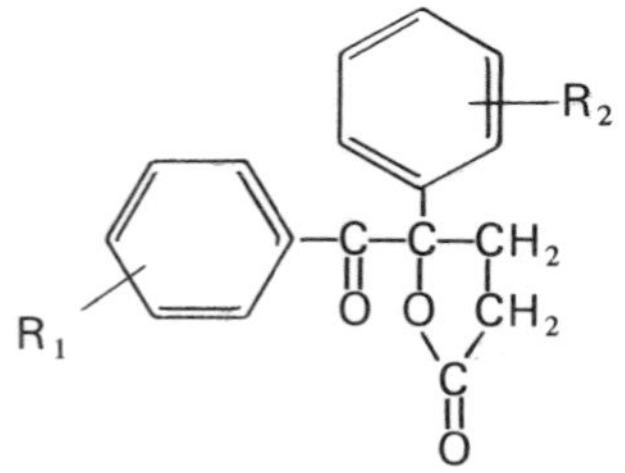

They are appropriately employed in amounts of from about 0.1 to 5% by weight, preferably of from about 0.5 to 2.5% by weight, either by themselves or mixed with one another or mixed with photosensitizers of a different nature. Suitable polymerizable compounds are all substances of which the carbon-carbon double bonds are activated by, for example, halogen atoms or carbonyl, cyanide, carboxyl, ester, amide, ether or aryl groups as well as carbon double bonds and carbon triple bonds.

By copolymerizable monomeric compounds there are to be understood the unsaturated compounds which are usual in polyester technology, having vinyl groups which are optionally substituted in the α-position, or allyl groups which are optionally substituted in the β-position, preferably styrene.

As radiation sources for carrying out the photopolymerization, it is possible to use natural sunlight or artificial radiation sources of which the emission lies in the range of about 250 to 500 mμ, preferably from about 300 to 400 mμ. Mercury vapor lamps, xenon lamps or tungsten lamps are for example suitable. The compositions also rapidly cure to give colorless molded and coated articles under the ultraviolet and visible radiation of low energy fluorescent lamps emitting rays of about 300 to 580 mμ.

Example 1: An unsaturated polyester manufactured by condensing 152 parts by weight of maleic anhydride, 141 parts by weight of phthalic anhydride and 195 parts by weight of propanediol-1,2 is mixed with 0.045 part by weight of hydroquinone and dissolved in styrene to give a 65% by weight solution.

Two parts by weight of two different known photosensitizers and three different photosensitizers according to the process are added to 100 parts by weight at a time of the resin solution and the mixtures are stored at 60°C, with exclusion of light, until they gel. The table below gives the sensitizers and the storage stability values at 60°C.

Sensitizer	Storage Stability at 60°C, days
Benzoin	<1
Benzoin ethyl ether	<1
γ-hydroxy-γ-phenyl-γ-benzoylbutyric acid	>10
γ-hydroxy-γ-phenyl-γ-benzoylbutyric acid-γ-lactone	>10
γ-hydroxy-γ-phenyl-γ-benzoylbutyric acid methyl ester	>10

Example 2: 20 parts by weight of styrene, 1 part by weight of a 10% by weight solution of paraffin (melting point 52° to 53°C) in toluene, as well as benzoin or photosensitizers according to the process in equimolar amounts, are mixed into 100 parts by weight at a time of the resin solution described in Example 1. The solutions thus obtained are applied to glass plates by means of a film spreader (500 μ) and are illuminated with radiation from a fluorescent lamp (Osram L 40W/70-1) at a distance of 5 cm.

The table on the following page gives the times for the paraffin to float to the surface and the times until a pencil hardness >6H is reached.

Sensitizer	Additive, parts by weight	Paraffin Surfacing Time, minutes	Pencil Hardness >6H, minutes
Benzoin	1.06	4.9	20.0
γ-hydroxy-γ-phenyl-γ-benzoylbutyric acid	1.42	1.9	16.0
γ-hydroxy-γ-phenyl-γ-benzoylbutyric acid methyl ester	1.49	2.6	19.0
α-(β-carbonamido-ethyl)benzoin	1.42	1.9	16.0

Benzoyl Derivatives of Diphenyl Sulfide plus Amine

A. Ravve, T. Jondahl, G. Pasternack and K.H. Brown; U.S. Patent 3,903,322; September 2, 1975; assigned to Continental Can Company, Inc. provide compositions sensitive to rapid polymerization by exposure to a source of ultraviolet radiation, the compositions being comprised of a photopolymerizable ethylenically unsaturated compound and a photoinitiating amount of a benzoyl derivative of diphenyl sulfide and an organic amine compound. The benzoyl derivatives of diphenyl sulfide useful as photoinitiators in the process are represented by the formula

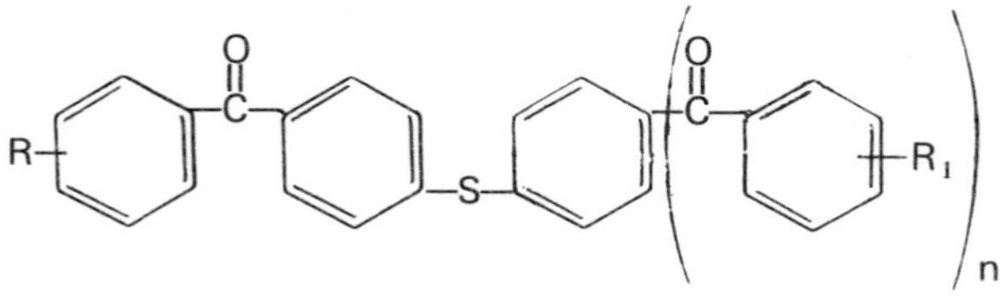

$$R-C_6H_4-\overset{O}{\overset{\|}{C}}-C_6H_4-S-C_6H_4\left(-\overset{O}{\overset{\|}{C}}-C_6H_4-R_1\right)_n$$

where each of R and R_1 is hydrogen, alkyl of 1 to 10 carbons, halogen, alkoxy, aryl, alkaryl arylalkyl, nitro or cyano groups and n is zero or 1. The benzoyl derivatives of diphenyl sulfide may be incorporated in the compositions at concentrations ranging from 0.1 to 10% by weight.

Organic amine compounds which may be used in combination with the benzoyl derivatives of diphenyl sulfide in the process are organic amine compounds having at least one alpha

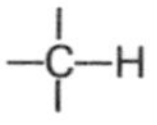

$$-\overset{|}{\underset{|}{C}}-H$$

group attached to the amino nitrogen and include primary, secondary and tertiary, aliphatic, heterocyclic and aromatic amines. The organic amine compounds are incorporated in the compositions at a concentration of about 0.1 to 10% by weight.

Example: A clear coating material composed of a photopolymerizable mixture of ethylenically unsaturated ester compositions was prepared and consisted of 272 parts of the diacrylate reaction product of bisphenol A-diglycidyl ether and acrylic acid, 190 parts pentaerythritol tetraacrylate, 28.7 parts of an unsaturated polyester prepared by reacting equal molar amounts of itaconic acid and bisphenol A-diglycidyl ether and 15.3 parts of tetraethylene glycol diacrylate.

A series of photopolymerizable coating compositions were prepared wherein 4.05×10^{-4} mol of a benzoyl derivative of diphenyl sulfide and 2.56% by weight triethanol amine were added to the ester mixture. These compounds and their concentrations in the ester mixture are listed in the table below. The benzoyl derivatives of diphenyl ether were prepared by following the procedure for the preparation of benzoyl diphenyl sulfide which is described immediately below.

In a one-liter round-bottom three-necked flask equipped with a reflux condenser, mechanical stirrer, thermometer and a dropping funnel were placed a reaction mixture of 0.5 mol (93 grams) diphenyl sulfide, 0.5 mol aluminum chloride (66.7 grams), and 200 ml tetrachloroethylene. The reaction mixture was stirred and cooled to 10°C. One-half mol of benzoyl chloride dissolved in 100 ml of tetrachloroethylene was added dropwise to the reaction mixture over a one-hour period.

After all the chloride salt had been added, the reaction mixture was stirred at room temperature, heated to 40°C to evolve HCl, and then stirred for an additional 2 hours. The reaction mixture was cooled and poured from the flask onto ice. The tetrachloroethylene solution was then sequentially washed with the dilute HCl solution, dilute Na_2CO_3 solution and water. The tetrachloroethylene was removed and the solid product obtained was purified by recrystallization.

For purposes of contrast, a series of coating compositions were prepared wherein 4.05×10^{-4} mol or more of a variety of aromatic carbonyl compounds outside the scope of the process were substituted for the benzoyl derivatives of diphenyl sulfide in the photopolymerizable ester mixture of the example. These aromatic carbonyl compounds and their concentrations in the ester mixture are also listed in the table and designated by the symbol C.

All of the benzoyl derivatives of diphenyl sulfide used in the example as well as the aromatic carbonyl compounds used in the comparative examples were purified by being chromatographed on a column of silica gel prior to use. Chlorothioxanthone was first chromatographed and then recrystallized to insure the purity of this compound.

The photopolymerizable compositions containing the photoinitiating compounds were then applied to steel plate of the type used in the manufacture of steel beverage containers using a number 10 draw bar which evenly distributed the compositions as thin film on the plate.

After application of the photopolymerizable compositions, the coated plates were placed under two low-pressure mercury lamps at a distance of about 2.75 inches from the lamp surface. The radiation emitted by the lamps was approximately 3 watts per inch of lamp surface. The coated plates were exposed to the ultraviolet radiation for 4 to 8 seconds to effect drying. The dryness of the irradiated coating was evaluated by rubbing the coating with one's fingers. The irradiated coatings received the following ratings:

Cured: Hard finger rubbing of coating surface does not rupture coating film.

Tack: Dry surface, but some rupture of coating film when rubbed with fingers, sl tack = slightly tacky, vsl tack = very slightly tacky.

Wet: Coating film has same physical state as existed prior to irradiation.

The results of the coating runs are recorded in the table below. In a comparative series of runs, the coating procedure was repeated with the exception that the coating formulations designated by the symbol C in the table were used. The results of the comparative coating runs are also recorded.

Composition No.	Photoinitiator	Concentration (% by weight)	Evaluation of Coating After Exposure to UV Radiation for 4 sec	6 sec	8 sec
1	Benzoyl diphenyl sulfide	1.14	vsl tack	vsl tack	cured
2	Benzoyl diphenyl sulfide	1.80*	vsl tack	cured	
3	Dibenzoyl diphenyl sulfide	1.53	vsl tack	cured	-
4	Di-o-toluoyl diphenyl sulfide	1.62	cured	-	-
5	Di-m-chlorobenzoyl diphenyl sulfide	1.80	sl tack	sl tack	sl tack
6	Di-o-chlorobenzoyl diphenyl sulfide	1.80	sl tack	vsl tack	vsl tack
C1	Dibenzoyl diphenyl ether	1.46	wet	wet	wet
C2	Dibenzoyl diphenyl ether	1.80*	wet	tack	tack
C3	Thiophenyl acetophenone	0.89	wet	wet	wet
C4	Thiophenyl acetophenone	1.80*	wet	wet	wet
C5	2,2-Diethoxyacetophenone	0.81	sl tack	sl tack	sl tack
C6	2,2-Diethoxyacetophenone	1.80*	sl tack	vsl tack	vsl tack
C7	2-Chlorothioxanthone	0.96	sl tack	vsl tack	cured
C8	2-Chlorothioxanthone	1.80*	vsl tack	cured	-

*Concentration is greater than 4.05×10^{-4} mol.

The data in the table indicate that ethylenically unsaturated polyester compositions photoinitiated in accordance with the process in most instances are photopolymerized to a dry state more quickly than the same ethylenically unsaturated polyester compositions photoinitiated with aromatic sulfur-containing ketone compounds outside the scope of the process when exposed to the same ultraviolet source, (compositions C1 to C6), chlorothioxanthone being the only exception.

Haloalkyl Benzoxazoles, Benzimidazoles and Benzothiazoles

The process of *J.G. Pacifici and C.A. Kelly; U.S. Patents 3,912,606; October 14, 1975; 3,962,055; June 8, 1976 and 3,962,056; June 8, 1976; all assigned to Eastman Kodak Company* relates to photopolymerizable polymeric compositions useful as coating and moldable compositions which are hardenable by ultraviolet radiation. They comprise mixtures of photopolymerizable or photocrosslinkable unsaturated compounds and at least one photoinitiator from the group consisting of haloalkyl benzoxazoles, benzimidazoles and benzothianzoles which has the formula below, where R_1, R_2 R_3 and R_4 are H, alkyl, alkoxy, carboxy, alkoxycarbonyl, Cl, Br, NO_2, or amino and can be the same or different.

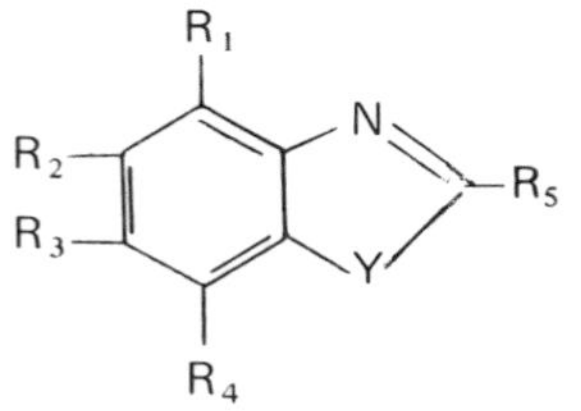

Example 1: 2-(p-α-chlorotolyl)benzoxazole is prepared according to the following procedure. To a stirred solution of 26 grams (0.12 mol) 2-p-tolylbenzoxazole and 17 grams (0.12 mol) of sulfuryl chloride in 150 ml chlorobenzene was added 0.1 gram of benzoyl peroxide. The reaction mixture was heated gently to reflux (134°C) and held for 2½ hours. An analysis of the reaction mixture by VPC showed the reaction to be about one-half complete. Additional sulfuryl chloride (17.0 grams) was added and the reaction continued for 1½ hours.

The solvent was removed by distillation and the residue dissolved in benzene, washed with water, 5% aqueous $NaHCO_3$ and finally with water. Hexane was added and the solution chilled. The crude product was collected by filtration, and recrystallized from ethanol to give 18 grams (63%) of an off-white solid, melting point 143° to 145°C. The identity and purity of the product was established by NMR analysis.

Example 2: 2-(p-α-bromotolyl)benzoxazole is prepared according to the following procedure. To a stirred mixture of 105 grams (0.5 mol) of 2-p-tolylbenzoxazole, 100 grams N-bromosuccinimide, and 1,000 ml of carbon tetrachloride was added 3.0 grams of benzoyl peroxide. The mixture was heated at reflux for 5 hours and the solvent removed on the steam bath. The residue was combined with 1,500 ml of water, heated to boiling, and the insoluble material collected by filtration. The filter cake was washed twice with 250 ml of hot water and recrystallized from ethyl acetate to give 89 grams (64%) of white plates, melting point 168° to 170°C. The identity and purity of the product were established by NMR analysis.

Example 3: The following compositions were prepared and films (1 mil) were cast on rolled steel plates with a Garner film casting knife. These samples were then exposed to a 1,200-watt, 17-inch Hanovia mercury arc at a distance of 6 inches from the arc. The tack-free time of the cured compositions was determined.

Composition Number	Ingredients	Parts by Weight	Tack-Free Time, seconds
1	Unsaturated polyester*	65	>2000 (no cure)
	Styrene	35	
2	Unsaturated polyester*	65	120
	Styrene	35	
	2-(p-α-chlorotolyl)benzoxazole	~1	
3	Unsaturated polyester*	65	86
	Styrene	35	
	2-(p-α-bromotolyl)benzoxazole	~1	
4	Unsaturated polyester**	65	74
	Styrene	35	
5	Unsaturated polyester**	65	96
	Styrene	35	
	Hydroquinone	100 ppm	
	2-(p-α-chlorotolyl)benzoxazole	~1	
6	Unsaturated polyester**	65	96
	Styrene	35	
	Hydroquinone	100 ppm	
	2-(p-α-bromotolyl)benzoxazole	~1	

(continued)

Composition Number	Ingredients	Parts by Weight	Tack-Free Time, seconds
7	Unsaturated polyester**	32	130
	Styrene	18	
	$CaCO_3$	25	
	Talc	25	
	2-(p-α-chlorotolyl)benzoxazole	~1	
8	Derakane 411-45 (acrylate bisphenol A epoxy resin)***	100	55
	2-(p-α-chlorotolyl)benzoxazole	~1	
9	Derakane 411-C-50†	100	55
	2-(p-α-chlorotolyl)benzoxazole	~2	

*The unsaturated polyester contains a mol ratio of maleic anhydride and isophthalic acid of 1:1 and has an acid number of 28 (ASTM D-1639-70).
**The unsaturated polyester contains a mol ratio of maleic anhydride and isophthalic acid of 1:1 and has an acid number of 32 (ASTM D-1639-70).
***45% styrene, viscosity, 500 cp.
†50% styrene, viscosity, 120 cp.

Halogenated Naphthalene Derivative

V.D. McGinniss; U.S. Patents 3,915,824; October 28, 1975 and 3,878,075; April 15, 1975; both assigned to SCM Corporation provides an improved process for laser or UV curing of pigmented-binder systems comprising ethylenically unsaturated polymers containing at least 0.05 weight part of opacifying pigment per weight part of binder. The pigmented-binder system includes as the improvement from 0.5 to 3% of a halogenated derivative of naphthalene in combination with 0.1 to 2% aromatic amino carbonyl photosensitizers and from 0.5 to 2% aromatic ketone or aldehyde photosensitizers. The pigmented-binder systems may be used for paint films having film thickness of less than 1 mil and preferably about 0.1 to 0.5 mil thick.

Example 1: A diacrylate (D.E.R. 332) was produced by reacting 2 mols of acrylic acid with 1 mol of diglycidyl ether of bisphenol A with 0.2% benzyldimethylamine at 120°C until the acid number of the reaction mixture was essentially zero. An ethylenically unsaturated binder was produced by mixing at room temperature about 30 parts 2-ethylhexyl acrylate, about 30 parts pentaerythritol triacrylate and about 30 parts diacrylate.

The foregoing binder was ground with rutile TiO_2 to produce a pigmented polymerizable binder having a pigment-binder ratio (P/B) of 0.9.

Example 2: About 1.5% of Michler's ketone and about 1.5% α-chloromethyl naphthalene were added to the pigmented binder of Example 1 based on the pigmented binder weight. The sample was drawn down on a steel panel to provide a 0.5 mil paint film and cured under nitrogen blanket on a plasma arc radiation system for 0.1 second which provided a complete cure. A similar paint film cured by 10 seconds' exposure to a conventional UV source with an inert atmosphere.

The UV source was two 4,000-watt mercury lamps placed 8 inches from the film. The inert atmosphere was created by placing a polyethylene film over the paint before curing to assimilate a nitrogen atmosphere.

Carboxy-Substituted Benzophenones

Compounds containing a benzophenone or a substituted benzophenone moiety (a) autophotopolymerizable, (b) photopolymerizable in compositions with another photoinitiator, or (c) photoinitiating in compositions with another photopolymerizable material are provided by *D.J. Carlick and R.H. Reiter; U.S. Patents 3,926,638 and 3,926,639 and G. Rosen; U.S. Patent 3,926,641; all December 16, 1975; all assigned to Sun Chemical Corporation.*

The compounds are prepared by reacting a polyfunctional polyethylenically unsaturated monomer or the like with a suitable carboxy-substituted benzophenone. Although the process will be illustrated by use of compounds prepared from benzophenone tetracarboxylic dianhydride (BTDA), it is to be understood that the process is equally applicable to compounds prepared from other carboxy-substituted benzophenones, their corresponding anhydrides, and substituted benzophenone mono- and polycarboxylic acids and anhydrides having the following formula:

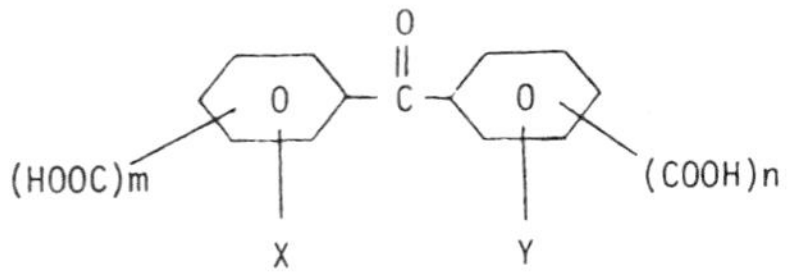

where m and n are each an integer from 0 to 3 and the sum of m and n is in the range of 1 to 6; and X and Y may each be 1 to 4 halogen atoms, e.g., chlorine, bromine, or iodine; dialkylamino groups having 1 to 4 carbons; or other groups which confer desirable properties to the product, such as for example mercaptan, disulfide, alkene, peroxy, alkoxy, carbonyl, amide, amine, nitro, hydroxy, ether, aryl, or the like; X and Y may be the same or different and either or both may be omitted.

When used as photosensitizers, the compounds of this process are used in a ratio to the polyethylenically unsaturated monomer of 1:99 to 90:10, and preferably from 30:70 to 70:30. When used in combination with a second initiator or sensitizer, 0.1 to 10 parts by weight of the secondary initiator per 100 parts of the carboxy-substituted benzophenone derivative are used.

Example: (a) A mixture of 747 parts of pentaerythritol-3,5-acrylate (1 equivalent OH) and 120 parts of benzophenone tetracarboxylic dianhydride (BTDA) was heated at 80° to 90°C in the presence of phosphoric acid as catalyst. The product was a half-ester adduct of the pentaerythritol-3,5-acrylate and BTDA.

(b) The product of part (a) was coated onto a glass slide at a wet film thickness of 0.3 micron and irradiated at a distance of 2 inches from a 6-inch 1,200-watt per inch mercury vapor lamp. The film dried in 0.95 second.

(c) To illustrate the use of the benzophenone-modified compound of part (a) as a photoinitiator, a solution of 30 parts of the product of part (a) in 70 parts of pentaerythritol tetraacrylate was applied in a thin film to corona-treated polyethylene film and laminated to vinylidene chloride-coated cellophane. The sample was exposed to a 200-watt/inch mercury vapor lamp for 0.2 second, causing complete cure of the adhesive and providing a laminate having excellent peel strength.

(d) For comparative purposes a mixture of 90 parts of pentaerythritol-3,5-acrylate and 10 parts of benzophenone was dried as in part (b) above. The film dried in 25 seconds.

Polyhaloacyl Aromatic Compounds

R.H. Reiter and G. Rosen; U.S. Patent 3,929,490; December 30, 1975; assigned to Sun Chemical Corporation have found that certain polyhaloacyl aromatic oligomers and higher polymers are effective initiators for the photopolymerization of ethylenically unsaturated compounds. The initiators are polyhaloacyl aromatic compounds which may be prepared by any known and convenient method.

Examples of the compounds of this process include, but are not limited to, poly(dichloroacetyl-α-methylstyrene), poly(dibromoacetyl-α-methylstyrene), poly(trichloroacetyl-α-methylstyrene), poly(tribromoacetylstyrene), poly(diiodoacetylstyrene), poly(trichloroacetylstyrene), poly(trichloroacetylphenyl oxide), poly(trichloroacetylphenyl glycidyl ether), poly(trichloroacetylvinyltoluene), poly(p-trichloroacetylphenyl acrylate), poly(trichloroacetylbenzyl), and trichloroacetyl-p-terphenyl. The ratio of the amount of monomeric compound to the amount of initiator is 50-99 to 1-50 and preferably 94-99 to 1-6.

Example 1: (a) To a flame-dried one-liter flask equipped with a stirrer, drying tube, addition funnel, and thermometer was charged 251 grams of aluminum chloride and 250 ml of carbon disulfide. Over a period of 80 minutes was added a solution of 218 grams of poly(α-methylstyrene), available as Dow resin 276-V2, in 158 grams of acetyl chloride while maintaining the temperature at –5° to 5°C.

The mixture was allowed to warm to 15°C over 40 minutes and then discharged into an ice-HCl mixture, washed until neutral, and taken up in benzene/methyl ethyl ketone. Residual water was removed azeotropically. The product was vacuum-stripped to yield 230 grams (77.5%) of a dark amber liquid having a Gardner viscosity of Z9 to Z10 (855 to 1066 poises). Analysis: theoretical, 10.00% O; found, 10.40% O. Its infrared spectrum showed a carbonyl absorption at 5.97 microns.

(b) A solution of 140 grams of the product of part (a) in 140 ml of benzene and 375 ml of acetic acid was charged to a one-liter flask fitted with a gas inlet tube, condenser, stirrer, and thermometer. Chlorine gas (238 grams) was added over one hour. The temperature was allowed to rise to 60°C and held until the reaction mixture showed a strong yellow-green color. Residual chlorine was swept out with nitrogen for one hour, and 225 grams of anhydrous sodium acetate was added. The temperature was raised to 95°C, and chlorine gas added at about half the previous rate.

The temperature was held at 95° to 99°C for one hour at which time 112 grams of chlorine had been charged. Nitrogen was sparged through the mixture for 40 minutes while a temperature of 90°C was maintained. After cooling to 70° to 75°C, the reaction mixture was poured with stirring into 18 grams of sodium sulfite dissolved in a mixture of 1,425 grams of water and 625 grams of ice.

After allowing the mixture to warm to room temperature, the lower organic layer was withdrawn and stripped under aspirator vacuum at 100°C for 2 hours. A yield of 232 grams of a dark brown tarry solid, poly(trichloroacetyl-α-methylstyrene) (PolyTCAP), was obtained, representing a weight gain corresponding to the reaction of 3.03 gram-atoms of chlorine per equivalent of aromatic ring. Analysis of $C_{11}H_9OCl_3$: theoretical, 40.5% Cl; found, 42.05% Cl. Its infrared spectrum showed a carboxyl absorption at 5.87 microns. The product was non-lachrymatory and had little color.

Example 2: The use of α,α-dichloroacetophenone (DCAP) and α,α,α-trichloroacetophenone (TCAP) as photoinitiators is known. These compounds, however, have limited commercial applicability. Because of its lachrymatory properties, the dichloro compound is unsuitable for use in inks and coatings. The trichloro compound is less irritating than the dichloro compound but it is somewhat irritating and has an offensive odor, precluding its use in inks and other thin-film applications. A comparison of the properties of these compounds and a product of this process (PolyTCAP prepared in Example 1) has been made.

One gram each of TCAP and PolyTCAP was placed on a watch crystal and kept in an oven at 50°C for 72 hours. The weight losses, due to evaporation, were as follows: 96.4% TCAP and 0.3% PolyTCAP. These data demonstrate the superiority of PolyTCAP over its analog TCAP in lack of volatility.

One gram each of TCAP and finely divided PolyTCAP was suspended and agitated in 100 cc of neutral distilled water for 18 hours. The aqueous layers were separated from the organic compounds, and the amounts of N/5 NaOH required to neutralize 50 cc aliquots of the aqueous liquids were measured. The amounts of base to reach a phenolphthalein end point were as follows: 0.32 ml TCAP and 0.03 ml PolyTCAP. These data illustrate the superiority of PolyTCAP over its analog TCAP in resistance to hydrolysis.

To demonstrate the relative cure speeds of mixtures of ethylenically unsaturated monomeric materials with DCAP, TCAP, and PolyTCAP as the initiators, runs were made with a variety of monomers with (a) no initiator, (b) DCAP, (c) TCAP, and (d) PolyTCAP; in (b), (c), and (d) the ratio of monomer:initiator was 90:10, except where additionally indicated for the isocyanate-modified pentaerythritol triacrylate. The compositions were exposed at a distance of 3 inches from a 200-watt/inch ultraviolet lamp.

Monomer	Cure speeds, seconds			
	no initiator (a)	DCAP (b)	TCAP (c)	PolyTCAP (d)
Pentaerythritol tetraacrylate	40	–	0.5	0.2
Trimethylolpropane triacrylate	90	–	5	3
Isocyanate-modified pentaerythritol triacrylate (as disclosed in U.S. patent No. 3,759,809)	28	2.3	0.7-0.8	0.5-0.6
+ 3% initiator		3.5	1.4-1.8	1.5-1.6
+ 5% initiator		2.5	1.0-1.2	1.0-1.1
1,6-Hexanediol diacrylate	145	–	6	4
Ethyl acrylate	180(evaporates)	–	45	40
Methyl methacrylate	180(evaporates)	–	25	20
Styrene	180(evaporates)	–	30	27
Divinyl benzene	180(evaporates)	–	45	36

From these data it can be seen that the compositions containing PolyTCAP as the initiator cure faster than the compositions containing either DCAP or TCAP as the initiator.

Three Component System

A.J. Bean; U.S. Patents 3,933,682; January 20, 1976; and 3,966,573; June 29, 1976; both assigned to Sun Chemical Corporation provide an improvement in the process for photopolymerizing a monomeric ethylenically unsaturated compound by irradiating the compound in the presence of a photoinitiator. The improvement comprises using as the photoinitiator a coinitiator system which consists essentially of (a) about 1 to 30 parts of at least one compound having the formula

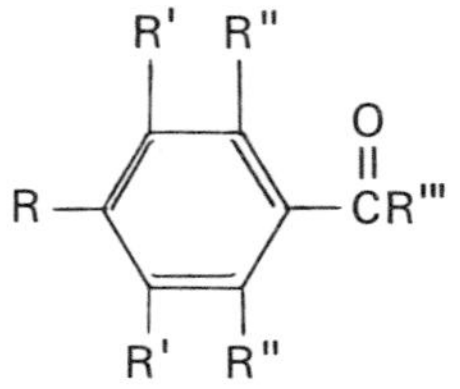

R' R''
R — (benzene ring) — $\overset{O}{\overset{\|}{C}}R'''$
R' R''

wherein R, R', and R'' are each hydrogen, C_{1-14} alkyl, aryl, alkylaryl, alkoxy, aryloxy, halogen, halogenated C_{1-14} alkyl, halogenated aryl, amino, or amino N-substituted with alkyl, aryl, β-alkanol, or a combination of these, and may be the same or different; and R''' is hydrogen, C_{1-14} alkyl, halogenated alkyl, aryl, —OZ where Z is C_{1-14} alkyl or aryl;

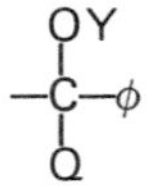

OY
—C—φ
Q

where Y is hydrogen, C_{1-14} alkyl, aryl, or alkylaryl and Q is hydrogen, C_{1-14} alkyl, aryl, alkylaryl, alkoxy, or aryloxy; or

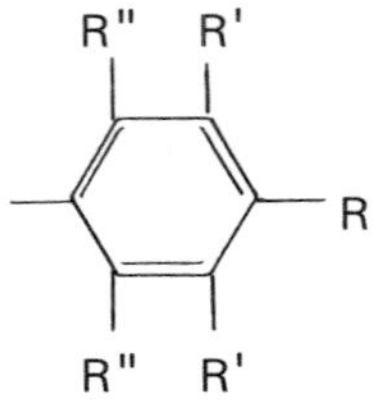

R'' R'
— (benzene ring) — R
R'' R'

wherein R, R', and R'' are each hydrogen, C_{1-14} alkyl, aryl, alkylaryl, alkoxy, aryloxy, halogen, halogenated C_{1-14} alkyl, halogenated aryl, amino, or amino N-substituted with alkyl, aryl, β-alkanol, or a combination of these, and may be the same or different; (b) about 1 to 30 parts of at least one compound having the formula

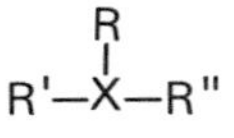

R
R'—X—R'' or R—X R'.

In the formulas X is nitrogen, phosphorus, arsenic, bismuth, or antimony and R, R', and R" are each alkyl, hydroxyalkyl, aryl, aralkyl, or alkaryl and may be the same or different and one or both of R and R" may be hydrogen; and (c) 1 to 30 parts of at least one halogenated hydrocarbon.

The ratio of the amounts of the components of the photoinitiator system (a):(b):(c) is 1-30 to 1-30 to 1-30, and preferably about 2-15 to 2-10 to 1-5. The ratio of the amount of the monomeric compound to the amount of the photoinitiator system is about 1-1,000 to 1, and preferably is about 10-50 to 1.

In the examples, the parts are given by weight unless otherwise specified. Unless otherwise indicated, when the ingredient is solid at room temperature, the mixture may be heated to melt the solid ingredient, but generally not above 100°C, or it may be used in a mixture with other liquid ingredients. The atmospheric and temperature conditions were ambient unless otherwise noted.

Examples 1 through 8: The samples were prepared as follows: 0.2 gram of the ester-photoinitiator composition was rolled onto a 2¼ x 8½ inch glass plate, using a Quick-Peak roller in order to form a uniform film. The wet film was then transferred to a 1 x 3 inch glass slide which was then exposed to consecutive 0.1-second flashes of a 100 watt/inch ultraviolet lamp until the composition was tack-free, as determined by rubbing the film with a finger.

Composition, % by weight	Example 1	2	3	4	5	6	7	8
Pentaerythritol tetraacrylate	100	95	95	95	95	95	95	95
Benzophenone	-	5	-	-	3	-	3	2
Triethanolamine	-	-	5	-	2	3	-	2
α-6-hexachloro-para-xylene	-	-	-	5	-	2	2	1
Cure speed, sec	30	15	4.5	1.5	2.5	3.0	10	0.7

From these data it can be seen that the cure speed of pentaerythritol tetraacrylate with a combination of benzophenone (a conventional photoinitiator), triethanolamine (an organic compound containing a Group V element), and α-6-hexachloro-para-xylene (a halogen-containing compound) is considerably faster (Example 8) than the ester alone (Example 1) or the ester with comparable quantities of benzophenone (Example 2), of triethanolamine (Example 3), of α-6-hexachloropara-xylene (Example 4), and of combinations of two of these (Examples 5, 6, and 7).

Activated Halogenated Azine

In accordance with the process of *V.D. McGinniss; U.S. Patent 3,959,100; May 25, 1976; assigned to SCM Corporation* polymerizable compositions comprising an ethylenically unsaturated vehicle selected from monomers, oligomers, and prepolymers, and photosensitizers for them are photopolymerized by exposure to a source of ultraviolet radiation having wavelengths of about 1800 to 4200 A; the photosensitizer consists essentially of an activated halogenated azine compound and is present from about 1 to 10% by weight of the vehicle. The activated halogenated azine photosensitizer is selected from the group consisting of activated halogenated benzazines, benzodiazines, diazines, and mixtures thereof;

the azine is characterized by having a substituted radical selected from chlorosulfonyl, α-haloalkyl, and α-haloalkylated aryls. Up to 20% by weight of the vehicle of an opacifying pigment may also be used.

Example: A number of acrylic resins and combinations thereof were utilized in evaluating the photoinitiator of the process. For convenience, the polymerizable binder composition comprises three acrylic resins in equal proportions. These resins were acrylic monomers, diacrylate, and triacrylate oligomers.

A polymerizable composition consisting of one-third 2-ethylhexyl acrylate, one-third ethylene glycol diacrylate, and one-third trimethylolpropane triacrylate was prepared with various photoinitiators in accordance with the process.

Each sensitized binder composition was poured over a pair of steel panels and drawn down with a No. 8 wound wire rod to a film thickness or coating of approximately 0.5 mil. The coated but wet panels were each exposed to a different source of ultraviolet light; one provided from a plasma arc radiation source (PARS) and the other from a conventional ultraviolet light, the light having two 4,000-watt mercury lamps. Exposure times were from 0.07 to 30 seconds, with the panels being placed approximately 5 inches from the ultraviolet source. These exposure times for the plasma arc are calculated from the speed of the conveyor belt on which the panels are placed. The photopolymerization was done in an inert atmosphere (protection of the sensitized binder was effected by a clear sheet of polyethylene). The table below shows the results of subjecting panels coated with the sensitized compositions to UV radiations from the PARS and Ash Dee units.

Photoinitiator, % by weight	UV Source	Time, seconds	Cure
2,3-bis(bromomethyl)-1,4-benzodiazine, 2%	PARS	0.2	Hard
2,3-bis(bromomethyl)-1,4-benzodiazine, 2%	Ash Dee	7	Hard
8-chlorosulfonyl-1-benzazine, 2%	PARS	0.2	Hard
8-chlorosulfonyl-1-benzazine, 2%	Ash Dee	7	Hard
3-bromomethyl-2-benzazine, 3%	PARS	0.2	Hard
3-bromomethyl-2-benzazine, 3%	Ash Dee	7	Hard

2-Methyl-Substituted Benzimidazole

In the process of *V.D. McGinniss; U.S. Patent 3,970,535; July 20, 1976; assigned to SCM Corporation* UV polymerization of a photopolymerizable ethylenically unsaturated vehicle is improved by incorporating into the vehicle about 0.5 to 10% by weight of a 2-methyl-substituted benzimidazole.

Example 1: A clear coating of one-third part melamine acrylate, one-third part hydroxyethyl acrylate, and one-third of the adduct described below was the vehicle for test curing using a plasma arc radiation unit. The special adduct is the reaction product of one mol of isophorone diisocyanate and 2 mols of hydroxyethyl acrylate.

The curing apparatus is an intense radiation torch (plasma arc) optically directed by a reflector system to irradiate a freshly painted flat aluminum workpiece passing below a rectangular irradiating window on an enclosed horizontal conveyor moving at various line speeds (providing about 0.2 second of irradiation at 100 feet a minute and 0.1 second of irradiation at 200 feet a minute). The atmosphere around the workpiece during its irradiation is kept essentially inert by purging it with nitrogen. Radiation energy supplied by such apparatus to the workpiece surface is about 35 kilowatts per square foot with slightly less than about 6 kilowatts per square foot thereof being in the UV spectrum. Such sort of torch is described in U.S. Patent 3,364,387.

The above clear coating was applied to an aluminum workpiece (4 x 8 inch size panel) as a film of about 0.4 mil thickness utilizing a wound wire rod No. 8. The coated panel was subjected to the radiation emitted by the apparatus described but the film did not cure completely even when exposure at line speed of 100 feet per minute was continued for 150 times (total actual exposure time in 30 seconds). Incomplete cure of the film is recognized by being tacky to the touch.

The vehicle was dosed with 2% by weight of 2-chloromethylbenzimidazole well mixed in. The sensitized coating was spread at about 0.4 mil thickness on the aluminum test panel and was cured at room temperature by subjecting it to the apparatus as above described. At 100 to 200 feet per minute line speed the vehicle cured tack-free and had a good scratch resistance to the fingernail, indicating good cure throughout the film depth.

Example 2: In this example the same apparatus, operation, and vehicle of Example 1 were used, except that the photosensitizer was replaced with 2-(trichloromethyl)-benzimidazole. The results of complete cure at line speeds of 100 to 200 feet per minute were obtained also.

Example 3: In this example the same apparatus, operation, and vehicle of Example 1 were used, except that 2% of 2-(α-chlorobenzyl) was used. Again, the same results as in Examples 1 and 2 were obtained.

Alcohol or Mercaptan Adducts of Triketone

N. Heap and J. Nemcek; U.S. Patent 3,974,052; August 10, 1976; assigned to Imperial Chemical Industries Limited, England provide a photopolymerizable composition comprising a polymerizable ethylenically unsaturated material and a photosensitizer. The photosensitizer is an addition product of an organic 1,2,3-triketone and an organic alcohol or organic mercaptan, the addition product having the formula

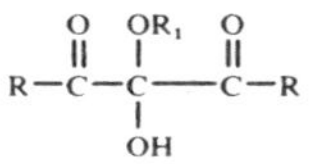

$$R{-}\overset{\overset{O}{\|}}{C}{-}\overset{\overset{OR_1}{|}}{\underset{\underset{OH}{|}}{C}}{-}\overset{\overset{O}{\|}}{C}{-}R \quad \text{or} \quad R{-}\overset{\overset{O}{\|}}{C}{-}\overset{\overset{SR_1}{|}}{\underset{\underset{OH}{|}}{C}}{-}\overset{\overset{O}{\|}}{C}{-}R$$

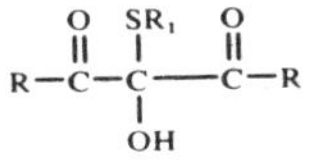

where the groups R and R_1 may be the same or different and each is a hydrocarbyl group or a substituted hydrocarbyl group or the groups R when taken together form part or the whole of a cyclic structure. The photosensitizer is present preferably in a concentration of 0.5 to 5% by weight of the ethylenically unsaturated material in the composition. A reducing agent may be present in the photopolymerizable composition to accelerate the rate of polymerization of the ethylenically unsaturated material. As the compositions are substantially stable such that little or no polymerization of the ethylenically unsaturated material takes place in the absence of radiation they form can-stable compositions which may be formed into a film, e.g., a paint film, and then caused or allowed to polymerize by exposure to light, e.g., by exposing the film to natural light, e.g., sunlight. When formed into a film and exposed to light the compositions rapidly polymerize.

Examples 1 through 20: These examples demonstrate the use as photosensitizers of adducts of 1,3-diphenyl propane trione with alcohols and mercaptans, i.e., adducts of formula

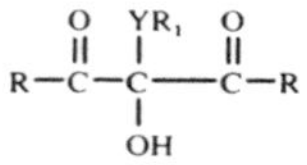

$$R-\overset{\overset{O}{\|}}{C}-\underset{\underset{OH}{|}}{\overset{\overset{YR_1}{|}}{C}}-\overset{\overset{O}{\|}}{C}-R$$

where R is the phenyl group, Y is O or S and YR_1 is as shown in the table below. Photopolymerizable compositions were prepared by mixing the photosensitizer, ethylenically unsaturated material and, if appropriate, dimethylaminoethyl methacrylate (DMAEM) in such proportions as to give compositions containing 0.2% by weight of the photosensitizer and if appropriate 3% by weight of dimethylaminoethyl methacrylate based on the weight of the ethylenically unsaturated material.

Example No.	Photosensitiser $-YR_1$	DMAEM (conc.-%)	Additive (3%)	Light Source Blue/Black	Resin	Gel-Time (minutes)
1	$-OCH_2CH_2O\overset{\overset{O}{\|}}{C}-\overset{\overset{CH_3}{|}}{C}=CH_2$	3	—	Black	HEMA	6
2	"	3	—	Blue	"	15
3	"	3	HEMA	Black	Crystic 199	7.5
4	$-OCH_2CH_3$	3	CH_3CH_2OH	Blue	"	15
5	$-OCH_3$	3	CH_3OH	Black	"	2
6	$-OCH_3$	—	CH_3OH	Black	"	4
7	$-OCH_3$	3	CH_3OH	Blue	"	15
8	$-OCH_3$	3	CH_3OH	Black	"	3
9	$-OCH_3$	—	CH_3OH	Black	"	4.5
10	$-OCH_2CH_2OCH_3$	3	$CH_3OCH_2CH_2OH$	Black	"	2.5
11	$-SC_8H_{17}$	3	$C_8H_{17}SH$	Black	"	4.5
12	$-SC_8H_{17}$	—	$C_8H_{17}SH$	Black	"	9.5
13	$-SC_8H_{17}$	3	$C_8H_{17}SH$	Blue	"	15
14	$-SC_8H_{17}$	—	$C_8H_{17}SH$	Blue	"	27
15	$-OCH_3$	3	—	Black	"	4.5
16	$-OCH_3$	—	—	Black	"	8
17	$-OCH_3$	3	CH_3OH	Black	"	2.5
18	$-OCH_3$	—	CH_3OH	Black	"	4
19	$-OCH_2CH_2OH$	4	—	Black	"	10
20	$-OCH_2CH_2OH$	—	—	Black	"	13

In Examples 1 and 2 the ethylenically unsaturated material was hydroxyethyl methacrylate (HEMA) which also was the alcohol used to form the photosensitizer; in these examples the photosensitizer was formed in situ by adding the appropriate amount of 1,3-diphenyl propane trione to the hydroxyethyl methacrylate. In Examples 3 through 14 the ethylenically unsaturated material was an unsaturated polyester resin known as Crystic 199; in these examples the photosensitizer was formed in situ by adding 1,3-diphenyl propane trione and the appropriate alcohol or thiol to the polyester resin.

In Examples 15 through 20 Crystic 199 again was used but in these examples the photosensitizer was added to the resin as a preformed compound. The procedure was as follows: A sample (3 to 3.5 grams) of the photopolymerizable composition was placed in a small glass vial which was then suspended at the center of a ring of eight fluorescent tubes (20 watt Thorn black-light tubes or 20 watt Thorn blue-light tubes. The Thorn black-light tubes emit radiation of wavelength 300 to 400 mμ and the Thorn blue-light tubes emit radiation of wavelength 320 to 600 mμ, with peak radiations of 350 and 425 mμ respectively. The diameter of the ring of tubes was eight inches.

N,N'-Oxalyl Indigo

J. Nemcek and N. Heap; U.S. Patent 3,974,053; August 10, 1976; assigned to Imperial Chemical Industries Limited, England provide photopolymerizable compositions consisting essentially of at least one polymerizable ethylenically unsaturated material and a photosensitive catalyst. The catalyst comprises (a) from 0.5 to 5% by weight based on the ethylenically unsaturated material of at least one photosensitizer having the structure

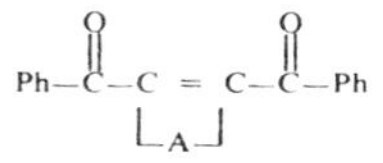

$$\mathrm{Ph{-}\overset{\overset{O}{\|}}{C}{-}C{=}C{-}\overset{\overset{O}{\|}}{C}{-}Ph}\quad(\text{the two central C atoms joined through } \mathrm{A})$$

where Ph is phenyl, halogen-substituted phenyl, phenylene or halogen-substituted phenylene and A is a cyclic hydrocarbyl group, a halogen-substituted cyclic hydrocarbyl group or a group of the formula

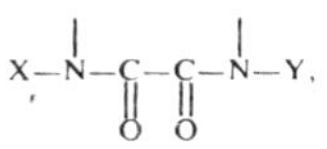

$$\mathrm{X{-}\overset{|}{N}{-}\underset{\underset{O}{\|}}{C}{-}\underset{\underset{O}{\|}}{C}{-}\overset{|}{N}{-}Y},$$

where X and Y each are hydrogen, a hydrocarbyl or a halogen-substituted hydrocarbyl group, and (b) from 1 to 5% by weight based on the ethylenically unsaturated material of a reducing agent capable of reducing the photosensitizer when the photosensitizer is in an excited state.

Where the composition contains a pigment a photosensitizer should be chosen which is excited by radiation having a wavelength which is not absorbed to an excessive extent by the pigment present in the composition. Preferably, the pigment should be transparent to radiation at the wavelength which excites the photosensitizer. Where the pigment absorbs ultraviolet radiation but absorbs little or no radiation in the visible region of the spectrum those photosensitizers of the process which are excited by visible light, for example N,N'-oxalyl indigo, are especially useful.

Example 1: Experiment 1 – 0.01 part of N,N'-oxalyl indigo and 4 parts of dimethylaminoethyl methacrylate were dissolved in 100 parts of a mixture of 38% by weight styrene and 62% by weight of an unsaturated polyester resin known as Crystic 199. The resulting solution was charged at room temperature to a Pyrex glass bottle which was then stoppered and irradiated with radiation from eight 20 watt blue-light fluorescent tubes disposed in a circle around the bottle at a distance of about 3 inches from the bottle. After 2¼ minutes the solution had gelled to an extent such that it could no longer be poured. The period of time required to gel the solution just to the extent that it could no longer be poured was recorded as the gel point of the solution.

Experiments 2, 3 and 4 – Three further experiments were carried out according to the above procedure except that 0.02, 0.05 and 0.005 part of N,N'-oxalyl indigo were used instead of 0.01 part. The gel points were 2, 1¾ and 2½ minutes respectively.

Experiment 5 – In a further experiment the procedure of Experiment 1 was repeated except that the 4 parts of dimethylaminoethyl methacrylate were omitted. The gel point had not been reached after 1 hour.

Experiments 6 and 7 – Experiments 2 and 4 were repeated except that eight 20 watt black-light fluorescent tubes were used instead of the blue-light tubes. The gel point was 1¾ minutes in each experiment.

Experiment 8 – Experiment 1 was repeated except that eight 20 watt green-light fluorescent tubes were used instead of the blue-light tubes. The gel point was 6¼ minutes.

Experiment 9 – Experiment 3 was repeated except that 0.05 part of 3,3',5,5'-tetrachloro-N,N'-oxalyl indigo were employed instead of the N,N'-oxalyl indigo. The gel point was 5½ minutes.

Example 2: 0.02 part of N,N'-oxalyl indigo and 3 parts of dimethylaminoethyl methacrylate were dissolved in 100 parts of hydroxymethyl methacrylate and the solution was charged to a pyrex bottle and irradiated as described in Example 1 with light from eight 20-watt black-light tubes. The gel point was reached in 11.5 minutes. For purposes of comparison the above procedure was repeated except that the dimethylaminoethyl methacrylate was excluded. The solution had not gelled after 30 minutes irradiation.

The compositions are suitable for the production of shaped articles of polymeric materials, for example sheets, and are particularly suitable for use in the preparation of polymeric materials in the form of films and in particular paint films. Thus, as the compositions are stable such that little or no polymerization of the ethylenically unsaturated material takes place in the absence of radiation they form can-stable compositions which may be formed into a film, e.g., a paint film, and then caused or allowed to polymerize by exposure to light, e.g., by exposing the film to natural light.

FOR IMAGING COMPOSITIONS

Complex of Porphyrin-SO_2

The process of *F.B. Erickson; U.S. Patent 3,897,255; July 29, 1975; assigned to Monsanto Company* concerns a photoimaging procedure in which the image is characterized by a differential concentration of sulfate groups, depending upon the degree of light exposure, and is developable by selective absorption of dyes and other procedures. The sulfate groups can be obtained by SO_2 treatment of hydroperoxy groups produced in a photooxidation imaging procedure.

In one aspect the process involves the use of sulfur dioxide for treatment of the image after exposure. In another aspect, the sulfur dioxide can be used to treat the coating or components thereof before exposure. The sulfur dioxide can, for example, be incorporated into a solution with a photosensitizer, apparently forming an adduct or complex with porphyrin sensitizer. If the sulfur dioxide is used to directly treat the photosensitive coating before exposure, it may complex with the sensitizer in such coating or otherwise remain in the coating to have a subsequent effect upon hydroperoxy groups. Materials which generate or provide SO_2, as well as SO_2 itself, can be used for the SO_2 treatments or reactions.

Example 1: A paperboard was coated with a styrene/butadiene 60/40 latex (Dow 636) to which a 2:1 pigment loading, based on polymer, had been added. The virtually colorless pigment was 9 parts clay to 1 part TiO_2. The coating was sensitized by wiping it with a solution of tetraphenylporphin in chloroform, chlorinated paraffins of 40 to 70% chlorine content (Chlorowax, grade 40) and a plasticizer, methylcarbityl benzyl phthalate. The coating was exposed to light through a positive transparency and developed by wiping with a kerosene solution of Sudan Brown dye. The nonexposed areas pick up the dye.

A one-minute exposure gave a satisfactory picture with good contrast, while a 15-second exposure gave poor contrast because the nonexposed areas picked up dye. When the light exposure was followed by five minutes treatment with SO_2 gas, a 15-to-30-second light exposure was sufficient to produce a picture with good contrast. With a one-minute light exposure and the SO_2 treatment, the resulting picture appeared overexposed. Substantially equivalent results were obtained when the SO_2 treatment preceded the light exposure.

Example 2: A styrene/butadiene block copolymer coating was sensitized with a tetraphenylporphin solution. The coating was exposed through a stencil for 10 to 15 minutes. The coating was then treated with sulfur dioxide. The light-exposed area was gray to colorless, while the nonexposed area was yellow-green. The coating was heated at about 120°C for 2 minutes to turn the exposed area black, while the nonexposed area was yellow-green. The coating was then contacted with ammonia to gradually lighten the yellow-green area until it disappeared, while the black area turned brown. An additional SO_2 contact produced the yellow-green color again, but this disappeared upon heating, while the brown color remained in the exposed area.

Another sample of the coating was similarly light-exposed and then heated for 1 to 2 minutes at 120°C. The coating was then contacted with SO_2, which turned the nonexposed area a yellow-green, while the exposed area remained colorless. Upon additional heating, the exposed area turned gray, while the nonexposed area remained yellow-green. Any moderate heat can be used in accelerating the effect of the sulfur dioxide, such as 50° to 150°C for a few seconds up to 5 minutes or more, with longer times generally being used as the heating temperature is lowered.

Uranyl Salt plus Aminimide

G.W. Brutchen and G.O. Fanger; U.S. Patent 3,898,087; August 5, 1975; assigned to Ball Corporation describe photopolymerizable compositions comprising an aqueous mixture of a water-insoluble resin such as polyvinyl acetate, polyvinyl acrylate, etc., a water-soluble binder such as polyvinyl alcohol, polyvinyl pyrrolidone, etc., a photopolymerization initiator selected from the group consisting essentially of uranyl nitrate, uranyl phosphate, uranyl chloride, uranyl carbonate, and uranyl dibutyl phosphate, and an aminimide represented by the general structural formula which follows.

$$R{-}N^{+}(CH_3)_2N^{-}{-}C(O)C(CH_3){=}CH_2$$

In the formula R is a radical selected from the group consisting of $-CH_3$, $-CH_2CH(OH)CH_3$, $-CH_2CH(OH)CH_2OH$, $-CH_2CH(OH)(CH_2)_5CH_3$ and also $-CH_2CH(OH)(CH_2)_7CH_3$. It has been found that in the presence of the aminimide compounds, the uranyl salts are very stable and do not reduce to the metal.

Example: A water-base emulsion containing 25 parts by weight of a water-base emulsion of polyvinyl acetate (about 55% solids) was thoroughly mixed with 6 parts by weight of a partially hydrolyzed polyvinyl acetate having a degree of hydrolysis of about 82 mol percent, average polymerization degree of about 500. To this aqueous mixture was added 0.50 part by weight of dimethyl-(2-hydroxylpropyl)-amine methacrylimide and about 0.5 part by weight of N,N'-methylenebisacrylamide, to which was added 2.25 parts by weight of a 50% aqueous solution of uranyl nitrate.

The aqueous mixture thus prepared was thoroughly mixed and a coating thereof applied onto an aluminum plate and allowed to dry. The coating after drying was about 15 mils in thickness and had a moisture content of about 6%. Thereafter, the coated plate was exposed using a 5 kv Ascor Addalux ultraviolet metal halide diazo lamp at a distance of about 24 inches for 1.75 minutes through a master negative, a graduated density step table. The unexposed portions were removed by allowing running water to play over the coating in the form of a pressurized spray which removed the unexposed portions therefrom.

There was excellent bonding between the metal surface and the polymeric material; there was a fine detail given throughout the treated plate reproducing in detail the master negative. A conventional printing ink was rolled onto the relief surface of the plate and rendered a splendid reproduction of the images of the original step table.

Diazo-N-Sulfonate plus Aromatic Hydroxy Compound

In the process of *S. Levinos; U.S. Patent 3,909,273; September 30, 1975; assigned to Keuffel & Esser Company* a latent polymerization initiator is provided which can be mixed with polymerizable ethylenically unsaturated monomers to yield compositions which will remain inactive for extended periods of time until exposed to light. Such an initiator comprises the combination of a diazo-N-sulfonate and an aromatic hydroxy compound.

The initiator has little effect in initiating polymerization of an ethylenically unsaturated compound, yet upon exposure to actinic radiation, such as ultraviolet light, the diazosulfonate readily dissociates to yield a diazonium cation which, in combination with the aromatic hydroxy compound, effects a redox reaction which generates polymerization-initiating free radical species.

The rate of polymerization generally increases with the amount of diazo-N-sulfonate included in the composition, and there has been found that an effective ratio of components in fluid compositions is one part of diazosulfonate to 50 to 100 parts of monomer on a dry weight basis. When employed in coated layers, an effective ratio of diazosulfonate to monomer and binder is in the range of about 2 to 4 parts for each 15 parts of the monomer mixture.

Example: A coated layer of polymerizable composition was prepared in the following manner:

Part A

Deionized water, ml	100
Gelatin (inert, high Bloom), g	5
Polyvinyl pyrrolidone (medium viscosity), g	1.5
Chlorohydroquinone, g	1.25
Denatured ethanol, ml	20
Deionized water, ml	20

The gelatin and polyvinyl pyrrolidone was dissolved in the 100 ml deionized water, followed by the addition of the chlorohydroquinone with stirring. This was followed by the addition of denatured ethanol admixed with deionized water in the quantities indicated. The pH of this mixture measured 5.8.

Part B

Deionized water, ml	120
Gelatin (inert, high Bloom), g	4
p-anisidine diazosulfonate, g	2.5
Acrylamide, g	4.2
N,N'-methylenebisacrylamide, g	0.8
Sodium dodecyl benzene sulfonate (20% solution), ml	5

Each of the ingredients was dissolved, with stirring, in the deionized water in the order given. The pH of this mixture was 6.7. Part B was then rapidly added to Part A with stirring. The final pH of this composite mix was 6.2.

The composition was coated at the rate of about 4 cm per second on a meniscus coater onto a presubbed polyester film support commonly used in the preparation of photographic materials. The coated film was allowed to air-dry in darkness at about 25°C. A section of the resulting coated film was exposed through a master stencil for 20 seconds to the light of a 300-watt Photoflood lamp positioned at a distance of about 12 cm.

Subsequent to exposure, the coated material was washed with water at a temperature of about 25°C with resulting removal of unexposed areas of the coated layer. The sheet was then immersed in an aqueous solution of methylene blue and rinsed with water to yield a positive blue image of the negative stencil.

Diazonium Salt with No Basic Groups in Cation

P.P. de Moira and J.P. Murphy; U.S. Patent 3,930,856; January 6, 1976; assigned to Ozalid Company Limited, England provide a material based on an epoxy resin which gives improved adhesion of the hardened resin to a metal support and such that little or no heating is required to harden the resin in the exposed areas of the coating.

The process is, in part, based on the consideration that amino and other basic groups should be avoided in the diazonium salt because, if such groups are present, the Lewis acid liberated upon decomposition of the diazonium salt tends to combine with such groups and no hardening of the resin can take place without heating the material to break up the resulting compound or complex. The process accordingly provides a light-sensitive material comprising a support carrying a coating of a photopolymerizable epoxy resin composition containing, as a photosensitive compound capable of catalyzing hardening of the epoxy resin, a diazonium salt soluble in organic solvents and whereof the cation is devoid of basic groups and the anion is selected from the following: difluorophosphate, phosphotungstate, phosphomolybdate, tungstogermanate, silicotungstate and molybdosilicate. The coating will normally contain 2 to 15%, and preferably 8 to 10% of diazonium salt based on the weight of epoxy resin.

Example: An aluminum support which had been roughened by brushing was thinly coated with the following composition:

2,5-dimethoxy-4-p-tolylmercaptobenzene diazonium phosphotungstate, g	2.0
Epikote 1007, g	20.0
Methyl cyclohexanone, ml	200.0

The coating was dried for two minutes at 100°C. The coated surface was then exposed to ultraviolet light for two minutes. The plate was then placed in an oven for a further 2 minutes at a temperature of 80°C to increase the rate of curing.

The plate was then washed with methyl ethyl ketone to remove the parts of the coating which had not been exposed to ultraviolet light. A printing plate was obtained which yielded long runs.

Aromatic Nitro Compounds as Epoxy Catalysts

The process of *S.I. Schlesinger; U.S. Patents 3,949,143; April 6, 1976; and 3,895,952; July 22, 1975; both assigned to American Can Company* relates to compositions comprising at least one substance capable of cationic polymerization in admixture with organic compounds which are phototropic and are converted to an acid catalyst upon exposure to radiation.

The composition comprises a monomeric or prepolymeric epoxide material, polymerizable to higher molecular weights by cationic polymerization, selected from the group consisting of glycidyl ethers of bisphenol A, epoxy-cycloalkanes, epoxy-cycloalkenes, epoxy-alkanes, glycidyl alkyl ethers, epoxy-novolacs and allyl glycidyl ether-glycidyl methacrylate copolymer containing as a photoactivatable initiator a phototropic aromatic compound containing at least one nitro radical in a position ortho to an alkyl substituent selected from compounds having the formula:

$$(R_1)(R_2)CH-C_6H_3(NO_2)-R_3$$

where R_1 and R_2 are selected from the group consisting of hydrogen, alkyl, aryl, carbalkoxy, pyridyl, carbazolyl, N-oxidopyridyl, nitroalkyl, nitroaryl, dinitrophenyl, alkaryl, aralkyl, haloalkyl and haloaryl radicals and R_3 is selected from the group consisting of hydrogen, alkyl, aryl, nitroalkyl, nitroaryl, alkaryl, aralkyl, haloalkyl, haloaryl radicals and electron-withdrawing groups selected from $-NO_2$, —CN, —COOR, —COR, $-NR_3^+$, —Br, —Cl, —F, —I, —RCOX, $-RX_3$, —NO, RCO— and R_2NCO— where X is halogen and R is alkyl.

Example 1: A mixture was prepared containing one gram of glycidyl methacrylate-allyl glycidyl ether copolymer, 0.3 gram of ethyl-bis-(2,4-dinitrophenyl) acetate, 1 ml acetonitrile and ½ ml acetone. Additional acetone was added until all of the catalyst dissolved forming a blue solution. The mixture was coated onto aluminum plate. An image of a key was made by exposing a coated plate sample with a key on the surface to a Gates Uviarc lamp for 6 minutes and then heating at 135°C for 30 minutes after which the exposed area had changed from blue to yellow.

The unexposed area was removed by washing with acetone leaving a negative image of the key in the form of a yellow polymer film in the exposed areas. A portion of this plate was dipped in dilute hydrochloric acid to determine the acid resistance of the coating. The coating was unaffected by the acid treatment.

A second coated plate sample was exposed on a portion of its surface for 15 minutes and heated for 15 minutes at 175°C. Washing with acetone then removed the exposed areas, leaving a yellow polymer film on the exposed areas.

Example 2: 30 grams of a 9% solution of glycidyl methacrylate-allyl glycidyl ether copolymer in a 6:1 butyronitrile-o-chlorotoluene solvent, was combined with 1.810 grams of 2,4-dinitrotoluene, whereupon a solution resulted which was spin-coated onto aluminum plate samples at 200 rpm spin speed. A portion of the surface of a coated sample was exposed to a 360-watt mercury arc at 20-cm distance for 2 minutes. The remainder of the surface was covered to exclude light.

Following the exposure, the sample was immersed in a developing solvent composed of three volumes of trichloroethylene to one volume of acetone. Only the unexposed coating washed off the aluminum plate. Another sample, which was similarly exposed for 3 minutes and developed the same way, had good resistance to methyl ethyl ketone, as well. However, a superior film in terms of gloss, thickness and smoothness was formed when another sample was exposed for 5 minutes.

Cyclic cis-α-Dicarbonyl Compounds

The process of *C.T.-L. Chang; U.S. Patent Reissue 28,789; April 27, 1976; assigned to E.I. du Pont de Nemours and Company* relates to photopolymerizable compositions comprising:

(a) At least one nongaseous ethylenically unsaturated compound having a boiling point above 100°C at normal atmospheric pressure and being capable of forming a high polymer by photoinitiated, addition polymerization;

(b) A cyclic cis-α-dicarbonyl compound, the excited state of which cannot react intramolecularly, the atoms adjacent to the two vicinal carbonyl groups being saturated when they are carbons;

(c) An actinic-radiation-absorbing compound capable of sensitizing the polymerization-initiating action of the cyclic cis-α-dicarbonyl compound, the actinic-radiation-absorbing compound being a nitrogen-containing aromatic or heterocyclic compound, having maximum absorption below 520 nm, selected from the class consisting of bis(p-aminophenyl-α-β-unsaturated)ketones; bis(alkylamino)acridine dyes; cyanine dyes containing two heterocyclic rings joined by a single methine group; styryl dye bases; 7-di(lower alkyl)amino-4-lower alkyl coumarins; p-aminophenyl ketones; p-dialkylaminophenyl unsaturated compounds; and 6-dialkylaminoquinaldines;

(d) Optionally, a free-radical-producing hydrogen or electron donor compound; and

(e) Optionally, a macromolecular organic polymer binder.

Constituents (b), (c) and (d) are employed in the respective amounts of 0.001 to 10, 0.001 to 10, and 0 to 10 parts by weight per 100 parts by weight of constituents (a) and (e). Constituents (a) and (e) are present in relative amounts from 3 to 100 and 0 to 97 parts by weight, respectively.

Compositions of the process have improved photoinitiation and photographic speed when used in imaging elements. Such elements may be self-supporting or comprise a support coated with the above-described composition. Imaging elements using the compositions displayed photographic speeds of at least about 2.5 times the speed of an element containing a cis-α-dicarbonyl compound without a radiation-absorbing sensitizer.

Example 1: A coating solution was prepared according to the following formula:

	Grams
Cellulose acetate (acetyl 40%, ASTM viscosity 25)	0.16
Cellulose acetate butyrate (butyrate 17%, ASTM viscosity 15)	0.25
Trimethylolpropane triacrylate	0.80
Acetone	6.9
Michler's ketone	0.0128
2,3-bornanedione	0.0080

The solution was coated with a doctor knife on a 0.001-inch-thick unsubbed polyethylene terephthalate web to a wet thickness of 0.002 inch. The coated element was dried and laminated at room temperature with a 0.001-inch-thick unsubbed polyethylene terephthalate cover sheet.

A sample of the element so prepared was exposed through a $\sqrt{2}$ step wedge with a rotary diazo printer (Blue Ray, average intensity 3.2 mf/cm^2) for 20 seconds. The cover sheet was removed and the coating was dusted with Quindo Magenta, CI Pigment Red 122. The steps up to step 16 inclusive had polymerized and did not accept pigment.

A control element made as above but without the 2,3-bornanedione polymerized only to step 6 inclusive when exposed as above.

Example 2: The following solution was prepared.

	Grams
Methyl methacrylate/methacrylic acid copolymer (90/10 mol ratio)	2
Diethylene glycol diacrylate	5
4,4'-bis(diethylamino)benzophenone	0.05
2,3-bornanedione	0.10
Ethyleneglycol monoethyl ether	8

This solution was dispersed in a solution of 10 grams of gelatin in 40 grams of water by mixing in a high-speed blender for 2 minutes. The dispersion was coated with a doctor knife on a 0.004-inch-thick polyethylene terephthalate web support to a wet thickness of 0.003 inch and dried.

A sample of the element so prepared was exposed through a process transparency using a 3,750-watt pulsed xenon arc at a distance of 18 inches for five seconds. The exposed element was immersed in a 2% aqueous solution of Rhodamine B Extra dye for 2 minutes, washed with cold water for 15 minutes, and dried. A positive, magenta-colored image was obtained.

The above element may also be used for preparing gravure printing plates. A method for forming a gravure printing plate, for example, comprises, in either order, (a) exposing to actinic radiation such a layer coated on a support, e.g., transparent film support, first through a gravure screen and then to a continuous tone image transparency which gives an image in the resist modulated by photopolymerization, and (b) adhering the surface of the layer to the surface to be imaged which has been moistened with water and peeling off the transparent film support, then (c) etching the surface to produce an intaglio printing surface with conventional etching solution, e.g., ferric chloride, (d) washing the etched surface, e.g., with hot water to remove the residual etching solution and the photopolymer resist.

The process may also be carried out by exposing the layer to the gravure screen through the transparent support and to the continuous tone image from the opposite side. In addition, the dispersion may be coated directly on the surface to be etched.

Acid Salt of an Indolinobenzospiropyran

Y. Noguchi, S. Watarai, C. Osada and H. Ono; U.S. Patent 3,933,509; January 20, 1976; assigned to Fuji Photo Film Co., Ltd., Japan describe a photopolymerizable composition comprising (a) a photopolymerization initiator comprising at least one inorganic or organic acid salt of a Lewis acid or proton acid effective as an initiator for cationic polymerization and an indolinobenzospiropyran represented by the following general formula:

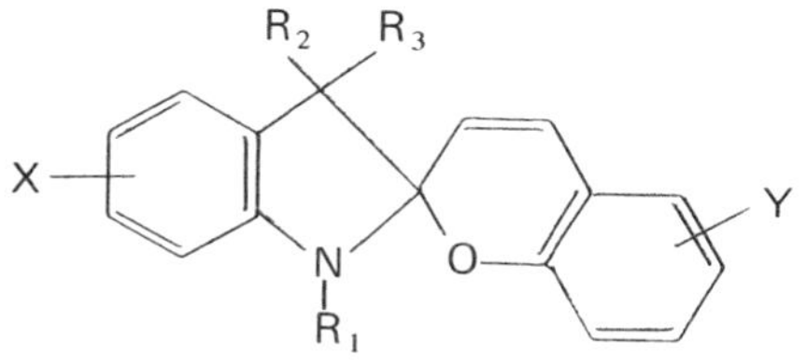

R_2 R_3 X N R_1 O Y

In the formula R_1 represents an alkyl group, substituted alkyl group or a phenyl group; R_2 and R_3 each represent an alkyl group or a phenyl group, or when taken together, R_2 and R_3 form a methylene chain $(CH_2)_n$ where n is an integer of 4 to 5; X represents a hydrogen atom, a nitro group, a halogen atom, a carboxy group, an alkoxycarbonyl group, an alkyl group or an alkoxy group; and Y represents one or more substituents selected from the group consisting of a hydrogen atom, a nitro group, a halogen atom, a formyl group and an alkoxy group; and (b) at least one cationically polymerizable substance having an ethylenically unsaturated bond or having a cyclic group capable of ring-opening for polymerization.

Upon being exposed to electromagnetic radiation of a wavelength of 250 to 700 mμ this composition photopolymerizes and, at the same time, undergoes a change in the optical density in the visible region.

The amount of the acid salt of indolinospirobenzopyran in the photopolymerizable composition is preferably 0.00001 to 15% by weight, especially preferably 0.001 to 5% by weight, based on the total amount of the cation-polymerizable compound(s).

The photopolymerizable composition of the process is colored yellow or orange before exposure but, when exposed, the color fades away to transparency. Therefore, when a plate for use in photo platemaking is prepared using this light-sensitive composition, a photo-faded image is produced simultaneously with the photo hardening. Thus, in the case of repeatedly exposing various portions of a plate, the plate of the process is extremely convenient since the exposed portions are discriminated from the unexposed portions.

Example 1: 2 grams of p-tert-butyl-phenoxyethyl vinyl ether was dissolved in 15 ml of a 10% acetone solution of cellulose acetate butyrate ester (Half Second Butyrate, molar percent: 37% butyryl and 13% acetyl). Further, a photopolymerization initiator solution prepared by dissolving 50 mg of 3,3-dimethyl-1-β-ethoxycarbonylethyl-6'-nitrospiro(indoline-2,2'-2'H-chromene) boron trifluoroetherate in 20 ml of acetone was added thereto.

The resulting light-sensitive composition was uniformly coated on a biaxially stretched polyethylene terephthalate film of about 100 microns in thickness using a coating rod. Thus, a yellowish orange light-sensitive sheet having coated thereon the light-sensitive composition in a dry thickness of about 2.5 microns was obtained. A positive line original for use in photographic platemaking was intimately superposed on the resulting light-sensitive sheet and the assembly was exposed for one minute to a 300-watt tungsten lamp for illumination spaced at a distance of 30 cm. The yellowish orange color faded and became almost clear at the exposed areas.

Subsequently, an art paper as an image-receiving paper was placed on a hot plate uniformly heated to 110°C, the above-described exposed sheet was pressed onto the image-receiving paper for about 5 seconds, then the sheet was removed. Areas corresponding to the unexposed areas were selectively transferred to the image-receiving paper. When toner powder was sprinkled over the surface of the image-receiving paper and the excess powder was removed, a visible copy in conformity with the original was obtained.

Example 2: When the same procedures as described in Example 1 were conducted except for using 2-(2,4,6-tribromophenoxy)ethyl vinyl ether in place of the vinyl ether used in Example 1, substantially similar results were obtained.

Diazine-Electron Donor Catalyst

In photopolymerization, the polymerization process is activated or initiated by radiation, for example, in the region of ultraviolet or visible waves. According to *N. Baumann and H.-P. Schlunke; U.S. Patent Application (Published) B 420,176; assigned to Ciba-Geigy AG, Switzerland* ethylenically unsaturated compounds can be polymerized advantageously with the aid of a catalyst system which consists of a diazine compound, preferably a quinoxaline and an electron donor, which together act as a photoredox pair.

Example: A cellulose triacetate film of size 234 cm^2 coated with hardened gelatin, is covered with 4 ml of the solution shown below (casting solution), to which, before casting, 10 mg of the quinoxaline compound of the formula

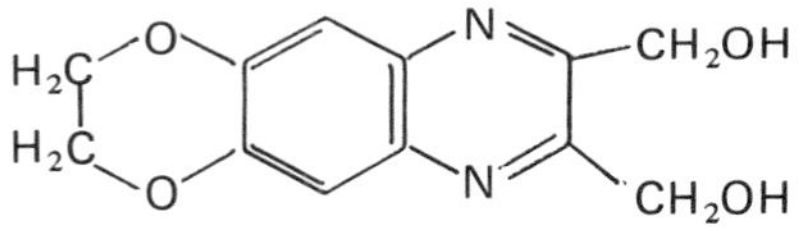

H_2C / H_2C (–O– … –O–) ; N ; N ; CH_2OH ; CH_2OH

in 2 ml of ethanol and 2 ml of an 0.016 molar aqueous sodium p-toluenesulfinate solution have also been added, and is dried.

Casting Solution – The casting solution consists of the following: 180 ml of 1.4 molar aqueous barium diacrylate solution; 60 ml of 1.6 molar aqueous acrylamide solution; 30 ml of 6% strength aqueous gelatin; and 30 ml of 0.25% strength aqueous solution of a nonionic polyfluorinated wetting agent (FC 170 from 3M).

The film treated in this way is exposed under a photographic step wedge (12 steps) for 30 seconds by the contact process to a 400-watt mercury high-pressure lamp at a distance of 40 cm. Thereafter the film is washed with water or rubbed down with moist cotton-wool and subsequently immersed for 10 seconds in a 2% strength aqueous solution of a dyestuff.

Numerous examples of applicable diazine compounds and electron donors are included in the published application.

α,ω-Diarylpolyene for Sulfonylazide Polymers

G.M. Peters, Jr., F.A. Stuber and H. Ulrich; U.S. Patent 3,947,337; March 30, 1976; assigned to The Upjohn Company describe compounds of the formula

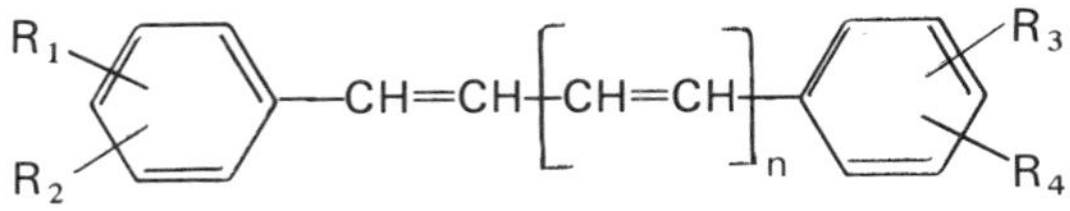

R_1, R_2 –CH=CH–[CH=CH]$_n$– R_3, R_4

where R_1 and R_2 are lower alkoxy or di(lower alkyl)amino and R_1 can be hydrogen and R_3 and R_4 are hydrogen, lower alkoxy or di(lower alkyl)amino and n is is an integer from 1 to 5. The compounds are sensitizers for photosensitive compounds particularly polymers containing sulfonazido groups as the light-sensitive moiety. Advantageously the sensitizer is employed in a proportion of about 10 to 20 parts by weight per 100 parts of light-sensitive polymer.

Example: The preparation of 1-(2,5-dimethoxyphenyl)-4-(4-dimethylaminophenyl)-butadiene is as follows. To a mixture of 5 grams (28.5 mmol) of 4-dimethylaminocinnamaldehyde and 8.2 grams (28.5 mmol) of O,O-diethyl 2,5-dimethoxybenzylphosphonate in approximately 30 ml of 1,2-dimethoxyethane was added 0.7 gram (29 mmol) of sodium hydride. The resulting mixture was heated under reflux for 90 minutes. The mixture was cooled to room temperature, the supernatant liquor was decanted from the orange gummy material which had settled out and the latter was washed with 50 ml of 1,2-dimethoxyethane.

The washings and the supernatant were combined and concentrated to about 60 ml by evaporation before being poured into 300 ml of water. The solid material which separated was isolated by filtration and recrystallized from ethanol. There was thus obtained 2.7 grams of 1-(2,5-dimethoxyphenyl)-4-(4-dimethylaminophenyl)-butadiene in the form of a crystalline solid having a melting point of 118° to 120°C. The UV spectrum of this material (2×10^{-5} molar solution in acetonitrile) exhibited a maximum at 380 nm (ϵ = 49,500). The nuclear magnetic resonance spectrum was also in accordance with the assigned structure.

In order to assess the efficiency of the above compound in extending the sensitivity of a sulfonylazide photosensitive polymer to radiation of wavelength of the order of 400 nm the following experiments were carried out.

A solution of 2.2 parts by weight of Photozid (a light-sensitive sulfonylazide polymer obtained by reacting a maleic anhydride-vinyl methyl ether copolymer with 2-hydroxyethyl 4-azidosulfonylcarbanilate) and 0.22 part by weight of the above butadiene in 100 parts of methyl ethyl ketone was used to whirler-cast films of light-sensitive polymer on strips of Mylar.

Individual coated Mylar strips were then exposed, via a negative image comprising a pattern of dots, to radiation of wavelengths 400 nm, 404 nm and 435 nm (produced by use of appropriate filters which excluded all other frequencies from the radiation produced by an Osram XBO 150-watt high-pressure xenon lamp. The exposure times were 30 seconds for 400 nm wavelength, 3 minutes for the 404 nm and 10 minutes for the 435 nm radiation. After exposure the irradiated Mylar strips were developed by immersion for 60 seconds in a bath containing a mixture of 5 parts of methyl ethyl ketone and 1 part of 4-methyl-2-pentanone.

The developed strips were then dyed by immersion for 60 seconds in a 1% w/w aqueous solution of Basic Blue 9. A clear blue image was achieved on all test strips. In contrast, strips which were prepared and irradiated in exactly the same manner but either omitting the butadiene from the initial polymer solution or replacing the butadiene by an equal weight of 1,4-diphenylbutadiene, showed no image thus indicating that, in the absence of the above butadiene sensitizer or when the butadiene was replaced by 1,4-diphenylbutadiene, the Photozid had not been activated by the radiation of the above wavelengths.

Part II. Electron Beam Curing

COATINGS

COATING COMPOSITIONS

Crosslinking of Polyvinylidene Fluoride for Wire Coating

R. Timmerman; U.S. Patent Reissue 25,904; November 16, 1965; assigned to Radiation Dynamics, Inc. found that polyvinylidene fluoride resins are not degraded by ionizing radiation to at least 100 megarads and, in fact, undergo crosslinking during irradiation thereby upgrading their thermal properties.

This result is quite contrary to the general polymer radiation rule which states that polymers with two hydrogen atoms substituted on the same carbon atom degrade when subjected to ionizing radiation. When polyvinylidene fluoride was irradiated in air at room temperature, its tensile strength increased, elongation decreased and deformation under load markedly increased at 200°C.

Moreover, tensile strength increased 14.2% up to 16 megarads, but only 10.5% up to 100 megarads. Elongation decreased 35.3% up to 8 megarads, but only 57.6% up to 100 megarads. Deformation under load at 200°C decreased 89% up to 8 megarads, but only 94.5% up to 32 megarads. Eight megarads appears to be a minimum dose necessary to obtain good structural and thermal properties. The values obtained from the deformation under load at 200°C indicate a life of at least one hour at that temperature. Continuous operation temperature limit is about 180°C, surpassing irradiated polyethylene.

Example: Vinylidene fluoride homopolymer (Kynar resin) was extruded as the insulation for a copper wire, then irradiated at a dosage of 16 megarads by high velocity electrons. One test specimen of the insulated wire exhibited about one hour life at 200°C.

Another test specimen of the wire appeared capable of indefinite service at an operating temperature of 170°C. Service in the presence of radiation did not degrade the insulated wire to the point of unserviceability, at least to the extent of the 100 megarads of radiation applied during this test.

Terminal Vinyl Ester Groups

L.S. Miller; U.S. Patent Reissue 27,656; June 5, 1973; assigned to Weyerhaeuser Company provides a process of coating a porous or nonporous substrate with a liquid polymerizable film and subjecting the coated substrate to ionizing radiation sufficient to impart to the coating composition a dose of from 1 to 10 megarads in one second or less.

The coating compositions, capable of substantially complete polymerization in less than one second, contain undiluted vinyl ester resins having terminal vinyl ester groups, or the above dissolved in vinyl monomers. The vinyl ester resins are made by reacting:

(1) A polyfunctional material selected from the group consisting of (a) dicarboxylic acids or acid chlorides having from 4 to 15 carbon atoms, and (b) polyfunctional isocyanates having terminal, reactive isocyanate groups with 2-hydroxyalkyl acrylates or methacrylates; or

(2) A half ester of a 2-hydroxyalkyl acrylate or methacrylate and a dibasic acid with a polyepoxide.

Example 1: A solution of 69.6 grams (0.6 mol) of 2-hydroxyethyl acrylate in 47.5 grams of pyridine and 150 ml of benzene was placed in a flask equipped with a stirrer, dropping funnel and condenser. A solution of 54.9 grams (0.3 mol) of adipoyl chloride in 75 ml of benzene was added with stirring and cooling. After the addition was complete, the mixture was heated one-half hour on a steam bath.

The clear solution was decanted from the precipitate and passed through a column of activated alumina. After removing the benzene on a steam bath, 82.8 grams of a clear residue, adipoyl bis(2-oxyethylacrylate) remained. A sample of wood and aluminum was given a 0.005 to 0.010 inch thick coating of the above resin composition and the coating covered with polyethylene terephthalate (Mylar) film 5 mils in thickness.

The covered products were then passed under a two million volt electron beam from a Van de Graaff accelerator in a way that the samples received 1.7 megarads in 0.3 second. After removing the Mylar cover, a clear hard coating remained on the substrates which did not appreciably soften in water and was not cracked or crazed by a sharp hammer blow.

A subsequent coating on aluminum of the same material was irradiated with a 500,000 volt electron beam giving a dose rate of two megarads in 0.5 second. Again a hard film resulted.

Example 2: In a glass stoppered flask, 20 parts of polypropylene glycol was mixed with 14.2 parts of toluene diisocyanate. The flask became warm from a mild exothermic reaction. After standing for 1 hour, 12.4 parts of 2-hydroxyethyl acrylate were added. After standing overnight, a clear viscous liquid remained.

A thin film of this resinous material was coated on aluminum, a 5 mil Mylar film placed over the coated substrate and the sample passed under a 300 kilovolt electron beam. The coating cured in 0.33 second to a very flexible, tough film which adhered strongly to the aluminum. A solution of 90% of the above reaction product in 10% n-butyl acrylate gave a harder, very flexible film on curing under the 300 kilovolt electron beam for 0.67 second.

Unsaturated Polyesters and N-3-Oxohydrocarbon-Substituted Acrylamides

The compositions of *T.C. Jennings; U.S. Patent Reissue 27,722; August 7, 1973; assigned to The Lubrizol Corporation* provides an unsaturated polyester and a crosslinking medium comprising (a) about 30 to 100 parts by weight of a monomeric N-3-oxohydrocarbon-substituted acrylamide having the formula

$$R^1-\overset{\overset{O}{\|}}{C}-\underset{R^3}{\overset{R^2}{\underset{|}{\overset{|}{C}}}}-\underset{R^5}{\overset{R^4}{\underset{|}{\overset{|}{C}}}}-\underset{H}{\underset{|}{N}}-\overset{\overset{O}{\|}}{C}-\underset{R^6}{\underset{|}{C}}=CH_2$$

wherein each of R^1, R^2, R^3, R^4 and R^5 is hydrogen or a hydrocarbon radical and R^6 is hydrogen or a lower alkyl radical, with (b) about 0 to 70 parts by weight of a second monomer capable of crosslinking the polyester, such as styrene diallyl phthalate.

An application of the curable compositions of this process is in radiation-cured coatings. Such coatings may be formed by spreading the polyester-monomer mixture over the surface to be coated, typically to a thickness of about 3 mils or less, and passing a beam of electrons or of radioactively produced particles (e.g., from a radioactive cobalt source) into the coating.

Halogen-Containing Olefinically Unsaturated Esters

M. Marx, A. Zosel, H. Spoor, H. Pohlemann and D. Heinze; U.S. Patent Reissue 28,173; September 24, 1974; assigned to Badische Anilin- & Soda-Fabrik AG, Germany provide a process for the production of crosslinked polymers by irradiation with doses of high energy radiation of from 1 to 30 Mrad at an irradiation period of from 0.5 second to 20 minutes. The system irradiated comprises (a) tetra-(allyloxymethyl)-acetylenediurea and (b) 10 to 90% by weight, with reference to the sum of (a) and (b), of esters of alpha,beta-unsaturated polymerizable carboxylic acids having three to six carbon atoms with chlorinated or brominated alcohols.

Example 1: A film having a thickness of 50 microns is drawn from a mixture of 7 parts of 2,3-dibromopropyl acrylate and 3 parts of tetra-(allyloxymethyl)-acetylenediurea by means of film drawing equipment onto a deep drawing metal sheet and irradiated with an electron beam at 2 megavolts accelerating potential with a dose of 12 Mrad at a dose rate of 19 Mrad per minute. A nontacky glossy hard coating is obtained which adheres very well. Its elasticity according to DIN 53,156 is 5.4.

Example 2: A deep-drawing metal sheet having a thickness of 0.7 mm is coated on both sides by dipping it in a mixture of 3 parts of 2,3-dibromopropyl acrylate and 3 parts of tetra-(allyloxymethyl)-acetylenediurea and then exposed in the air at 30°C to x-ray irradiation (anode: gold; potential: 120 kilovolts). At a dose rate of 0.1 Mrad per minute an irradiation period of 20 minutes is necessary to obtain both on the side of the sheet facing the radiation source and on the opposite side a hard, glossy, elastic and well-adhering solvent-resistant coating.

Hot Melts with Reduced Mixing Times

Hot melt compositions are produced by *J.D. Domine and R.H. Schaufelberger; U.S. Patent 3,884,786; May 20, 1975; assigned to Union Carbide Corp.* by preparing a concentrate of wax and copolymer, irradiating the concentrate to produce an irradiated concentrate and then diluting the irradiated concentrate with additional quantities of wax. The compositions are used as coatings.

The improved three step method permits the incorporation of copolymers having higher molecular weights in the hot melt composition; it was also found that the copolymers are more readily dissolved in the wax. The copolymers used are the ethylene-vinyl acylate and ethylene-alkyl acrylate copolymers where the vinyl acylate or alkyl acrylate content is from 5 to 40 weight percent. The copolymer has an initial melt index of from 1.5 to 100 dg/min. High energy ionizing radiation to an absorbed dose of from 0.01 to 5 megareps is used.

Example: A series of three concentrates was produced by initially compounding mixtures of wax and ethylene-vinyl acetate (EVA) copolymer in a Banbury mill at 70°C for 10 minutes. The copolymer used has an initial relatively high melt index of 20 dg/min and a vinyl acetate content of 28 weight percent.

The three concentrates were irradiated in air at about 20°C with high energy electrons by exposure to the radiation from a two mev Van de Graaff electron accelerator. The irradiated copolymer in the irradiated concentrates now had a low melt index calculated to be about 1.2 dg/min. This value is estimated on the basis that the melt viscosities of these irradiated concentrates were similar to the melt viscosity of the control at the same concentration and temperature; the control was produced using a copolymer having a melt index of 1.2 dg/min.

The dose was adjusted so that a hot melt blend of 40 weight percent irradiated copolymer and 60 weight percent wax would have about the same melt viscosity as the comparative control blend hereinafter described. When all other factors are equal, e.g., concentration, temperature, etc., melt viscosity can be used as a measure of molecular weight. The three irradiated concentrates were diluted with wax to yield hot melt coating and adhesive compositions containing 40 weight percent copolymer. This dilution was carried out in a standard commercial type laboratory mixer at 120° to 125°C.

The comparative control blend was prepared by mixing 40 parts of ethylene-vinyl acetate copolymer having an initial melt index of 1.2 dg/min and a vinyl acetate content of 28 weight percent with 60 parts of wax in a commercial type laboratory mixer. The details and results are summarized in the table on the following page.

The data show that the total mixing time for the preparation of Run A was only 44% of the mixing time required for the preparation of the control; for the preparation of Run B only 50% and for the preparation of Run C only 56%. These reductions in time are of great commercial significance.

Run	A	B	C	Control
Concentrate, parts by weight				
EVA (MI = 20 dg/min)	80	85	90	0
Wax	20	15	10	0
Irradiated concentrate				
Dosage, megarep	1.5	1.3	0.9	0
Hot melt composition, parts				
Irradiated concentrate A	50	0	0	0
Irradiated concentrate B	0	47.1	0	0
Irradiated concentrate C	0	0	44.4	0
EVA (MI = 1.2 dg/min)	-	-	-	40
Wax	50.0	52.9	55.6	60
Mixing time, minutes	115	130	145	260
Viscosity at 121.1°C, cp	42,300	39,900	41,800	42,800

Alkyd Resin

An alkyd resin composition is prepared by *T. Watanabe, K. Murata and T. Maruyama; U.S. Patent 3,882,006; May 6, 1975; assigned to Kansai Paint Company, Ltd., Japan* by dissolving a polymerizable alkyd resin having from 10 to 50% of oil length into a vinyl monomer.

The polymerizable alkyd resin is obtained by a half-esterification reaction of an acid anhydride having a polymerizable unsaturated group and an alkyd resin modified with conjugated unsaturated fatty acid and/or conjugated unsaturated oil having at least one reactive hydroxyl group per one molecule. The alkyd resin composition thus obtained is coated on an article, and ionizing radiation is applied on the article to cure the coated film thereon.

Example: Into a reactor provided with a thermometer, stirrer, pipe for introducing inert gas, and water separator, 384 parts of phthalic anhydride, 115 parts of pentaerythritol, 233 parts of trimethylolethane, 288 parts of tung oil fatty acids, 49 parts of p-tertiary-butyl benzoic acid, and 60 parts of xylene (mixed xylene consisting of 30% of o-xylene, 30% of m-xylene, and 40% of p-xylene) as azeotropic solvent were charged.

Nitrogen gas was introduced into the resulting mixture, the temperature of the contents in the reactor was elevated to 200°C while stirring, and the reaction was continued for about 10 hours at this temperature thereby to condense the resin thus obtained to an acid value of 12 (oil length being 28.8% and hydroxyl value about 155). Thereafter, the temperature of the contents was decreased to 110°C, 271 parts of maleic anhydride and 0.6 part of hydroquinone were added thereto, and the stirring was continued for 1 hour at 110°C thereby to complete the reaction.

By this reaction operation, an alkyd resin modified by a conjugated unsaturated fatty acid and obtained by reacting free hydroxyl groups with acid anhydride having polymerizable unsaturated group was obtained.

Immediately after the half-esterification reaction, 500 parts of styrene was added to 500 parts of the resulting alkyd resin, nitrogen gas was introduced thereinto, and the resulting mixture was sufficiently stirred to dissolve the resin while heating the mixture at a temperature of 80°C. After dissolving the mixture, it was cooled to room temperature and as a result, a styrene monomer solution having a resin content of 50% was prepared.

A coating material prepared by blending well 1.0 part of 60% toluene solution of cobalt naphthenate with 100 parts of the above described alkyd resin composition was coated on one side of a glass plate having a thickness of 1.5 mm so as to obtain a film having a thickness of 20 microns.

The glass plate thus coated was placed on a conveyor device for electron ray radiation and was subjected to a total radiation dose of 5 Mrad by employing a transformer type electron accelerator under such conditions of acceleration energy being 300 kv and electron rays current 25 ma, respectively. The radiation time was about 1 second.

The results of pencil scratch hardness and gelation percentage obtained by measuring the cured film are indicated in the table below. For comparison, another coating material prepared by blending 2.0 parts of 50% diethyl fumarate solution of methyl ethyl ketone peroxide and 1.0 part of 60% toluene solution of cobalt naphthenate with 100 parts of the above alkyd resin composition was coated on two glass plates so as to obtain a film having a thickness of 20 microns.

	Curing Conditions		
Test Items	**Curing at 25°C for 8 hours**	**Baking at 100°C for 5 minutes**	**Radiation by 5 Mrad**
Pencil scratch hardness	B	HB	F-H
Gelation percentage	68.1	70.8	78.3

One plate was baked in a hot blast stove at a temperature of 100°C for 5 minutes, and the other was left in a thermostatic chamber at a temperature of 25°C for 8 hours. The measured results of thus prepared films are similarly indicated in the table above.

Pencil scratch hardness was determined according to JIS K5652 (5.15) (1957). To determine gelation percentage each sample film was peeled off from the glass plate, the film was placed in a wire gauze of stainless steel of 300 mesh, and extracted in a Soxhlet extractor by the use of acetone solvent for 20 hours. Then, the resulting extract was dried at a temperature of 110°C for 3 hours. The thus dried extract was weighed and the gelation percentage was calculated in accordance with the following equation: (weight of residue/weight of sample) x 100 = gelation percentage.

In accordance with the value of gelation percentage, the degree of crosslinkage can be relatively estimated.

Modified Polyester and Blends

The process of *S.F. Hudak; U.S. Patent 3,882,189; May 6, 1975; assigned to Ashland Oil, Inc.* relates to the modification of two distinct classes of polyesters listed on the following page.

(1) saturated oil free polyesters dispersed in organic solvents or water and
(2) unsaturated polyesters

The process concept is identical in both cases, although the products are quite different because of the differences in the nature of the polyester portion of the polymer molecule. Both polyester resins are prepared in a two step reaction sequence comprising the steps of:

(1) forming the polyester by condensation of polyol and polycarboxylic acid, and
(2) introduction of urethane linkages by reaction of an excess of the polyester of step 1 with an organic polyisocyanate.

The first step of polyester formation is well known in the art. It is advantageously conducted under reflux in an inert solvent such as xylene at a temperature from about 200° to 450°F for a period sufficient to lower the acid number to a point where substantially complete reaction has occurred.

This point will be reached in a period from about 5 minutes to 10 hours. The acid number of the saturated polyester after the first step and prior to reaction with polyisocyanate is preferably in the range about 4 to 20 for the organic solvent cuts and from 30 to 70 for the water-dispersible types. The acid number of the unsaturated polyester may have a value up to about 60.

The second step of the process, the reaction of polyester with organic polyisocyanate, is conducted at a temperature in the range of about 120°F to about 400°F. The polyesters that are cut in organic solvents are ordinarily processed under reflux. The water-dispersible types are processed close to 100% solids.

The polyisocyanate is added dropwise over a period of about 5 minutes to about 2 hours. The reaction mixture may then be diluted with inert solvent in the case of the organic solvent types and maintained at an elevated temperature of about 150° to 400°F for a period of about one-half hour to about 4 hours to obtain a constant viscosity and to insure complete reaction of the polyisocyanate. The water-dispersible type is processed in like manner except that no organic solvent is added during the reaction.

The ratios of the various reactants depend on the nature of the polyester and whether it is saturated, oil free, or unsaturated. In the preparation of the solvent cut, saturated urethane modified polyester, the ingredients are chosen so that the ratio of equivalents of hydroxyl/equivalents of (carboxyl + isocyanate) is greater than 1 and up to about 2.0, preferably about 1.15 to 1.35, and the ratio of carboxyl/isocyanate being greater than 1 and up to about 30, preferably about 4 to 8.

In the preparation of water-dispersible, saturated, urethane modified polyesters, the ingredients are chosen so that the ratio of equivalents of hydroxyl/equivalents of (carboxyl + isocyanate) is 0.5 to 2.0, preferably about 1.1 to 1.5, and the ratio of carboxyl/isocyanate being greater than 1 and up to about 30, preferably about 4 to 8.

In the preparation of the urethane modified unsaturated polyester, the ratio of equivalents of hydroxyl/equivalents of (carboxyl + isocyanate) is from 0.5 to 2.0, the ratio of carboxyl/isocyanate being greater than 1 and up to about 30. Although it is operable for the isocyanate to be in excess of the carboxyl, it is not preferred. The preferred ratios are 1.1 to 1.3 for hydroxyl/equivalents of (carboxyl + isocyanate) and 5 to 15 for carboxyl/isocyanate. The reactants used in the preparation of the polymeric materials of this process are those conventionally used in the polyester art and the polyurethane art.

The urethane modified saturated polyesters of this process are suited for blending with a variety of resins, particularly aminoplasts. The organic solvent cut, saturated polyesters are further suited for blending with free isocyanate containing adducts or prepolymers, to prepare coating compositions. The aminoplast should comprise 5 to 50%, and preferably 15 to 30% of the composition. The urethane modified unsaturated polyesters also are intended to be blended with copolymerizable ethylenically unsaturated monomers. The cure of the compositions can be accelerated by exposure to high energy electrons (100 kilo electron volts to 1.5 million electron volts, for example) by γ-radiation, x-ray radiation or ultraviolet radiation.

Nonglossy Coatings

E.A. Hahn; U.S. Patent 3,918,393; November 11, 1975; assigned to PPG Industries, Inc. found that a strong film or coating may be achieved with a flattened effect (low gloss) by first subjecting a radiation-sensitive material to ionizing irradiation or actinic light in an atmosphere containing at least about 5,000 parts per million of oxygen or more and subsequently subjecting the material to ionizing irradiation in an inert atmosphere containing less than about 1,000 parts per million of oxygen. The radiation-sensitive material to be subjected to ionizing irradiation or actinic light may be any substantially solventless actinic light or irradiation-sensitive, curable, organic material. The most useful organic materials to be used are polyester resins and acrylic resins and monomers. While solvents are not necessary for this flatting effect they may be used to produce other properties. The amount of ionizing irradiation employed in the first step is from about 0.2 Mrad to about 20 Mrad. The second radiation step requires about the same dosage as the first.

Example 1: A panel of aluminum was coated with hexahydrophthalic ethylene glycol diacrylate coating formulation. The coated panel was subjected to ionizing irradiation from an electron beam in air. The coating received a total dosage of 3 Mrad. The coated panel was then subjected to ionizing irradiation by an electron beam in a nitrogen atmosphere having 80 parts per million of oxygen. The total dosage of this second treatment was 3 Mrad.

The above coated panel was compared to a similar panel coated with hexahydrophthalic ethylene glycol diacrylate and subjected to 3 Mrad in nitrogen only. The physical strength of the two coatings were equivalent but the gloss of the coating subjected to only one pass under the electron beam was 50% as measured by the 60° glossmeter and the gloss of the coating subjected to electron beam in air and then in nitrogen was only 5%.

Example 2: A steel panel was coated with phthalic ethylene glycol diacrylate coating formulation and irradiated with a total dosage of 3 Mrad by an electron beam in air. The panel was then irradiated with a total dosage of 3 Mrad in a N_2 atmosphere having 80 ppm of O_2.

The above coated panel was compared to a similar panel coated with phthalic ethylene glycol diacrylate and subjected to 3 megarads in nitrogen only. The physical strength of the two coatings were equivalent but the gloss of the coating subjected to only one pass under the electron beam was 50 as measured by the 60° glossmeter and the gloss of the coating subjected to electron beam in air and then in nitrogen was only five.

Unsaturated Polyesters

T. Maruyama and K. Murata; U.S. Patent 3,919,063; November 11, 1975; assigned to Kansai Paint Company, Ltd., Japan provides a method for curing unsaturated polyester compositions which comprises applying electron beam at a total dose of 0.5 to 50 Mrad and a dose rate of 0.1 to 30 Mrad per second to an unsaturated polyester composition.

The composition comprises a vinyl monomer and unsaturated polyester dissolved therein, the vinyl monomer being a mixture of 40 to 70 weight percent of at least one of acrylic and methacrylic monomers and 30 to 60 weight percent of at least one of styrene and vinyl toluene, and the unsaturated polyester having an acid value of 4 to 11 and being a condensation product of 1.0 mol of a dicarboxylic acid and 1.0 to 1.2 mols of a polyhydric alcohol.

The dicarboxylic acid is a mixture of (a) 30 to 45 mol percent of at least one of endo-cis-bicyclo[2.2.1]-5-heptene-2,3-dicarboxylic acid and anhydride thereof, (b) 20 to 40 mol percent of at least one of unsaturated dicarboxylic acids and anhydrides thereof and (c) 15 to 55 mol percent of at least one of aromatic dicarboxylic acids, saturated aliphatic dicarboxylic acids and anhydrides thereof.

The process provides a method which makes it possible to obtain a coating film of an unsaturated polyester resin having excellent weather resistance.

Example 1: Three mols of fumaric acid, 2 mols of phthalic anhydride, 2 mols of adipic acid, 3 mols of endo-cis-bicyclo[2.2.1]-5-heptene-2,3-dicarboxylic acid, 10 mols of neopentyl glycol and 1 mol of trimethylol propane were subjected to condensation reaction in a nitrogen gas atmosphere at 200°C for 17 hours, whereby unsaturated polyester having an acid value of 8.5 was obtained.

The unsaturated polyester was then dissolved to a polymer concentration of 70 weight percent in a monomer mixture consisting of styrene, methyl methacrylate and butyl acrylate in the ratio by weight of 50:8:42, whereby varnish (Varnish A) was obtained.

To 100 parts of the varnish prepared as above were added 100 parts of titanium dioxide of the rutile type and 25 parts of the above monomer mixture, and the resultant mixture was dispersed in a paint conditioner to obtain white enamel. The resultant enamel was applied to a plate of ABS resin (acrylonitrile-butadiene-styrene resin) to a thickness of about 25 microns. The coating was irradiated and cured with electron beam by a Van de Graaff accelerator at a dose of about 12 Mrad and a dose rate of about 0.5 Mrad per second using electron beam energy of 800 kv and electron beam current of 150 μm.

Comparative Example 1: White enamel was prepared from the Varnish A in the same manner as in Example 1. One part of dimethyl phthalate solution containing 60 weight percent methyl ethyl ketone peroxide and 0.5 part of cobalt naphthenate (xylene solution containing 6.25 weight percent of cobalt) were added to 100 parts of the white enamel obtained as above. The resulting mixture was then applied to an ABS plate in the same manner as in Example 1 and the coating was left to stand at 20°C for four days for curing.

Comparative Example 2: White enamel was prepared from the Varnish A in the same manner as in Example 1. One part of benzoin ethyl ether and 0.05 part of solid paraffin having a melting point of 62°C were added to 100 parts of the white enamel obtained as above. The resulting mixture was then applied to an ABS plate in the same manner as in Example 1 and the coating was irradiated with a high pressure mercury lamp of 400 w for 3 minutes at 40 cm distant from the coating.

Acrylated Epoxidized Soybean Oil Amine

The amine derivatives of acrylated epoxidized soybean oil, which is the reaction product of epoxidized soybean oil with acrylic acid or methacrylic acid are produced by the reaction of acrylated epoxidized soybean oil with an organic amine. They are useful alone, or in conjunction with a photosensitizer and/or a pigment as inks and coatings. The compositions can be cured by radiation as described by *D.J. Trecker, G.W. Borden and O.W. Smith; U.S. Patents 3,931,075; January 6, 1976; 3,931,071; January 6, 1976; 3,979,270; September 7, 1976 and 3,878,077; April 15, 1975; all assigned to Union Carbide Corporation.*

An epoxidized soybean oil having an average molecular weight of about 1,000, an oxirane content of about 7% by weight and being the epoxide of the triester of glycerol with soybean oil was used herein.

Two liters of epoxidized soybean oil, 8 liters of acrylic acid, 8 grams of hydroquinone, 2 grams of p-methoxyphenol and 0.5 gram of phenothiazine were charged to a reaction flask and reacted at 100° to 110°C for 5 hours while continuously purging dry air through the mixture.

The unreacted acrylic acid was distilled under vacuum on a rotary film evaporator and the residual product was diluted with diethyl ether. The solution was passed through a column of amine containing ion exchange resin, Rohm & Haas A-21 and then distilled to remove the diethyl ether. The acrylated epoxidized soybean oil was a straw yellow color and was free of unreacted acrylic acid.

Epoxy Ester-Saturated Alkyd

E. Takiyama, S. Hokamura and T. Hanyuda; U.S. Patent 3,882,187; May 6, 1975; assigned to Showa High Polymer Company, Ltd., Japan describe an unsaturated epoxy-ester resinous composition curable under irradiation by an electron beam to form a coating film exhibiting excellent dryness and adhesiveness on the surface of a coated metal. The epoxy-ester resinous composition comprises a reaction product of:

(a) one equivalent weight of an epoxy compound having at least two epoxy radicals in the molecule,

(b) 0.5 equivalent weight of an unsaturated monocarboxylic acid, and
(c) 0.5 equivalent weight of a saturated alkyd which has a terminal carboxyl radical in the molecule and is produced by heating one mol of a glycol with 1.33 to 2.0 mols of a saturated dicarboxylic acid, the reaction product being produced under heating and in the presence of an esterification catalyst and with or without a polymerization inhibitor, solvent, polymerizable monomer or additional solvent.

Example: A saturated alkyd having the terminal carboxyl radicals and an acid value of 108 was prepared by heating an admixture of 159 grams (1.5 mols) of diethylene glycol and 332 grams (2.0 mols) of tetrahydrophthalic acid anhydride in a carbon dioxide gas stream at 180°C for about 6 hours.

Then 510 grams (one equivalent) of the saturated alkyd, 374 grams (two equivalents) of Shodine-508 which is a diglycidyl phthalate and has an epoxy equivalent of 187, 72 grams (one equivalent) of acrylic acid, 3 grams of diethylamine hydrochloride (an esterification catalyst) and 0.4 gram of hydroquinone were charged into a flask equipped with a stirrer, a refluxing condenser and a thermometer, and then the resultant mixture was heated at 120°C for about 180 minutes under agitation to produce a reaction product having an acid value of 25. The reaction product was cooled and then it was mixed with styrene to produce an unsaturated epoxy ester resinous composition A containing 35 weight percent of styrene.

A steel plate was coated with the unsaturated epoxy ester resinous composition A to have a layer having a 50 micron thickness and the layer was exposed in air to a radiation of 5 Mrad, 10 Mrad, or 15 Mrad by using a Van de Graaff electron accelerator. It was found that the layer lost its tackiness and has a pencil hardness of H to 2 H.

Also the layer was tested for adhesiveness in accordance with the specification of the National Coil Coater's Association. The test results are given in the following table

	5 Mrad	10 Mrad	15 Mrad
Pencil hardness	H	2H	H
Test on squares	100/100	100/100	100/100
Bending test at an angle of 180°	Pass	Pass	Pass
Cross-cut Erichsen test	above 8 mm	above 8 mm	above 8 mm

Chlorinated Paraffin and Acrylate

P.G. Garratt and J. Hoigne; U.S. Patent 3,892,884; July 1, 1975; assigned to Lonza Ltd., Switzerland provide a method for the hardening of synthetic resins, particularly in the shape of thin layers or coatings, by means of ionizing radiation where synthetic resin mixtures are used which contain (a) chlorinated polyolefins as prepolymers and (b) reactive, olefinically unsaturated monomers.

The amount of prepolymer in the synthetic resin mixture is advantageously 20 to 80% by weight, preferably 30 to 50% by weight. The hardening of the synthetic resin mixture can be effected with any ionizing radiation, preferably with an energy rich electromagnetic radiation, for instance, with x-rays or gamma rays and with accelerated electrons such as accelerated electrons having an average electron energy of 50 to 4,000 kev. When hardening thin layers of the synthetic resin mixture, for instance, thin coatings, an average electron energy of 50 to 600 kev is used.

Example 1: 30 to 120 micron thick films of a synthetic resin mixture which contains one part of a chlorinated polypropylene (Alprodur 646 J) mixed with one part n-butyl acrylate, were deposited on electrolytically pretreated steel sheets. After approximately 1½ to 2½ minutes these films were exposed to accelerated electrons with an average energy of 400 kev, by moving the sheets in the longitudinal direction through an electron beam.

These tests were made at room temperature and without any particular steps to exclude oxygen. The films were then immediately tested for stickiness of the surface and their surface hardness was tested with a steel blade. The resistance of the synthetic resin films against chemical solvents was tested by application of a drop of toluene.

The degree of crosslinking was determined by measuring the insoluble residue in a Soxhlet extraction apparatus after 24 hour extraction with toluene. The hardening dose is the smallest radiation dose which was required in order to obtain a nonsticky film with satisfactory surface hardness. The table below lists the variation of the hardening dose with the thickness of the film.

Film Thickness in microns	Hardening Dose (Intensity of Radiation 12 Mrad/sec)	Percent Insoluble Portion if Hardening Dose Is Applied
30	3.5	72
60	4.0	74
90	4.25	68
120	4.5	73

Example 2: 90 micron thick films of various synthetic resin mixtures were applied, irradiated and tested in the same manner as that described in Example 1. The irradiation took place in a nitrogen atmosphere, however. The table below, shows the films hardened by irradiation with electrons, the hardening dose required and the insoluble portion in percent.

Prepolymer	Monomer	Weight Ratio Prepolymer: Monomer	Intensity of Irradiation (Mrad/sec)	Hardening Dose (Mrad)	Insoluble Part at the Hardening Dose (percent)
Chlorinated polypropylene (Alprodur 646 J)	n-Butyl acrylate	1:1	12	1.75	71
Chlorinated polyethylene (Alloprene CPE 20)	n-Butyl acrylate	1:1	12	1.75	65
Chlorinated rubber (Alloprene R 20)	n-Butyl acrylate	1:1	12	2.0	27

Acrylic Acid Solvent for Radiation Curable Oil

T.F. Huemmer and R.J. Plooy; U.S. Patent 3,912,670; October 14, 1975; assigned to The O'Brien Corp. describe radiation curable coating compositions for metal containers comprising a radiation curable oil having at least two radiation sensitive sites per molecule, a flow control additive and a reactive solvent selected from the group consisting of acrylic acid and mixtures of acrylic acid with another acrylic monomer.

The three essential components are present in the following proportions: radiation curable oil, 10 to 78% by weight; flow control additive, 2 to 40%; and solvent, 20 to 80% by weight.

Example 1: An acrylic acid-epoxy adduct was formed by mixing the following materials in a 1,000 ml three neck flask equipped with a stirrer, thermometer and condenser.

	Grams
Acrylic acid	360
Epoxidized linseed oil (Epoxol 9-5)	442.5
Monomethyl ether of hydroquinone	0.5
Triethylamine	4.0

The mixture was heated to 100°C for a period of 12 hours. The epoxy number of the resulting acrylic acid epoxy adduct was then determined and found to be 0.15.

The following were mixed in a 1,000 ml three neck flask equipped with a stirrer, condenser and thermometer.

	Grams
Acrylic acid-epoxy adduct, above	130
Acrylic acid	255
2-Ethylhexyl acrylate	348
Hydrocarbon resin modifier (Modaflow)	80

Films were drawn down on aluminum substrate at the 0.5 mil thickness with a drawdown rod and cured at the 5 Mrad dose level under nitrogen. The data appear below.

	Result
Initial viscosity, sec, No. 4 Ford cup	35
Adhesion, cross hatch	100% pass
Adhesion, burr edge	100% pass
Surface cure at 5 Mrad (N_2)	Tack free
Wedge bend	100% pass
Leveling and flow appearance	good

Example 2: Into a 1,000 ml three neck flask equipped with a stirrer, thermometer, Dean-Stark trap and condenser were added the following.

	Grams
Hydrogenated castor oil (Castorwax)	467
Acrylic acid	216
Monomethyl ether of hydroquinone	0.5
Concentrated sulfuric acid	18
Benzene	50

The mixture was heated to the reflux for a period of 10 hours. After the removal of 20 ml of water, the mixture was sparged with air to remove the benzene solvent. The above product was then mixed accordingly.

	Grams
Hydrogenated castor oil-acrylic acid adduct	130
Acrylic acid	500
Resin modifier (Gantez AN 8194)	35

The composition was tested as a 1.0 mil drawdown on "as received" aluminum substrate and found to cure tack-free at the 5 Mrad dose level under nitrogen. Cross hatch adhesion to the aluminum substrate immediately upon emerging from the beam was 100% pass. The coating was able to withstand 20 inch-pounds of indirect impact and pass a water pasteurization test of 100°C for 10 minutes with no blushing or adhesion failure of the coating.

Radiation Curable Epoxy

N.S. Marans and A. Glueckmann; U.S. Patents 3,912,608; October 14, 1975; and 3,929,927; December 30, 1975; both assigned to W.R. Grace & Company provide a radiation curable epoxy composition. The composition consists of an intimate mixture of 40 to 95 parts of an epoxy compound having the formula

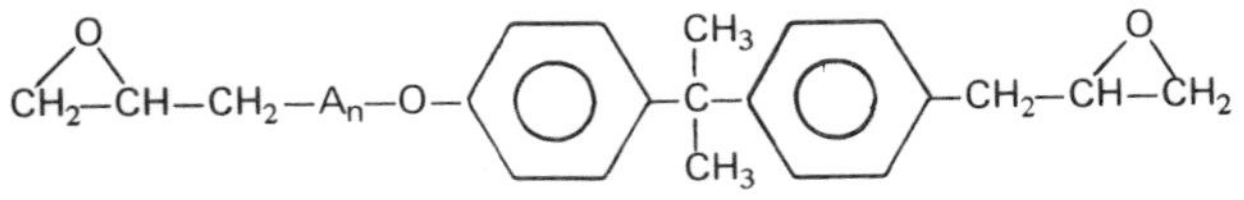

$$\overset{O}{\overbrace{CH_2-CH}}-CH_2-A_n-O-C_6H_4-\underset{CH_3}{\overset{CH_3}{\underset{|}{\overset{|}{C}}}}-C_6H_4-CH_2-\overset{O}{\overbrace{CH-CH_2}}$$

in which A is

$$-O-C_6H_4-\underset{CH_3}{\overset{CH_3}{\underset{|}{\overset{|}{C}}}}-C_6H_4-O-CH_2-\overset{OH}{\overset{|}{C}H}-CH_2-$$

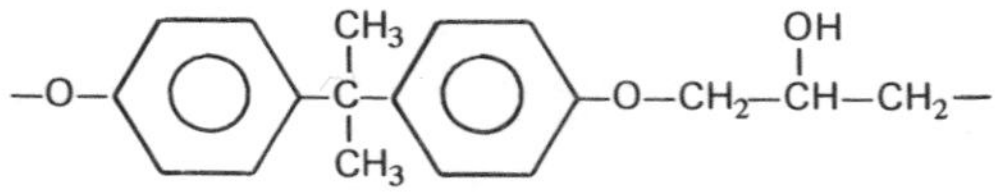

and n has an average value of 0 to 13 and 5 to 60 parts of an ethylenically unsaturated compound having the formula

$$CH_2{=}\underset{R_1}{\underset{|}{C}}COO-R_2-N\begin{matrix}R_3\\R_4\end{matrix}$$

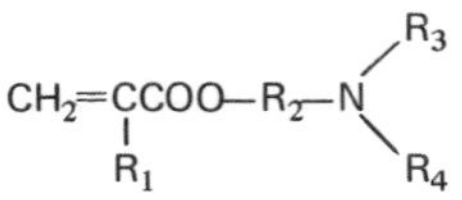

in which R_1 is hydrogen or a lower alkyl group, R_3 and R_4 are lower alkyl groups, and R_2 is an alkylene group having 2 to 8 carbon atoms.

In a preferred embodiment the composition is cured by subjecting it to high energy ionizing radiation including x-rays, gamma rays, an electron beam, a neutron beam, or a proton beam. The radiation dosage required to produce a good cure is about 0.1 to 5.0, preferably about 0.5 to 2.5 Mrads.

This result (the curing of the composition) is surprising because a similar dosage of radiation failed to cure the epoxy component of the composition where the epoxy component was irradiated in the absence of the ethylenically unsaturated compound.

Example 1: 75 parts of a commercially available epoxy compound (Epon 828) was admixed with 25 parts of dimethylaminoethyl methacrylate. A portion of the above composition was applied to a glass plate and irradiated with a high energy electron beam generated by a Van de Graaff generator. The radiation dosage was 1.2 Mrad. The composition cured to produce a hard film firmly bonded to the glass plate.

Example 2: The general procedure of Example 1 was repeated. However, in this instance 90 parts of Epon 828 was admixed with 10 parts of dimethylaminoethyl methacrylate. A coating of this composition when applied to a glass plate and irradiated with an electron beam (1.25 Mrad) formed a hard coating which adhered firmly to the glass plate.

In related work *N.S. Marans and A. Gluecksmann; U.S. Patent 3,926,755; December 16, 1975; assigned to W.R. Grace & Company* provide a composition consisting essentially of an intimate mixture of (a) 35 to 85 parts of polyester of an ethylenically unsaturated polycarboxylic acid and a polyhydric alcohol and (b) 15 to 65 parts of an amino compound having the formula

$$CH_2{=}\underset{\displaystyle R_1}{\underset{|}{C}}COO{-}R_2{-}N\begin{matrix}\diagup R_3\\ \diagdown R_4\end{matrix}$$

in which R_1 is hydrogen or a lower alkyl group, R_2 is alkylene group having 2 to 8 carbon atoms, and R_3 and R_4 are lower alkyl groups, the polyester plus the amino compound totaling 100 parts.

A preferred embodiment is directed to a radiation curable composition consisting essentially of an intimate mixture of (a) 5 to 85 parts of a polyester of an ethylenically unsaturated polycarboxylic acid and a polyhydric alcohol (b) 5 to 50 parts of the above described amino compound and (c) 5 to 45 parts of styrene or diallyl phthalate, the polyester, plus the amino compound, plus the styrene or diallyl phthalate totaling 100 parts. This composition is cured (polymerized) by irradiation with 0.01 to 10 Mrad (more preferably 0.1 to 2.5 Mrad) of high energy ionizing radiation.

Example 1: The polyester used in this example was a commercially available polyester of an ethylenically unsaturated polycarboxylic acid, a saturated polycarboxylic acid, and a polyhydric alcohol. The polyester was prepared by the reaction of phthalic anhydride and maleic anhydride with ethylene glycol. As purchased the polyester was a mixture of the aforementioned polyester and styrene. The proportions of polyester to styrene were about 70 parts of polyester and 30 parts of styrene.

95 parts of the polyester-styrene mixture (PSM) were admixed with 5 parts of a dimethylaminoethyl methacrylate (DMAEMA) to form a mixture, a portion of which was applied to a glass plate and irradiated with a high energy electron beam generated by a Van de Graaff generator. The radiation dosage was 0.25 Mrad. The composition cured to produce a hard film of cured polymer bonded to the glass plate.

Example 2: The general procedure of Example 1 was repeated however, in this instance 90 parts of the PSM described in Example 1 was admixed with 10 parts of DMAEMA. A coating of this composition was applied to a glass plate and irradiated with an electron beam, the dose being less than 0.25 Mrad. The irradiated composition formed a hard coating of cured polymer which firmly adhered to the glass plate.

Polyol-Carbamate with Unsaturated Monomer

K. Jellinek and R. Oellig; U.S. Patent 3,928,287; December 23, 1975; assigned to Rutgerswerke AG, Germany provide a compound comprising the reaction product of at least one half of the hydroxyl groups in a polyol of the formula

$$\text{(I)}\qquad R\left[O-CH_2-\underset{\displaystyle OH}{\underset{|}{CH}}-CH_2-N\begin{matrix}\diagup \overset{\displaystyle R'}{\overset{|}{CH}}-\overset{\displaystyle R'}{\overset{|}{CH}}-OH\\ \diagdown \underset{\displaystyle R'}{\underset{|}{CH}}-\underset{\displaystyle R'}{\underset{|}{CH}}-OH\end{matrix}\right]_n$$

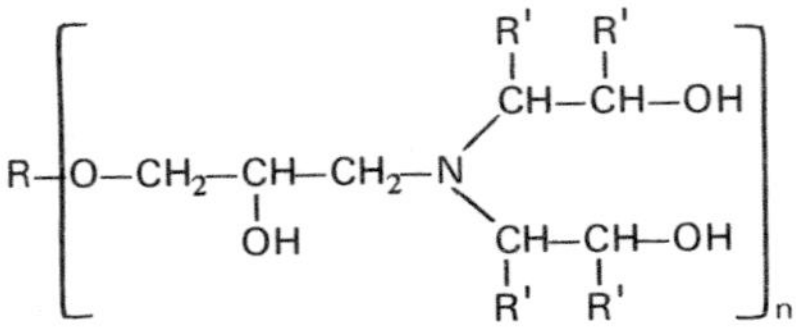

the polyol being the reaction product of an epoxy compound and a dialkanolamine in a stoichiometric ratio between amine hydrogen and epoxy groups, and a monoisocyanatoallyl carbamate or monoisocyanatomethallyl carbamate of the formula

$$\text{(II)}\qquad O{=}C{=}N-R''-NH-\overset{\displaystyle O}{\overset{\|}{C}}-O-\overset{\displaystyle R'''}{\overset{|}{CH}}-\overset{\displaystyle R'''}{\overset{|}{C}}{=}CH_2$$

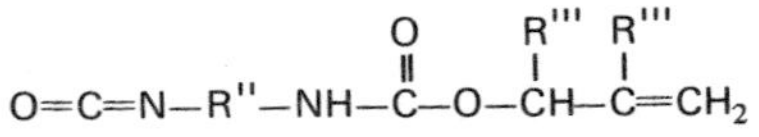

or

$$\text{(III)}\qquad O{=}C{=}N-R''''\begin{matrix}\diagup NH-\overset{\displaystyle O}{\overset{\|}{C}}-O-\overset{\displaystyle R'''}{\overset{|}{CH}}-\overset{\displaystyle R'''}{\overset{|}{C}}{=}CH_2\\ \diagdown NH-\underset{\displaystyle O}{\underset{\|}{C}}-O-\underset{\displaystyle R'''}{\underset{|}{CH}}-\underset{\displaystyle R'''}{\underset{|}{C}}{=}CH_2\end{matrix}$$

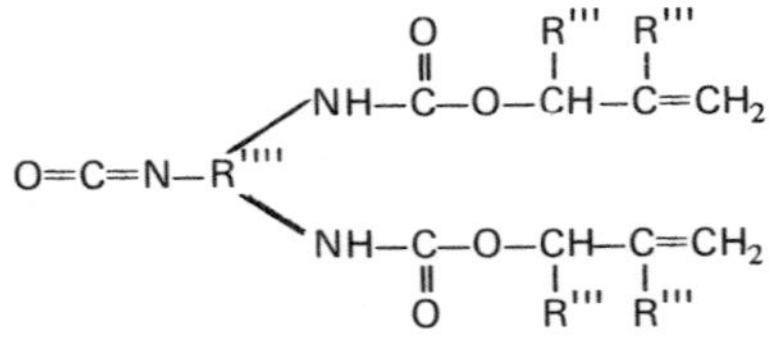

where R is a C_1-C_{10} alkyl, cycloalkyl, aryl or aralkyl radical having a valency of from 1 to 10; each R' is independently selected from hydrogen, methyl and ethyl; R'' is a C_1-C_{10} alkylene, cycloalkylene, arylene or aralkylene group; R'''' is a trivalent aryl, carbamate or carbamide group; each R''' is independently selected from hydrogen and methyl and n is a number from 1 to 10.

The compound of this process is useful in compositions which are polymerizable by ionizing radiation, wherein the composition comprises a mixture of the compound of this process and at least one vinyl or allyl monomer or mixture of such monomers.

The radiation hardenable composition may be used as a coating composition. The use comprises the process of applying a layer of the composition to a substrate and irradiating the layer to thereby harden the composition, wherein the composition is such that a 50 micron thick layer thereof is capable of being hardened when subjected to a radiation dosage of $\leqslant$ 2 Mrad.

Unsaturated Poly(Amide-Esters) and Reactive Monomer

The unsaturated poly(amide-esters) described by *D.A. Tomalia and B.P. Thill; U.S. Patent 3,928,499; December 23, 1975; assigned to The Dow Chemical Company* are linear, toluene-soluble vinyl addition polymers whose backbones comprise a plurality of units corresponding to the formula

$$\begin{array}{l} -(\overset{}{C}R_1CH_2)- \\ \quad\ | \\ O{=}C{-}NH{-}CR_2R_3{-}(CH_2)_n{-}CR_4R_5{-}O{-}C(O){-}CR_6{=}CH_2 \end{array}$$

where R_1 is hydrogen or methyl; R_2 to R_5 are hydrogen or lower alkyl; R_6 is hydrogen or methyl; and n is 0 or 1. The polymers may be cast as coatings and be crosslinked by conventional techniques. Additionally, the polymers can be dissolved in a reactive monomer diluent, such as methyl methacrylate, and this curable mixture used to form coatings or shaped articles.

The polymers are prepared by reacting a vinyl addition prepolymer having pendant oxazoline or 5,6-dihydro-4H-1,3-oxazine groups with acrylic or methacrylic acid.

Example: Several of the solutions of unsaturated poly(amide-esters) in various liquid vinyl monomers were coated to 1 mil thickness on bonderized field test panels and subjected to a 2 million electron volt electron beam at various total dosages. The results of this series of runs are summarized below.

Run No.	Reactive* Monomer Diluent Added	Unsaturated Poly(Amide-Ester) in Solution (percent)	Radiation Dosage (Mrad)	Pencil Hardness	Tape Test (percent)
1	None	50	1	6H	87
2	None	50	2	6H	75
3	None	50	1	>6H	Not measured
4	None	50	2	>6H	100
5	BA	40	2	2-3H	69
6	HPA	40	1	2H	100
7	HPA	40	3	3H	100

*BA is butyl acrylate and HPA is 2-hydroxypropyl acrylate.

In runs 1 to 4 the curable reaction mixture of poly(amide-ester) was dissolved in a 56/44 mixture of ethyl acrylate/methacrylic acid. In runs 5 to 7, the curable reaction mixture of poly(amide-ester) dissolved in methyl methacrylate was further blended with either butyl acrylate or 2-hydroxypropyl acrylate to form other curable compositions.

These new compositions had a concentration of 40 weight percent, total weight basis, of unsaturated poly(amide-ester) in the combination of reactive monomer diluents. The cured coatings were very hard and insoluble in water and methyl ethyl ketone. Additionally, the cured coating adhered strongly to the substrate, as evidenced by the Tape Test.

In this test, the coating was scored in a cross hatch fashion forming squares one-sixteenth inch on a side and Magic Tape was pressed firmly to the scored surface and ripped at 90° angle from the test panel. The percent of the coating remaining on the test panel is given above.

High Boiling Polyvinyl Monomer

In accordance with the process of *O. Isozaki and S. Iwase; U.S. Patent 3,933,939; January 20, 1976; assigned to Kansai Paint Company, Ltd., Japan* a liquid unsaturated resin composition comprises 60 to 95% by weight of liquid resin and 5 to 40% by weight of polyvinyl monomer having a high boiling point. The liquid resin is produced by adding 1.5 to 2 molecules, as an average, of a monomer having both an epoxy group and a polymerizable unsaturated group to each molecule of a linear polyester of about 500 to about 5,000 preferably, 1,000 to 3,000 in number average molecular weight and having terminal carboxy groups.

The linear polyester is produced by condensation of a composition consisting essentially of (a) 10 to 25% by weight of aromatic dibasic acid or its anhydride and (b) 75 to 90% by weight of aliphatic saturated compounds consisting essentially of saturated aliphatic dibasic acid or its anhydride and saturated aliphatic glycol, where 10 to 60% by weight of the aliphatic saturated compounds is alicyclic saturated dibasic acid or its acid anhydride.

Example: A polyester having terminal carboxyl group was prepared by condensing 90 grams of hexahydrophthalic anhydride, 202 grams of adipic acid, 148 grams of phthalic anhydride, 104 grams of neopentyl glycol and 90 grams of 1,3-butylene glycol.

The peak molecular weight in gel permeation chromatography of the polyester was about 600. Then, 220 grams of glycidyl acrylate was added to the polyester and caused to react for 5 hours at 140°C. Further, 85 grams of thus obtained resin was mixed with 15 grams of 1,6-hexanediol diacrylate. Hereinafter this mixture is referred to as Composition 1.

Subsequently 30 grams of titanium white was added to 50 grams of the Composition 1 and mixed well to prepare a coating material. This coating material was applied to the surface of an iron plate forming a film of 15 microns in thickness, and irradiated with 10 Mrad of electron beam to obtain a cured coating which had excellent water and acid resistance and an impact resistance of 40 cm (Du Pont Impact Tester).

Polyester Type Oligo(Meth)Acrylate

T. Ogasawara, Y. Kobayashi, K. Mizutani, T. Nakagawa, N. Hisanaga and H. Tatemichi; U.S. Patent 3,935,173; January 27, 1976; assigned to Toagosei Chemical Industrial Company, Ltd., Japan provide a curable, solvent-free and liquid composition having a low content of volatile components and comprising both a polyester type oligo(meth)acrylate having a plurality of (meth)acryloyl groups and a polyol(meth)acrylate.

The term (meth)acryloyl group means acryloyl and methacryloyl groups; (meth)-acrylate means acrylate and methacrylate; and (meth)acrylic acid means acrylic and methacrylic acids.

The composition comprises (a) an oligo(meth)acrylate having a (meth)acryloyl group equivalent of not more than 1,000 and is prepared by esterifying (1) a polyhydric alcohol, (2) a polycarboxylic compound selected from the group consisting of a polycarboxylic acid and the anhydride thereof, the polyhydric alcohol and polycarboxylic compound not being a dihydric alcohol and dicarboxylic compound respectively at the same time, and (3) (meth)acrylic acid to produce the oligo(meth)acrylate and (b) a polyol poly(meth)acrylate prepared by esterifying a polyhydric alcohol and (meth)acrylic acid to produce the polyol poly(meth)acrylate.

Example 1: Synthesis of Polyacrylate of Polyester Polyol of Tetrahydrophthalic Acid, Diethylene Glycol and Pentaerythritol — A reactor provided with an agitator, thermometer and water separator was charged with 53 grams (0.5 mol) of diethylene glycol, 152 grams (1 mol) of tetrahydrophthalic anhydride and 0.2 gram of trimethylbenzylammonium chloride to form a mixture which was then heated to about 100°C for three hours.

The mixture so heated was incorporated with 136 grams (1 mol) of pentaerythritol, 216 grams (3 mols) of acrylic acid, 1,500 cc of toluene, 14 grams of sulfuric acid and 0.2 gram of phenothiazine, and the whole mass was esterified to produce an oligoacrylate. The oligoacrylate so produced was a light brown colored, viscous liquid which contained 0.9% by weight of toluene and had a viscosity of 90,000 cp (21°C) and an acryloyl group equivalent of 162.

Example 2: 60 parts by weight of the oligoacrylate obtained in Example 1 and 40 parts by weight of trimethylolpropane triacrylate were mixed together and the resulting mixture was then applied to a cold rolled steel plate (degreased with trichloroethylene, JIS-G-3141, stain-finish, Bt 144 treatment) to a thickness of about 25 microns by the use of a bar coater.

The coat thus formed on the plate was then cured by the irradiation of electron beams of Mrad at a current of 10 ma by the use of an electron beam accelerator in the air. For comparison, coating and curing were conducted in the same manner as above with the exception that the trimethylolpropane triacrylate was substituted by 40 parts by weight of styrene (comparative example). The results are indicated in the following table.

	Volatilization Loss*	Pencil Hardness	Appearance of Coat
Example 2	2.1%	4H	Colorless, transparent and smooth
Comparative Example 2	23.6%	3H	Colorless, transparent and smooth

*Decrease in weight of the coating during the time from the end of coating to the end of curing.

Linear Copolymers of Glycidol

V.L. Stevens, A.R. Sexton and F.P. Corson; U.S. Patent Application (Published) B 444,078; March 23, 1976; assigned to The Dow Chemical Co. provide linear copolymers of glycidol, glycidyl esters of unsaturated fatty acids and, optionally, alkylene oxides made by the reaction of an unsaturated fatty acid with a polymer or copolymer of tert-butyl glycidyl ether in the presence of an acid catalyst. The products are polymerizable and copolymerizable with vinyl monomers and are useful as curable resins.

The compounds of the process are polymers which range from oily liquids to solids depending on molecular weight, the nature of the initiator moiety and the identity, proportions and arrangement of the various other moieties present. Those compounds that are initially liquid can be converted to solid form by polymerization or copolymerization through the polymerizable double bond of the α,β-unsaturated acid.

These materials are useful as curable resins that can be formed into coatings or shaped articles which can then be cured by exposure to heat, radiation or source of free radicals, thus being rendered harder and more resistant to heat and solvents. Numerous examples of esters of the process and details of their preparation are included in the patent.

To show the utility of the esters, they alone or in admixture with another polymerizable monomer, were applied to Bonderite 37 treated cold-rolled 20 gauge steel plates as a film approximately 0.001 inch thick. The sheets were then exposed to 1 to 3 Mrad of radiation in the form of a two million volt electron beam. The resulting films were found to be harder and more water-resistant after irradiation than before.

Aromatic Polysulfones with Increased Flow Resistance

P.A. Staniland and G. Jarrett; U.S. Patents 3,975,249; August 17, 1976 and 3,959,101; May 25, 1976; both assigned to Imperial Chemical Industries Ltd., England provide polymeric materials useful for heat-resistant films, foams, coatings, insulation and especially for coating noncooking surfaces of cookware.

Aromatic polymers having increased molecular weight and resistance to flow are produced by the exposure to a total dose of 10 to 150 Mrad of ionizing radiation selected from electron beam and β-ray at temperatures of 100°C to 400°C of at least one aromatic polysulfone containing repeat units $—Ar—SO_2—$ (where Ar is a bivalent aromatic residue which may vary from unit to unit in the polymer chain and at least some of the Ar units have an aromatic ether or thioether group in the polymer chain ortho or para to at least one $—SO_2—$ group).

Low doses of electron beam/β-ray radiation may increase the molecular weight of the polysulfone while leaving it thermoplastic; the higher doses result in crosslinking of the polymer chains so that the polymer becomes insoluble in solvents and is no longer thermoplastic. The effect of irradiation is enhanced if the irradiated aromatic polysulfone is heated after irradiation at temperatures of 200° to 400°C.

Aromatic polymers having increased molecular weight and resistance to flow are also produced by irradiation with γ-rays at temperatures up to 400°C followed by heating at 200° to 400°C.

Example: A sample of polysulfone having repeat units of the following formula prepared in a manner similar to that described in Example 3 of British Patent 1,153,035 and having a reduced viscosity of 0.5 was extruded into film having a thickness of 250 μm using an extruder having screw diameter 40 mm and barrel temperature 350°C.

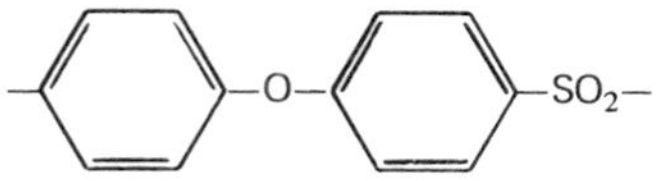

Portions (A) to (F) of the film were irradiated with electron beam produced on a linear accelerator having an arc voltage of 4 mev and providing a dose rate of 3.5 x 10^6 rads/min. Portion (G) was not irradiated and served as a control sample and portion (H) was heated for 10 minutes at 280°C but not irradiated. The results shown in the accompanying table show that irradiation produced films have increased resistance to flow, higher cut-through temperature, superior resistance to solvents and lower elongation at breaks while the yield strength remained essentially unchanged.

Properties (after irradiation)	A	B	C	D	E	F	G	H
	Exposure Time, minutes							
	3	5	10	10	10	15	0	0
	Exposure Temperature, °C							
	295	350	160	275	330	300	–	280
Immersion in xylene	very slight crazing after 2.5 hr		slight crazing after 1 hr		no crazing after 2.5 hr		immediate craze	-
Time to flow at 270°C	>24 hr	>24 hr	40 min	40 min	>24 hr	>24 hr	10 min	-
Yield strength (MN/m²)	85	83.5				88.3	86.4	80.1
Break strength (MN/m²)	71.5	69.5				73.6	68.7	66.7
Elongation at break (%)	120	65				30	160	110
Thickness (μm)	54	56				54	55	62
Cut-through temperature (°C)	280	290				300	250	

Stain-Resistant Diacrylates

R. Demajistre; U.S. Patent 3,979,426; September 7, 1976; assigned to PPG Industries, Inc. prepares diacrylates by reacting a monohydroxylated acrylic monomer with a polyisocyanate. The reaction product may be polymerized by subjection to ionizing irradiation, actinic light or to free radical catalysts to form a useful coating material. The diacrylates may also be copolymerized with other radiation sensitive materials.

The monomers will polymerize to hard, mar-resistant and stain-resistant films at a total dosage of less than 4 Mrad. The preferable dosage used is from about 0.5 Mrad to about 10 Mrad.

Example 1: A reactor was charged with 36 grams of acrylic acid and heated to 100°C. Over a 1 hour period, 122.5 grams of Cardura E ester were added and the reactants were kept at that temperature until the acid value was less than 5. To 317 grams of the above monomer were added dropwise a mixture of 111 grams of 1-isocyanomethyl-5-isocyano-1,3,3-trimethylcyclohexane and one drop of dibutyltin acetate over a one hour period and the reactants were kept at 95° to 100°C for an additional period of about 2 hours. The product had an NCO equivalence of 20,000 and a Gardner-Holdt viscosity of Z7-8.

Example 2: To 90 parts of the composition of Example 1 were added 10 parts of butyl acrylate and the mixture was coated onto an aluminum panel and subjected to electron beam bombardment to a total of 2.6 Mrad in a nitrogen atmosphere. The coating was stained with ink, mustard and merthiolate and was found to be resistant to all stains.

Unsaturated Olefin plus Unsaturated Olefin with Urethane Groups

D. Chaudhari, E. Häring, A. Dobbelstein and W. Hoselmann; U.S. Patent 3,948,739; April 6, 1976; assigned to BASF Farben & Fasern AG, Germany provide binding systems which may be hardened by ionizing radiation and which will provide a higher speed of hardening and therefore will allow industrial application of the coating process.

The coating compositions comprise as binding agents a mixture of (a) at least 1 unsaturated olefin compound containing urethane groups and (b) at least 1 further unsaturated olefin compound that may be copolymerized. The unsaturated olefin compound (a) containing the urethane groups is a reaction product of (1) a compound of the general formula:

$$\left(CHR_1{=}CR_2{-}\overset{\displaystyle O}{\overset{\|}{C}}{-}O{-}CH_2{-}\underset{\displaystyle OH}{\underset{|}{CH}}{-}CH_2{-}O{-}\overset{\displaystyle O}{\overset{\|}{C}}{-}\right)_n R$$

where n is 1 or 2, where R stands for a straight chain or branched alkyl group of valence n, and where R_1 is hydrogen, methyl or the group

$$-\overset{\displaystyle O}{\overset{\|}{C}}{-}O{-}CH_2{-}\underset{\displaystyle OH}{\underset{|}{CH}}{-}CH_2{-}O{-}\overset{\displaystyle O}{\overset{\|}{C}}{-}R_3$$

where R_3 is a monovalent alkyl residue and where R_2 is hydrogen or methyl, and (2) a compound containing at least 1 isocyanate group. The mixture of (a) and (b) may contain conventional additives of the lacquer industry.

Example: 250 parts of a glycidyl ester of a mixture of alpha branched, tertiary carboxylic acids with 9 to 11 carbon atoms (for instance Cardura E) and 72 parts of acrylic acid with 0.1 part hydroquinone are heated to 40°C in a four necked flask. 0.25 part triethylamine are added as catalyst and the temperature is raised to 100°C.

At acid no. 0, the preparation is cooled to room temperature and subsequently 87 parts of toluylene diisocyanate are slowly added drop by drop in a nitrogen atmosphere with constant stirring so that the temperature does not exceed 30°C.

When being reacted with the diisocyanate, a catalyst may be added in the form of triethylenediamine-1,4-diazobicyclo[2.2.2] octane (Dabco). The reaction product so obtained will be divided into two parts and each will be dissolved to 75% in (a) butanediol-1,4-diacrylate and (b) 2-hydroxyethyl acrylate.

These solutions prepared according to (a) and (b) are deposited at a thickness of approximately 60 microns on a pretreated steel plate and are exposed in an inert atmosphere to a 300 kv electron beam system with 50 ma current. The speed of the band on which the test samples are being conveyed through the irradiation room may be raised to 60 meters per minute. Hard, adhesion-free and scratchproof coatings characterized by good adhesion, elasticity and resistance to solvents are obtained.

RESISTS

Polymers Containing Epoxy Groups

T. Hirai, Y. Hatano, S. Nonogaki and T. Kobayashi; U.S. Patent 3,885,060; May 20, 1975; assigned to Hitachi, Ltd., Japan describe organic polymer compositions containing epoxy groups in the main chain or subchain or both thereof, for example, polyglycidyl methacrylate, which have excellent cathode ray-sensitive properties.

The compositions provide insolubilized films suitably used as a resist composition in place of a photoresist composition and as a memory medium for a high density memory.

The organic polymer is selected from the group consisting of a diene polymer selected from the group consisting of epoxidized 1,4-polybutadiene, epoxidized 1,4-polyisoprene, epoxidized 1,4-polychloroprene, and epoxidized 1,2-polybutadiene and an epoxy-containing vinyl polymer selected from the group consisting of polyglycidyl methacrylate, polyglycidyl acrylate, a glycidyl methacrylate-acrylonitrile copolymer and a glycidyl acrylate-acrylonitrile copolymer, the organic polymer having a molecular weight of from 100,000 to 10,000,000 and the diene polymer having a degree of epoxidation of from 8 to 70 percent.

Example 1: An example of processing an SiO_2 layer formed on a semiconductor substrate in accordance with the process is described below. At first 4% methyl ethyl ketone (MEK) solution of epoxidized 1,2-polybutadiene (molecular weight: 300,000; the degree of epoxidation: 45%) was rotary coated on an SiO_2 layer of 5000 A thickness formed on an Si wafer of 0.3 mm thickness at the rate of about 3,000 rpm so as to form a homogeneous coating film on the SiO_2 layer.

A cathode ray beam of 15 kv of acceleration voltage was applied onto the thus formed coating film in an irradiation amount of 2×10^{-8} coulomb/cm^2. Thereafter, the nonirradiated portion was washed with MEK to be dissolved and flowed out, and thus the developing was completed. Thus obtained developed surface was heat-treated at 110°C for 1 hour, and dipped in a solution mixture of 46% aqueous solution of HF and 41% aqueous solution of NH_4F for 10 minutes to be etched and to remove the portion of SiO_2 layer the coating film was not formed on. Thereafter by dissolving and removing the abovementioned residual coating film, a Si wafer having a SiO_2 layer in the desired position was obtained.

Example 2: A 5% MEK solution of polyglycidyl methacrylate, molecular weight 1,700,000, degree of epoxidation 100%, was rotary coated on an SiO_2 wafer in accordance with the same method as in Example 1 to form a coating film of about 0.15 micron. This polymer-coated wafer is placed in a cathode ray processing apparatus and a narrow cathode ray beam of 10^{-9} amp beam current and of 20 kv acceleration volt was scanned on the coating film at 1 micron width and 1 micron distance.

This processed wafer was taken out of the cathode ray processing apparatus. By washing this wafer with MEK, the nonscanned portion of the coating film was dissolved and removed. Thus, memory of 1 micron width and 1 micron distance was obtained.

Olefin-SiO_2 Copolymers

M. Kaplan and E.B. Davidson; U.S. Patent 3,893,127; July 1, 1975; assigned to RCA Corp. have found that certain copolymer films of SO_2 and olefins are excellent electron beam resists, having high sensitivity. Films of these copolymers on suitable supports provide improved electron beam recording media. Some of these polymers do not require solvent development.

Comonomers suitable for preparing the electron beam resists form film-forming, soluble polymers and include straight chain olefins, branched-chain olefins, cycloaliphatic olefins, aryl-substituted olefins, olefins substituted with groups such as hydroxyl groups, and heterocyclic unsaturated compounds such as 2-isopropenylthiophene. Compounds having vinyl unsaturation are preferred.

Example: A series of SO_2-olefin copolymers were prepared by adding about one mol of the olefin to about one mol of cold, liquified SO_2 containing about 0.1 gram of azobisisobutyronitrile and exposing to a 200 watt mercury lamp for about 4 hours at a temperature below -10°C. The polymers were precipitated from methanol.

The purified polymer was dissolved in a solvent to make a 2 to 6% by weight solution and spun onto ½" x ½" glass plates coated with a 200 A thick layer of chromium and a 2000 A thick layer of nickel. The films were exposed to the beam of a scanning electron microscope at an acclerating potential of 5 kv and a beam current of 3 na.

An approximately Gaussian shaped beam having a full width at half amplitude of about 0.5 micron was scanned to describe rasters on the surface of the films at various speeds, thereby varying the total exposure of the films to the beam. The samples were developed in various solutions as set forth below by immersing the exposed fllm for about up to 30 seconds.

Comonomer Dodecene-1 – This copolymer was applied to the substrate from a 3% by weight solution in toluene. Direct print-out rasters were observed at very fast scan speeds, up to 3,400 cm/sec, although no direct print-out trough went through to the substrate. When the trough obtained by scanning at 50 cm/sec was developed in acetone for 30 seconds, the trough width was 0.8 micron and was through to the substrate.

Comonomer 2-Methylpentene-1 — This copolymer was applied to the substrate from a 5% by weight solution in chlorobenzene. Direct print-out rasters were observed at scan speeds up to 125 cm/sec which were through to the substrate and were about 0.5 micron wide.

Olefin-SO_2 copolymers could not be employed when thicker resist films, on the order of 1 micron or more, were required, e.g., when the films were not to be developed through to the substrate and etched, because these films cracked during solvent development. These cracks distorted and even nullified the information in the developing relief pattern and limited the use of olefin-SO_2 copolymer resists to applications requiring only thin films.

In related work *E.S. Poliniak, H.G. Scheible and R.J. Himics; U.S. Patent 3,935,331; January 27, 1976; assigned to RCA Corp.* found that cracking during development of exposed olefin-SO_2 copolymer films can be prevented by filtering the copolymer-solvent solution prior to applying to a substrate; drying to remove the solvent under a high vacuum, i.e., over 10^{-6} torr, after applying to the substrate; and storing the film in a dry atmosphere until it is to be exposed to an electron beam.

In related work *E.S. Poliniak and R.J. Himics; U.S. Patent 3,935,332; January 27, 1976; assigned to RCA Corp.* found that three developer solvents, 2-methylcyclohexanone, 3-methylcyclohexanone and a mixture of acetonyl acetate and acetone, improve the resolution of electron beam exposed films of poly(1-methyl-1-cyclopentene-SO_2) copolymers.

1,4-Diphenyl-1,3-Butadiene to Increase Sensitivity

The use of a scanning electron beam to generate a pattern in a negative photoresist is known. Electron beam equipment can be made which is capable of scanning very quickly, but a standard negative photoresist such as partially cyclized cis-polyisoprene requires such a large flux of electrons for proper exposure that the scanning equipment must be operated at speeds substantially slower than the capability of the equipment.

According to *B. Broyde; U.S. Patent 3,894,163; July 8, 1975; assigned to Western Electric Company, Inc.* by adding 1,4-diphenyl-1,3-butadiene which dissociates readily into free radicals, to the photoresist, the sensitivity or speed of the photoresist is effectively increased. As a result, the electron beam can scan at a higher rate.

Example 1: To establish a basis for comparison, tests were first made with partially cyclized cis-polyisoprene photoresist without any additives. A partially cyclized cis-polyisoprene photoresist-solvent solution was mixed with a thinner (mixed xylenes) in a 1 to 3 ratio. The resist-solvent solution was commercially obtained and comprised partially cyclized cis-polyisoprene (averaging one double bond per 10 carbon atoms; number average molecular weight of 65,000 ± 5,000; weight average molecular weight of about 120,000) dissolved in 12% ethylbenzene, 82% mixed xylenes and 6% methyl Cellosolve.

The mixture was applied to a chromium-coated glass plate (or, alternatively, to a silicon slice onto which a 18000 A SiO_2 layer had been grown), and then spun to a thickness of 8000 A. After baking at 150°C for 10 minutes, the film was 6000 A thick. The coated plates were then put into a vacuum chamber and radiated with 15 kv electrons.

The plate was developed with a partially cyclized cis-polyisoprene developer, commercially obtained, and a partially cyclized cis-polyisoprene rinse, commercially obtained and baked at 150°C for 10 minutes. The following results were found:

(a) Flux needed to record image (gel dose) = 0.9×10^{-6} coul/cm^2
(b) Flux needed to form 3000 A film = 4×10^{-6} coul/cm^2
(c) Flux needed to form maximum (4000 A) thickness = 7.5×10^{-6} coul/cm^2

Under identical conditions, but with 5 kv electrons, the dose densities required to expose partially cyclized cis-polyisoprene films were:

(a) 0.5×10^{-6} coul/cm^2
(b) 0.75×10^{-6} coul/cm^2
(c) 2×10^{-6} coul/cm^2

Example 2: The procedure of Example 1 was repeated except that a 0.1 weight percent solution of 1,4-diphenyl-1,3 butadiene in the thinned partially cyclized cis-polyisoprene photoresist-solvent solution was prepared and employed. Dose densities for the three tests were as follows:

(a) 0.5×10^{-6} coul/cm^2
(b) 1.5×10^{-6} coul/cm^2
(c) 1.5×10^{-6} coul/cm^2

Crack-Resistant Terpolymers

In the process of *E. Gipstein and W.A. Hewett; U.S. Patent 3,898,350; August 5, 1975; assigned to International Business Machines Corp.* electron beam positive resists are formed from terpolymers of (a) an alpha olefin, (b) sulfur dioxide, and (c) a compound selected from the group consisting of cyclopentene, bicycloheptene and methyl methacrylate. The terpolymers have the particular unexpected advantage of being resistant to cracking of the films.

Example: Synthesis of Polysulfone Terpolymers — When two olefins can each copolymerize with SO_2 in a 1:1 ratio, the three component system also behaves as a 1:1 ratio (total vinyl monomers:SO_2). Poly(cyclopentene sulfone-co-hexene-1-sulfone) was prepared as follows.

A mixture of 13.6 grams (0.2 mol) cyclopentene, 33.6 grams (0.3 mol) hexene-1 and 0.36 gram (4×10^{-3} mol) t-butyl hydroperoxide initiator dissolved in 250 milliliters dry toluene was polymerized at -20°C with 48 grams of SO_2 (0.75 mol) added dropwise to the stirred solution.

After 1 hour, the viscous solution was poured into 2 liters of cold MeOH to precipitate a white polymer. The polymer was purified by dissolution in $CHCl_3$ and reprecipitation in MeOH. After drying 48 hours at 45°C under vacuum 56.7 grams (100%) of product was obtained. Other terpolymers were prepared as described and films spun onto SiO_2 wafers.

Cyclopentene/butene-1-SO_2 films 4000 to 9100 A thick did not crack or craze and could be successfully processed to give excellent images after exposure. For example, 6000 to 9000 A thick films spun from 7 to 10% solutions of the polymer in CH_3NO_2 on SiO_2 wafers precoated with bis-trimethylsilyl acetamide, an adhesion promoter, were prebaked for 1 hour at 100°C under vacuum. A pattern was written with an E-beam at 110 nsec exposure (4×10^{-6} coul/cm^2) and images developed with a solvent mixture of cycloheptanone and cyclohexanone (80/20).

The developed wafers were postbaked at 165° to 200°C for 30 minutes to 1 hour and then etched with HF for 5 minutes. Excellent images of high definition and fidelity with fine line geometry remained.

Methyl Methacrylate-Acrylonitrile Copolymers with Improved Sensitivity

A positive type cathode ray-sensitive coating film is provided by *H. Morishita and S. Nonogaki; U.S. Patent 3,914,462; October 21, 1975; assigned to Hitachi, Ltd., Japan.* It consists essentially of a copolymer of methyl methacrylate with a comonomer selected from the group consisting of acrylonitrile, methacrylonitrile and maleic anhydride.

A coating film consisting of the copolymer described above can obtain therein fine cathode ray images of the positive type in about 1/10 the amount of irradiation of that required for conventional positive type cathode ray-sensitive coating films. It is preferable that the mol ratio of the comonomer be from about 0.5 to about 10 mol percent.

Example: Methyl methacrylate-acrylonitrile copolymer is obtained by the steps of maintaining water (passed through ion-exchange resin) in a reaction vessel; dissolving an emulsifying agent comprising sodium lauryl sulfate in the water; controlling the atmosphere in the reaction vessel to provide an oxygen-free atmosphere, such as by using a nitrogen atmosphere, introducing a mixture of acrylonitrile and methyl methacrylate whose mol ratio is controlled into the water, adding small amounts of potassium persulfate and sodium bisulfate into the water, and maintaining the reaction vessel at about 45°C for about 4 hours with stirring of the contents.

Methyl methacrylate-acrylonitrile copolymers having a mol ratio of acrylonitrile of 0.5 mol percent, 5 mol percent, 10 mol percent, 50 mol percent and 70 mol percent, respectively, are obtained by controlling the ratio of quantities of methyl methacrylate and acrylonitrile. In these runs the amount of water used is eight times the total amount of monomers charged to the vessel wherein the amount of sodium lauryl sulfate, potassium persulfate and sodium bisulfate, respectively is 0.8 times the total monomer charge.

A coating film of the methyl methacrylate-acrylonitrile copolymer having a thickness of about 0.5 micron is formed on a chrome film coated on a glass substrate by the steps of dissolving the copolymer in methyl Cellosolve acetate, applying the dissolved copolymer on the chrome film, and drying the applied copolymer. The table shows irradiation amounts of cathode ray and developers for obtaining positive type fine cathode ray images on respective coating films of the copolymers.

In order to develop the coating film it is immersed in developer solvent for 2 minutes and the developer solvent used is then blown against immersed coating film by a spray gun for 30 seconds.

Mol Ratio of Acrylonitrile (mol percent)	Amount of Irradiation by Cathode Ray (coulomb/cm^2)	Developer Solvent (parts by volume)
0.5	1×10^{-5}	Methylisobutylketone + isopropyl alcohol (2:3)
5	5×10^{-6}	Methylisobutylketone
10	5×10^{-6}	Methylisobutylketone + isopropyl alcohol (2:3)
40	1×10^{-5}	Methylisobutylketone + isopropyl alcohol (1:1)
50	-	-
70	-	-
0	5×10^{-5}	Methylisobutylketone + isopropyl alcohol (2:3)

As is apparent from the table, positive type fine cathode ray images can be obtained when the mol ratio of acrylonitrile is 0.5% by about one-half the amount of irradiation as compared with that of conventional polymethyl methacrylate and when the mol ratio of acrylonitrile is 5 to 10%, by about one-tenth the irradiation as compared with that of conventional polymethyl methacrylate. All images obtained in accordance with this process are similar in clarity, and irradiation values given are the minimum values required to provide soluble film.

However, when the mol ratio of acrylonitrile of the copolymer of the process is over 50%, positive type fine cathode ray images cannot be obtained.

Scanning Speed Increase Using Epoxy-Polymer Mixture

T.L. Brewer; U.S. Patent 3,916,035; October 28, 1975; assigned to Texas Instruments Inc. has found that by adding an epoxy to a polymer that acts as a negative electron beam resist, the electron beam scanning speed of the polymer is increased without affecting any of the other characteristics of the polymer as a negative electron beam resist.

When a mixture of 10% ERRA-4090 (a cyclohexylepoxy) and 90% polystyrene is processed to form a negative electron beam resist, the scanning speed of the epoxy and polystyrene resist, as compared to the pure polystyrene resist is increased by a factor of 3. When a negative electron beam resist, comprising a mixture of 10% ERRA and 90% styrene-butadiene is prepared, the speed of the epoxy and styrene-butadiene resist, as compared to the pure styrene-butadiene resist, is increased by a factor of 50%.

The scanning speed of a negative electron beam resist made from a mixture of

10% ERRA and 90% polydimethylsiloxane, as compared to the pure siloxane resist, was increased by a factor of 3.

An interesting effect of the addition of epoxy to a polymer is that the epoxy seems to increase the speed of the slower scanning speed polymers more than the faster speed polymers. In other words, the faster the polymer crosslinks by itself, the less effect the addition of the epoxy has on increasing the crosslinkage speed.

Another interesting aspect of the epoxy addition is that the molecular weight of the polymer has no relationship to the increase in speed due to the epoxy additive. For example, if an epoxy added to a polymer with a molecular weight of 30,000 increases the crosslinking speed of the polymer by a factor of 3, the epoxy added to the same polymer having a molecular weight of 90,000 increases the crosslinkage speed of the higher weight polymer also by only a factor of 3.

The epoxy compound, ERRA-4090 is added as a solid to an aromatic solvent, such as xylene or toluene to form a 0.5% concentration. Solid polystyrene is added also to the same solvent to form a 5% solution. The two solutions are mixed to form a solution of polystyrene and epoxy. The ratio of epoxy to polystyrene, by weight, will range from a low of about 5% to a high of about 30% with the optimum amount being about 10%. The epoxy and polystyrene solution can vary from approximately 2 to 10% by weight.

A small amount of the epoxy and polystyrene solution is applied to the chrome support and the chrome support with the covering epoxy and polystyrene solution is spun at a speed of approximately 3,000 rpm, in order to form a uniform layer of epoxy and polystyrene on the support as a thin film.

The thin film is baked to remove the solvent, if at all, at a temperature below 40°C; at higher baking temperatures, the epoxy will tend to decompose. The chrome substrate with the thin film of epoxy and polystyrene is then placed in an electron beam irradiator and electron beam is allowed to scan the surface of the thin film in a predetermined pattern.

The epoxy-polystyrene resist is developed by spraying or dipping the thin film covered chrome support in an aromatic solvent for approximately 30 seconds. To harden the crosslinked pattern remaining on the chrome support, the resist covered support is baked at a temperature of between 80° and 180°C in any atmosphere, preferably air for convenience, for approximately 30 minutes.

For use as a mask, the chrome support with its patterned resist is subjected to a chrome etch for a period of time sufficient to remove the chrome exposed by the openings in the resist. Finally, the resist is removed by dipping the resist covered chrome support in diethyl phthalate at 170°C for 30 minutes, or by spraying with a hot dioxane-pyrrolidone solution. The patterned chrome support is ready to be used to form an image on a photoresist formed on a semiconductor wafer. The specific temperatures and times given are not critical.

Polydialdehydes

In the process of *J. Fech, Jr. and E.S. Poliniak; U.S. Patent 3,940,507; February 24, 1976; assigned to RCA Corp.* polymers found useful as electron beam resists are homopolymers of dialdehydes having recurring units of the formula

$$\left[-\overset{\displaystyle X}{HC\quad CH}-O- \right] \quad \text{(HC and CH bridged by X above and by O below)}$$

where X can be a straight or branched chain alkylene group having 1 to 10 carbon atoms or a saturated or unsaturated cycloalkylene or arylene group having up to 10 carbon atoms. The polymers can be acetylated for increased stability.

Example: Preparation of Polymer — 13.4 grams (0.1 mol) of phthalaldehyde was polymerized in 134 ml of methylene chloride at -78°C using 0.076 ml (0.06 mol) of boron trifluoroetherate as catalyst. After reacting for 1 hour, the solution was quenched with 0.5 ml of pyridine. The polymer was precipitated with 800 ml of diethyl ether, filtered, washed with 500 ml of ether and vacuum dried at room temperature.

The polymer was obtained in 92.5% yield and had a melting point of 123° to 123.5°C. The structure was confirmed by nuclear magnetic resonance spectroscopy. The polymer product was reprecipitated from methylene chloride with methanol, filtered and dried. A yield of 82% was obtained of the polymer now having a melting point of 142° to 143°C.

Termination of Polymer — 3 grams of polymer product were reacted with 60 ml of acetic anhydride at 35°C for 3 days. The solution was poured into ice water, the precipitated polymer filtered, washed with water, then methanol, and dried under vacuum.

The polymer product was purified by dissolving in 50 ml of tetrahydrofuran, reprecipitated with 800 ml of methanol, filtered and dried. A yield of about 70% of a polymer having the structure

$$CH_3-\overset{O}{\overset{\|}{C}}-O\left[-HC\quad CH-O- \right]\overset{O}{\overset{\|}{C}}-CH_3 \quad \text{(HC and CH bridged by O and fused to a benzene ring)}$$

was obtained having a melting point of 162.5° to 164°C. The number average molecular weight of the polymer was 48,000 and the weight average molecular weight was 71,600.

Electron Beam Recording — A solution containing 1 part of the polymer in 9 parts of Cellosolve acetate was made and spun coated onto nickel coated glass plates fitted with silver paste electrodes. The film thickness was about 300 nanometers. The films were given line exposures to the beam of a scanning electron microscope at an accelerating potential of 5 kv and a beam current of 3 na. The gaussian-shaped beam, having a width at one-half amplitude of 0.42 micron, was scanned at several speeds to describe rasters on the surface of the films, thereby varying the total exposure of the films to the beam.

The exposed films were developed by immersing in a 65:35 mixture of cyclohexyl acetate and butyl Cellosolve for two minutes and 15 seconds. After drying, the films were measured for erosion of the unexposed area; about 60 nanometers of erosion had occurred.

A layer of gold about 200 A thick was evaporated onto the developed films which were then examined using a scanning electron microscope. The width and depth of the raster lines were measured. The data are summarized below.

Sample	Electron Beam Velocity, cm/sec	Exposure μcoulombs/cm^2	Line Widths, microns
1	2.5	26	1
2	10	6.7	0.925
3	25	2.6	0.6

Photomicrographs of the above samples show that the trenches have an excellent geometry, that is, a clearly defined, straight-walled trench.

Adding Acetate to Poly(Methyl Methacrylate)

E. Gipstein, W.M. Moreau and O.U. Need III; U.S. Patent 3,931,435; January 6, 1976; assigned to International Business Machines Corp. found that the speed of poly(methyl methacrylate) electron beam positive resists can be enhanced by the addition of acetate polymers. The acetates useful in the process are vinyl acetate and isopropenyl acetate in amounts of 1 to 20% by weight.

Example: In separate experiments, poly(vinyl acetate) and poly(isopropenyl acetate) were each added to poly(methyl methacrylate) in a solvent at 5% by weight and spin coated with a thickness of 7000 A. A control of poly(methyl methacrylate) without additive was coated to the same thickness.

The films were exposed to 25 kv electron beam dosage. The films were then developed using amyl acetate. The thickness of unexposed and developed areas were measured for a determination of the ratio S/S_0. (S/S_0 is called the solubility ratio. It is defined as the solubility of the resist after irradiation divided by the solubility before irradiation.) The higher solubility ratio obtained for the poly(methyl methacrylate) with additives proves a speed enhancement of from 8 to 10 times. The results are reported in the table below.

Additive	S/S_0	Dose (25 kv)	S/S_0	Dose (25 kv)
None	4	12 μcoul/cm^2	1.5	3 μcoul/cm^2
5% PVA	32	12 μcoul/cm^2	8.0	3 μcoul/cm^2
5% PIPA	40	12 μcoul/cm^2	9.0	3 μcoul/cm^2

Plural Layers for Metal Lift-Off

W.M. Moreau and C.H. Ting; U.S. Patent 3,934,057; January 20, 1976; assigned to International Business Machines Corp. provides a high sensitivity resist layer structure for high energy radiation exposure formed by coating plural layers of radiation degradable polymers on a substrate which layers are successively slower dissolving in the resist developer. Upon exposure and solvent development, a resist edge profile is obtained which is particularly useful for metal lift-off.

The resists useful in the process are those which are degraded under high energy radiation at dosage levels above about 1 x 10^{-6} coulomb per square centimeter and which are conventionally employed in positive high energy exposure resist systems.

The plural layers of resist can be made up of the same polymer or different polymers so long as the underlayers are faster dissolving in developer solution than the overlying layer. Two or more layers can be employed.

The solubility rate differences of the layers can be established by using one or a combination of known chemical and structural differences in the resist layer compositions. A convenient way of achieving the required solubility rate differences is by varying the molecular weight of the polymer in each layer since lower average molecular weight materials of a family of polymers are faster dissolving.

Another way of establishing the required solubility rate differences is by isomerism such as the tacticity of the polymers, with polymers of the same molecular weight but different spatial arrangement of monomer units having different solubility rates. Copolymers with varying ratios of monomers can also be employed as well as layers of polymers having functional groups of varying polarity.

Example: A dual layer of polymethyl methacrylate was employed as described below. As illustrated in Figures 5.1a through 5.1e, onto wafer **21** was spin coated at 14000 A thick layer **23** of polymethyl methacrylate (M_w 82,400, M_n 41,560) from an 18% chlorobenzene solution.

FIGURE 5.1: DUAL LAYER METAL LIFT-OFF

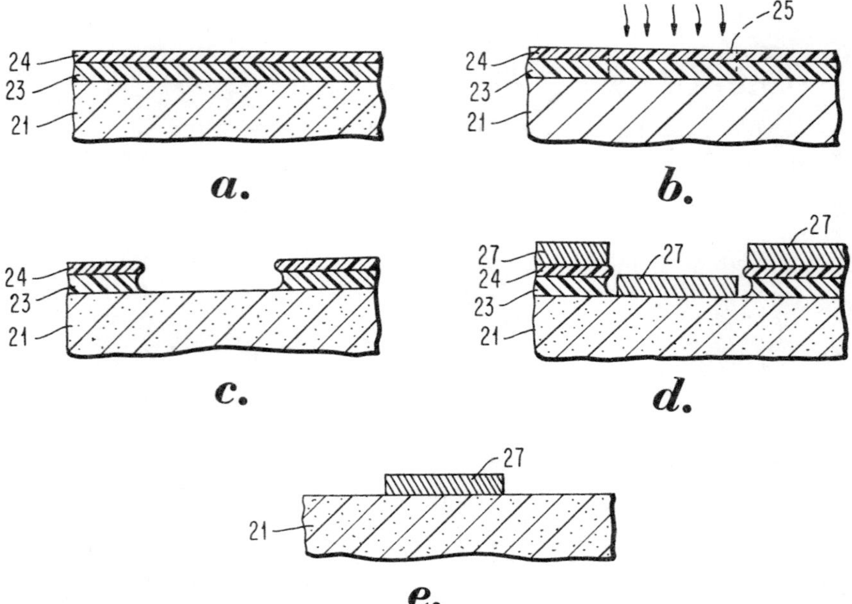

Source: U.S. Patent 3,934,057

Layer **23** was prebaked at about 160°C for about 1 hour. On top of layer **23** was formed a second layer **24** comprising a 7000 A thick film of polymethyl methacrylate of a higher molecular weight (M_w 750,450; M_n 246,190) from an 8% by weight solution of polymer in chlorobenzene.

The dual layer film was prebaked for 1 hour at 160°C and then exposed to a scanning 1 micron beam of 25 kv electrons at a dosage of 7.5×10^{-6} coulomb per square centimeter. The exposure produced an exposed portion **25** in the layers as illustrated in Figure 5.1b. The solubility rate, (S_0) of the unexposed top layer polymer was determined to be 75 A per minute.

The dual resist films were developed for about 50 minutes in methyl isobutyl ketone developer at a temperature of about 21°C. The unexposed portion of layer **24** lost about 3750 A during development. The net remaining dual film thickness then was determined to be 17250 A. The undercut developed resist profile illustrated in Figure 5.1c was obtained as shown by a cross sectional, scanning electron microscope photograph of the developed resist layers at a magnification of 7,000 times.

The developed resist film was aluminized by vapor deposition with a 10000 A thick layer **27** of aluminum. The resist layers and overlying aluminum was lifted off as illustrated in Figure 5.1d. The resist thickness to metal thickness ratio was 1.7 to 1 and the image resolution of aluminum stripes was found to be a width to height ratio of about 1.3 to 1.0 (i.e., 13000 A width image 10000 A in height).

The minimum spacing between aluminum stripes obtainable is about one-half the net resist thickness or about 8500 A. It was found that the minimum dosage needed to obtain lift-off with a two layer process and a 10000 A thick aluminum layer was only one-sixth the dosage needed for a single layer film.

Thermally Stable Polycarbonates

E. Gipstein, W.M. Moreau and O.U. Need III; U.S. Patent 3,961,099; June 1, 1976; assigned to International Business Machines Corp. describe positive electron beam resists which are made of polycarbonates and which possess the advantage of being thermally stable.

Polycarbonates that are particularly useful in the process are materials having the formula

$$(-O-R-O-\underset{\underset{O}{\|}}{C}-)_n$$

where R may be either alkyl or aryl. By the process it is possible to form resists which are sensitive to electron beam radiation and are thermally stable up to about 400°C.

Example: The most preferred polycarbonate, poly(2,2-propanebis-4-phenyl carbonate), (1) was tested for its sensitivity. It required only a minimum dose of 3×10^{-6} coul/cm^2 at 15 kv. The high radiation sensitivity of this polymer is surprising, since aromatic groups in a polymer main chain or as side chain are inhibitors of polymer degradation. The susceptibility to radiation may be due to localized scission of the carbonate linkage with elimination of CO_2. The cast film of (1) was exposed with a 1 μ spot at a decreasing series of dose at 15 kv. The film was developed in 1,6-dichlorohexane. One-half of the unexposed film (3500 A) remained after development. The image was stable to 260°C heat for 3 hours.

Polymer Containing Dimethylglutarimide Units

According to *J. Bargon, E. Gipstein and H. Hiraoka; U.S. Patent 3,964,908; June 22, 1976; assigned to International Business Machines Corp.* images are formed using as a positive resist a thin film of a polymeric composition comprising dimethylglutarimide units, having the structure

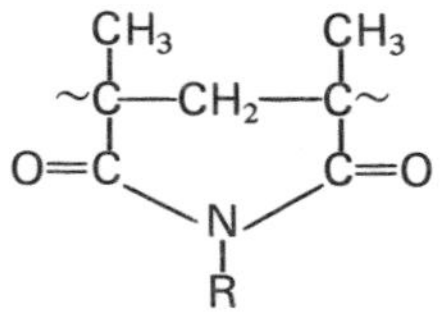

$$\begin{array}{ccccc} & CH_3 & & CH_3 & \\ & | & & | & \\ \sim & C & -CH_2- & C & \sim \\ & | & & | & \\ O= & C & & C & =O \\ & \diagdown & N & \diagup & \\ & & | & & \\ & & R & & \end{array}$$

where R is H, CH_3 or C_2H_5. The dimethylglutarimide unit containing polymers make possible resists simultaneously having very fine spatial resolution capabilities, high glass transition temperatures and high temperature stability, up to 300°C.

Still an additional advantage is that the homopolymers and copolymers of the process generate far fewer bubbles than found in the nonmodified polymers. They are very suitable for use as lithographic resists in the positive mode and are sensitive to both electron beam and light radiation, for example, light of about 2537 A in wavelength. The resists are particularly useful in micro circuitry processings.

In one preferred embodiment the resist film is formed in situ by heating in the presence of gaseous ammonia, methylamine or ethylamine, a resist film containing dimethacrylic acid units.

When the dimethacrylic acid containing polymer is treated with ammonia or an amine in the vapor state, the corresponding imide is formed in situ. Such an in situ method has the advantage of not requiring film casting, an advantage particularly useful in regard to the dimethylglutarimide polymers of the process since they have very little solubility in common nonreacting solvents.

Another advantage of the in situ preparation is that it is possible to combine the heating with ammonia vapor with the conventional prebaking process to which many polymer resist films are subjected before treatment with developer solvents in order to eliminate cracking.

Example 1: Poly(Dimethacrylamide) – Thin films (a few microns thick) of poly(methacrylic acid) were cast from an 8% (w/v) 2-methoxy ethanol solution on a silicon wafer, and on a sodium chloride plate. They were placed in a vessel which was then pumped down to 10^{-3} torr to remove traces of the solvent. After introducing ammonia to 15 cm Hg, the entire vessel was placed in an oil bath held at 200°C for 1 hour and then again pumped down to 10^{-3} torr at room temperature. The polymer film thus treated on a silicon wafer was used for electron beam exposure, while the polymer film on the sodium chloride plate was used for IR measurement. The polymer film thus obtained has exactly the same IR spectrum as the one reported in the literature.

The reported method for the preparation of poly(dimethacrylimide), however, requires multiple processes including a high pressure operation, while the present method is a single one step process.

Example 2: Poly(N-Methyl Dimethacrylimide) – The same procedure described above for poly(dimethacrylimide) in Example 1 was used except monomethylamine or dimethylamine replaced ammonia. The polymer film thus obtained has exactly the same IR spectrum as that reported in the literature for poly(N-methyl dimethacrylimide).

Example 3: The same procedure described above for Example 1 was used except monoethylamine replaced ammonia. In this way poly(N-ethyl dimethacrylimide) was obtained.

Example 4: Electron Beam Exposure – About 1 micron thick films of poly(dimethacrylimide), poly(N-methyl dimethacrylimide), and poly(N-ethyl dimethacrylimide) were exposed to scanning electron beam at 23 kev with a dosage of 10^{-5} coulomb per square centimeter. All these thin polymer films showed positive electron beam characteristics, that is, degradation of the film where the electron beam scanned, after development in 2-ethoxy ethanol (ethyl Cellosolve) or in an aqueous ethanol.

For comparison purposes, films of the same thickness of the copolymer of methacrylic acid and methyl methacrylate and their ammonia modified copolymers were exposed to 23 kev scanning electron beam at a dosage of 10^{-5} coulomb per square centimeter. These films were developed in 2-ethoxy ethyl acetate. The ammonia and amine modified films gave clearer and sharper images than the nonmodified copolymer.

Nitrocellulose

Very sensitive electron beam positive resists have been obtained using films of nitrocellulose containing 10.5 to 12% nitrogen.

E. Gipstein, W.M. Moreau and O.U. Need III; U.S. Patent 3,985,915; October 12, 1976; assigned to International Business Machines Corp. found that electron beam positive resists are formed using cellulose derivatives which have been rendered soluble by including in them a substituent group. Suitable solubilizing substituent groups include, for example, the nitro group, ester groups such as acetate or butyrate and ether groups such as methoxy groups.

Solubilized cellulose derivatives developed in appropriate solvents are high speed electron beam positive resists.

Examples 1 through 5: Films of these materials were coated from organic solvents, dried at 100°C and exposed to 25 kv electron beams with decreasing doses to 1 x 10^{-6} coul/cm². A minimum solubility ratio of final thickness to initial thickness after development of 2.0 was used as a minimum criterion of resist sensitivity. The films exhibited good adhesion to silicon dioxide during etching of 0.5 micron lines. In comparison, U.S. Patent 3,535,137 requires a minimum dose of 100 x 10^{-6} coul/cm² using polymethyl methacrylate as the resist material. The results obtained according to the process are shown below.

Example	Cellulose Derivative	Developer	Minimum Dose, μcoul/cm²
1	Nitrocellulose, N 10.5 to 12%	Pinacolone (di-tert-butyl ketone)	1.5
2	Cellulose triacetate	1,5-Dichloropentane	2.5
3	Cellulose acetate butyrate	1,5-Dichloropentane	2.5
4	Cellulose acetate propionate	1,5-Dichloropentane	2.5
5	Methylcellulose	Water (pH 1-6)	2.0

ADDITIVES

Amines as Dosage Reducing Agents

According to *I. Mani; U.S. Patents 3,882,003; May 6, 1975 and 3,901,779; August 26, 1975; both assigned to The Dow Chemical Company* the curing dosage of ionizing radiation required to cure in an inert atmosphere a mixture of certain vinyl monomers and a polymerizable vinyl ester resin is reduced by adding to the mixture at least 0.3 weight percent of certain nitrogen-containing materials. The nitrogen materials include 2-oxazolines, guanidines and certain amines.

Amines which may be employed have the formula $R_1R_2R_3N$ wherein R_1 may be an alkyl or an aralkyl group, R_2 and R_3 each may be hydrogen or an alkyl group or R_1 and R_2 together may be a cyclic alkylene radical or an oxydialkylene radical.

Example: A vinyl ester resin was prepared by reacting methacrylic acid (31%) with a glycidyl polyether of bisphenol A having an epoxide equivalent weight of 186 to 192. The vinyl ester resin was then mixed with n-butyl acrylate in the proportions of 2:1 and to this resin was then added 3% of various nitrogen-containing compounds, based on weight of resin and monomer.

A film was cast with a 7 mil draw-down bar on a Q-panel (4 x 12 x 0.03 inch) and covered with a 2 mil sheet of Mylar (polyester) film to exclude air. The coated polished steel Q panel was then passed through a 2 mev electron beam from a Van de Graaff accelerator filtered with 0.33 gram per square centimeter Al. A 50 microampere beam current and a conveyor speed of 3.4 centimeter per second delivered a dose of 0.1 Mrad for each pass through the beam.

Without any accelerator the resin required 0.5 Mrad to cure. While this dose level is already low, the addition of di-n-butylamine reduced the dosage required to 0.1 to 0.2 Mrad. Similarly, when tetramethylguanidine and 2-oxazoline were employed as the accelerators the dosage was reduced to 0.2 and 0.4 Mrad, respectively. Tests made with styrene in place of the butyl acrylate failed to evidence the reduction in curing dosage.

Example 2: When the resin of the previous example was mixed with 2-hydroxyethyl acrylate (50:50), the mixture required 0.3 Mrad to cure. Even at this low dosage, the addition of 3% by weight of dibutylamine reduced the dosage to cure to 0.15 Mrad.

In related work *I. Mani; U.S. Patent 3,882,004; May 6, 1975; assigned to The Dow Chemical Company* uses the same nitrogen-containing materials as above to reduce the dosage level necessary to cure a thermosettable mixture of an alkenyl aromatic monomer such as styrene and a polymerizable vinyl ester resin. In this case, however, a considerably larger dosage is necessary.

Example: A vinyl ester resin was prepared by reacting 2-hydroxypropyl acrylate (37.5%) with maleic anhydride (34.3%) to form a half ester which was subsequently reacted with 1,4-butanediol diglycidyl ether (28.2%) according to the procedure of U.S. Patent 3,367,992. The resin was then mixed with styrene (2 parts to 1 part). To the mixture was then added 3% of a nitrogen-containing compound. A film was cast and exposed to ionizing radiation as described above.

The curing dose in Mrad for each of three different nitrogen compounds is shown below. At least about 0.3% of the nitrogen compound is necessary and a minimum curing dose is found at about 1.5 to 5 weight percent. No advantage is found with amounts above 10%.

Nitrogen Compound	Curing Dose, Mrad
None	2.8-3.6
Di-n-butylamine	1.4
Tetramethyl guanidine	1.3-1.4
2-Oxazoline	1.5-1.6

Thermoplastic Vinyl Polymer to Improve Adhesion

J.R. Deamud and C.L. Hickson; U.S. Patent 3,895,171; July 15, 1975; assigned to Ford Motor Company provide an improvement in radiation polymerizable paints which comprise a film-forming binder solution of an alpha,beta olefinically unsaturated organic resin and a compound which is polymerizable with the resin.

The improvement comprises including in the paint from about 2 to 7 weight percent based on the total formula weight of the paint of a saturated, thermoplastic, vinyl polymer selected from the group consisting of homopolymers and copolymers of vinyl hydrocarbon monomers and acrylic monomers and having a molecular weight within the range of from about 10,000 to about 50,000.

Example 1: An alpha,beta olefinically unsaturated resin is prepared as follows.

Material	Parts by Weight
(a) Methyl methacrylate	328
(b) Ethyl acrylate	549

(continued)

Material	Parts by Weight
(c) Glycidyl methacrylate	123
(d) Azobisisobutyronitrile	10
(e) Hydroquinone	0.4
(f) Tetraethylammonium bromide	3.6
(g) Methacrylic acid	74.5
(h) Xylene (solvent)	1,000

To a reaction vessel equipped with a stirrer, condenser, thermometer and nitrogen inlet and exit tubes is added solvent (h) which is then heated to reflux temperature (138°C). To this refluxing xylene is added a mixture of components (a), (b), (c) and (d) over a period of 2 hours. Heating is continued at 135°C for 4 hours and the reaction mixture is cooled to 90°C.

Components (e), (g) and (f) are then added in that order. The temperature is then raised to 135°C and reaction continued until an acid number of less than one is obtained. Finally the xylene is removed by vacuum distillation.

The following materials are mixed together in the weight percentages indicated to form a radiation polymerizable coating containing thermoplastic polymer:

Material	Weight Percent
Radiation curable resin from above	13.2
Hydroxypropyl acrylate	10.8
Hydroxyethyl acrylate	13.0
Isobutyl acrylate	14.8
2-Ethylhexyl acrylate	7.6
TiO_2	28.3
SiO_2 (silica flattener)	7.3
Acryloid B-82*	5.0

*This is a commercially available methyl methacrylate copolymer; a 40% solution of which in toluene exhibits a Brookfield viscosity at 25°C of 400 to 700 centipoises; polymer exhibits a T_g of 35°C and gives an ultimate Tukon hardness of 10 to 11.

Example 2: A radiation curable coating formulation is prepared according to Example 1 except that instead of the Acryloid B-82, the formulation includes 5.0 weight percent of a thermoplastic polymer prepared as follows.

To a 1 liter, three-necked flask fitted with a sealed stirrer, thermometer and condenser is added 300 ml of water containing 1% by weight sodium polymethacrylate, 0.10 gram of monosodium phosphate in 10 grams of water as a buffer solution, and a mixture of 1.0 gram of benzoyl peroxide with 50 grams of methyl methacrylate and 50 grams of ethyl acrylate monomers.

This mixture is stirred rapidly and heated in a water bath at 80°C for 1 hour to accomplish polymerization of the monomers. The solid polymer granules can be collected on a filter, thoroughly washed with water and dried. The polymer is analyzed to have a molecular weight of 28,000.

The coating formulation containing the above thermoplastic is mixed, sprayed on substrates of metal, wood and plastic and cured by electron beam irradiation.

Vinyl Phosphate as Anticorrosive

According to *M. Gotoda, K. Yokoyama, Y. Kono, S. Toyonishi, K. Hiwano and T. Shimoyama; U.S. Patent 3,909,379; September 30, 1975; assigned to Japan Atomic Energy Research Institute, Japan* a composition comprising diallyl phthalate prepolymer and acrylonitrile or a mixture of acrylonitrile and an alkyl acrylate is cured by means of an electron beam as well as other ionizing radiations to form a crosslinked resin having excellent properties.

A vinyl phosphate incorporated in the composition makes the cured coating anticorrosive and adhesive to metal surface. The composition is useful as a coating or adhesion agent, or for the manufacture of decorative plywood, colored coiled steel sheet, etc., and for preparing an anticorrosive paint.

The diallyl phthalate prepolymer used is obtained by polymerizing diallyl phthalate by the conventional method, stopping the polymerization before gelling takes place and separating the unreacted monomer. A prepolymer having softening point in the range of 50° to 110°C, iodine value in the range of 45 to 65, viscosity (as 50% methyl ethyl ketone solution) in the range of 35 to 110 cp (30°C) is preferable.

Hardening is retarded as the concentration of vinyl monomer increases, and therefore the proportion of vinyl monomer is most suitable for use in the range of 30 to 70% by weight of the mixture.

The composition is curable by a radiation of high dose rate (e.g., an electron beam from a high power accelerator) and, therefore, curing is effected in an extremely short period of time. This is a great advantage in industry, especially in such fields as the manufacture of plywood and the continuous coating of coiled steel sheets.

Free radical initiators such as benzoyl peroxide can be added to the resin-forming mixture to reduce the radiation dose and increase the degree of curing.

The curing is satisfactorily effected when irradiation is carried out in air, but the radiation dose can be reduced further if the irradiation is carried out in an inert gas such as nitrogen or in an atmosphere maintained at low oxygen concentration.

When a small amount of a phosphate having a reactive vinyl group or groups is added to the above resin-forming composition, the adhesion and anticorrosive properties of the cured coating thereof are remarkably improved.

Example 1: A resin-forming composition was obtained by dissolving 100 parts by weight of diallyl phthalate prepolymer (iodine value 56, softening point 80° to 95°C) and 4 parts by weight of commercially available benzoyl peroxide in a mixture of 60 parts by weight of commercially available acrylonitrile and 20 parts by weight of commercially available methacrylate to give a homogenous mixture.

This composition was poured on the surface of a plywood board laminated with printed paper at the rate of 180 grams per square meter. The coated surface was covered with a mold release sheet, which was pressed and stretched well to expel bubbles. The composite was subjected to electron beam irradiation of 6 Mrad from a Van de Graaff electron accelerator (1.5 mev 100 μa, 0.5 Mrad per second).

A stable and durable decorative surface coating with beautiful luster was obtained. The obtained product passed all the tests carried out in accordance with JAS and JIS. The tests include Test for Resistance to Chemicals (passed tests with 5% acetic acid, 1% sodium carbonate solution, methanol, toluene, acetone, carbon tetrachloride, chloroform); Repeated Heating and Cooling Test (cycle of heating at 100°C for 2 hours and cooling at -20°C for 2 hours repeated five times and no change observed); Soiling Test (passed test with mercurochrome and black shoe polish, slightly stained with iodine tincture); Test Against Cigarette (passed); Test for Machineability with Electric Saw (passed).

Example 2: A coating compound was prepared by dissolving 50 parts of diallyl phthalate prepolymer (iodine value 56, softening point 90° to 95°C), 40 parts of n-butyl acrylate, 10 parts of acrylonitrile and 3 parts of phosphate of 2-hydroxyethyl methacrylate (monoester about 80%, diester about 20%), formulating 15 parts of black iron oxide pigment by means of a ball mill for 24 hours, and adding 2 parts of benzoyl peroxide.

A coating 30 microns thick was applied to the surface of 70 x 150 x 0.8 mm mild steel plates (JIS-G-3310) with a bar coater. The thus coated plates were given 6 Mrad of electron beam irradiation under nitrogen atmosphere from a Van de Graaff type accelerator (1.5 mev and 100 μa). Excellent results were obtained in the various tests made on the cured coatings as shown below.

Cross-Cut Adhesion Test — The cured coating was cross-cut into 100 pieces of 1 mm squares by the steel needle for a gramophone. A cellophane tape was put thereon, and then pulled off at an angle of 45° against the coating surface. None of the coating peeled off.

Impact Test — Neither peeling nor cracking was exhibited under the Du Pont type testing (load 1 kg, diameter of hammer one-half inch, 50 cm fall).

Xylene Immersion Test — After being immersed in xylene for 5 days at 20°C, neither blistering nor peeling occurred, although a trace of swelling was exhibited.

Silicon Carbide as Dosage Reducing Agent

T. Maruyama and K. Yamashita; U.S. Patent 3,924,021; December 2, 1975; assigned to Kansai Paint Company, Ltd., Japan found that by incorporating silicon carbide and/or corundum in conventional coating materials for curing by electron beams, the dose of electron beam necessary for curing is greatly lowered and the productivity in the coating line is also greatly improved.

The electron beam curable coating composition is prepared by combining 5 to 30% by weight of silicon carbide and/or corundum to the coated film forming components of the coating material.

Coating materials useful in the process are those of which the main components are resins having alpha,beta olefinically unsaturated bonds such as unsaturated polyester, unsaturated acrylic copolymers and unsaturated polyurethane.

Example 1: 440 grams of maleic anhydride, 740 grams of phthalic anhydride, 936 grams of neopentyl glycol and 135 grams of diethylene glycol were taken in a flask provided with a stirrer, thermometer and reflux condenser, and condensation reaction was carried out in nitrogen stream under heating and conventional esterifying conditions to obtain an unsaturated polyester with an acid value of 38.

To 55 parts of the polyester were added 30 parts of styrene and 15 parts of methyl methacrylate to dissolve, and a coating material 1 was obtained. To 93 parts of the coating material 1 was subsequently added 7 parts of silicon carbide and full dispersing and mixing were carried out to make a coating material 1'. In the same manner, to 94 parts of coating material 1 was added 6 parts of corundum to obtain a coating material 1''.

Example 2: Each one of coating materials 1, 1', 1'' from Example 1 was coated on a lauan plywood (12 mm thick) pretreated with a filler in a film thickness of 40 μ. Each of the coated films was exposed to an irradiation of a dose of 8 Mrad of electron beam under a nitrogen atmosphere, an electron beam energy of 300 kev, and electron beam current of 25 ma. The tests results of the coated films cured thus are tabulated as follows.

Coating Material Test Subject	1	1'	1''
Pencil hardness	2H	4H	4H
Adhesion	90/100	100/100	100/100
Abrasion resistance (Taber)	125 times	673 times	654 times

Unsaturated Phosphoric Ester

The coating compositions of *R.A. Dickie and J.C. Cassatta; U.S. Patent 3,957,918; May 18, 1976; assigned to Ford Motor Company* on a nonpolymerizable solvent, pigment initiator and particulate filler-free basis consist essentially of a binder solution of: (1) between about 90 and about 10 parts of a saturated, thermoplastic vinyl polymer prepared from at least about 85 weight percent of monofunctional vinyl monomers; (2) between about 10 and about 90 parts of vinyl solvent monomers for the vinyl polymer, at least about 10 weight percent, preferably at least about 30 weight percent of the solvent monomers being selected from the group consisting of divinyl monomers, trivinyl monomers, tetravinyl monomers and mixtures thereof; and (3) between about 1.0 and about 15.0 parts per 100 parts of the total of the thermoplastic, vinyl polymer and the vinyl solvent monomers of a triester of phosphoric acid bearing one or more sites of vinyl unsaturation and having the formula

$$\left[H_2C{=}\overset{R}{\overset{|}{C}}-\overset{O}{\overset{\|}{C}}-O-A-O\right]_m-\overset{O}{\overset{\|}{P}}-[OR']_{3-m}$$

where R is H, Cl or CH_3; A is C_nH_{2n}, n is 2 to 6; m is 1 to 3; R' is C_1-C_4

chloro- or bromoalkyl. These coating compositions, which are radiation polymerizable, are preferably cured by exposure to ionizing radiation or ultraviolet radiation. The coating compositions provide an excellent protective surface which adheres well to a variety of substrates, in particular metals and vapor deposited metals, and, thus, can be employed in the preparation of a wide variety of articles.

One such preferred article or material is useful as a substitute for metal plated materials used for trim or brightwork on the exterior of automobiles.

Example: A radiation polymerizable coating is prepared from the following materials in the manner hereinafter set forth.

Step 1: Preparation of Paint Binder Resin

	Parts by Weight
(1) Water	150
(2) Triton X200	5.2
(3) 1% aqueous $K_2S_2O_8$	30
(4) Methyl methacrylate	300
(5) Water	270
(6) Triton X200	3.5
(7) Triton X305	10.7
(8) $K_2S_2O_8$	1.2
(9) Octanethiol	2.1

Items 1 and 2 are charged to a reactor provided with a condenser, a thermometer, an agitator, and a dropping funnel. The mixture is boiled to remove dissolved oxygen, and cooled slightly to 90°C. Item 3 is added. A mixture of the remaining ingredients is then added slowly over a period of about 40 minutes while maintaining the reaction mixture at reflux. Following the monomer addition, the mixture is maintained at reflux for an additional 2 hours.

The latex so obtained is cooled and coagulated by adding it dropwise to three volumes of rapidly stirred methanol heated to about 40°C. The polymeric precipitant is isolated by filtration, washed with methanol, dried in vacuo and used in the subsequent preparation of coating materials. The polymer molecular weight is about 10,000.

Step 2: Formulation of Coating

	Parts by Weight
(1) Polymer from Step 1	24.2
(2) Neopentylglycol diacrylate	36.4
(3) 2-Ethylhexyl acrylate	39.4
(4) Tris(methacryloyloxyethyl) phosphate	4.0
(5) Butyl acetate	40
(6) Toluene	40
(7) Methyl ethyl ketone	10
(8) Isopropanol	10

A solution of polymer is prepared using the above listed monomers and solvents.

The film-forming solution so obtained is applied to a plastic substrate bearing a vacuum deposited aluminum surface and cured thereon under an inert atmosphere by electron beam irradiation using a total dosage of 9 Mrad (voltage 275 kv, current 40 ma).

The coating so obtained displays no softening or color change, and the underlying metal is similarly unaffected by 240 hours exposure to water at 90°F. No failure is observed in 168 hours 5% salt spray exposure. By way of comparison, a similarly prepared coating from which the phosphate additive is omitted, when subjected to similar test conditions, is observed to allow attack on and virtually complete removal of the underlying metal adjacent to a scribed line.

Metal Oxide or Hydroxide

L.S. Miller; U.S. Patent 3,963,798; June 15, 1976; assigned to The Dow Chemical Company provides thixotropic coating compositions capable of being rapidly cured by exposure to high energy radiation. These compositions are produced by incorporating into a liquid, hydrophobic, essentially solvent-free coating vehicle curable by exposure to high energy radiation and essentially free of reactive carboxylic acid groups (a) an additive selected from the group consisting of metal oxides and metal hydroxides, the additive being dispersible in the vehicle in a finely divided state and (b) an acid selected from the group consisting of acrylic acid and methacrylic acid, the additive being reactive with the acid to form a metal salt thereof and water. The amounts of additive and acid are sufficient to render the composition thixotropic. Since the resulting water solution of the metal salt is substantially insoluble in the hydrophobic vehicle, there results a two-phase dispersion having markedly different viscosity properties than the vehicle alone.

The thixotropy exhibited by these compositions renders them especially useful for application to porous substrates. Because of their nondrip, low flow characteristics, they are particularly well suited for application to vertical surfaces. The high viscosity also minimizes oxygen absorption into the covering surface, and as a consequence, little or no oxygen inhibition of curing occurs at the surface of the coating when high energy ionizing radiation is used. Moreover, as compared to the coating vehicle alone, these compositions are more easily curable by exposure to high energy radiation.

Preferred additives are ZnO, MgO, CaO, SrO, BaO, HgO, PbO, Pb_3O_4 and the hydroxides corresponding thereto. The use of at least 3% by weight of additive will generally be preferred. The acrylic and methacrylic acids used in the compositions of this process are most effective when used in a proportion of one mol of acid per equivalent of additive.

Example: (a) A pigmented base coating composition is formulated from 17.5 parts of a resin (A) produced by reacting the half ester condensation product of 2 mols of 2-hydroxyethyl acrylate and 2 mols maleic anhydride with 1 mol bisphenol A diglycidyl ether; 17.5 parts of the resinous condensation product (B) produced by reacting the half ester condensation product of 2 mols 2-hydroxypropyl acrylate and 2 mols of maleic anhydride with 1 mol n-butanediol diglycidyl ether; 15 parts of n-butylacrylate; 31 parts of titanium dioxide and 19 parts of calcium carbonate.

Similar coatings are prepared from (1) the above formulation with 10% zinc oxide added, (2) the above formulation with 10% acrylic acid added and (3) the above formulation with 10% zinc oxide and 10% acrylic acid added (approximately 0.56 mol acrylic acid per equivalent of zinc oxide). Formulation (3) becomes very thick after approximately 5 minutes, while the viscosities of formulations (1) and (2) remain essentially unchanged even upon prolonged standing.

Each of these formulations is spread onto a separate plywood panel with a draw bar to a thickness of 0.005 inch. The coated panels are then cured in air by passing them at 100 feet per minute under the beam of an electron accelerator operating at 300 kilovolts and 20 milliamps.

A dose of 1.7 Mrad per pass is absorbed. After each pass, the degree of coating cure is estimated by moderate scraping with the edge of a coin. Cure is considered incomplete if the coating is removed. The base formulation and formulations (1) and (2) each require three passes or 5.1 Mrad to attain a hard coating while formulation (3) is hard after only one pass.

(b) Formulation (3) is coated on plywood as described above and is then irradiated in air at 300 kv and 10 milliamps and with one pass at 100 feet per minute (0.85 Mrad). The resulting coating is sufficiently cured to resist the coin scratch test.

PLASTICS

ELASTOMERS

Heat-Flowable Material in Elastomer

Heat recoverable articles of manufacture made from organic polymeric compositions are described by *P.M. Cook; U.S. Patent Reissue 28,688; January 20, 1976; assigned to Raychem Corporation.* The compositions comprise a mixture of a thermoplastic resin material, or other organic, normally solid heat flowable material, in an elastomeric material. A heat recoverable article obtained from these compositions is elastomeric in both its heat unstable and heat stable states.

Mixing and Molding: The preferred method is as follows. A quantity of the resin is placed on a 2-roll mixer operating at a temperature sufficient to soften or melt the resin. The resin is milled until completely softened and then an equal amount of elastomer is slowly added to the resin. Mixing is continued until a homogeneous composition is secured. This is removed from the rolls and cooled. The balance of the mixture is done by placing a requisite amount of mixture on a cold 2-roll mill and adding the additional elastomer along with antioxidants, accelerators, fillers, plasticizers, etc., as required. Molded slabs were prepared using a 6" x 6" x 0.062" rubber mold as described in ASTM D-15.

Irradiation Technique: Samples were irradiated using a resonant transformer operating effectively at 850 kv and 5 milliamps. The samples were cross fired to insure uniform irradiation through the sample, and the irradiation dose was predetermined by Faraday cage measurements. A well grounded thermocouple in contact with the sample served to measure the temperature of the sample.

This example illustrates the production of heat recoverable silicone elastomeric articles in accordance with the process. Test specimens having the compositions set forth below were made as described above and irradiated at a dose level of 10 megarads with the following results.

Specimen 1

	Percent
Silastic 82U	90
6001 polyethylene	10
Memory	62
Retraction	94

Specimen 2

	Percent
Silastic 82U	90
Styron 666	10
Memory	87
Retraction	100

Acrylonitrile-Butadiene

An acrylonitrile-butadiene (NBR) elastomeric composition having improved elongation properties is provided by *R.J. Eldred; U.S. Patent 3,950,238; April 13, 1976; assigned to General Motors Corporation*. The composition comprises the high energy electron beam radiation induced reaction product of, by weight:

(a) one hundred parts of any acrylonitrile-butadiene random copolymer consisting of from 30 to 45% acrylonitrile,

(b) from 15 to 80 parts of carbon black,

(c) from 4 to 16 parts of trimethylolpropane trimethacrylate (TMPT), and

(d) a polymerizable noncrosslinking monomer selected from the group consisting of

(1) $CH_2{=}C(CH_3){-}C({=}O){-}O{-}CH_2{-}CH{-}CH_2$ (epoxide: O bridging CH and CH_2)

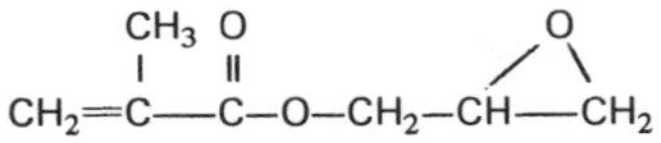

(2) $CH_2{=}C(R_1){-}C({=}O){-}O{-}CH_2{-}CH(OH){-}CH_3$

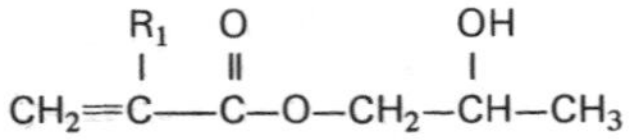

(3) $CH_2{=}C(R_1){-}C({=}O){-}O{-}CH_2{-}CH_2{-}O{-}R_2$

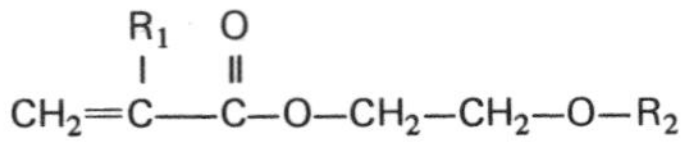

wherein R_1 may be —H or —CH_3, and R_2 may be —H or an alkyl group having up to 10 carbons. The mol ratio of noncrosslinking monomer to trimethylolpropane trimethacylate is within the range of 1.0 to 1.4 and the trimethylolpropane trimethacrylate noncrosslinking monomer and carbon black are initially uniformly dispersed throughout the copolymer. After the ingredients are blended into a relatively uniform dispersion, the mixture is then subjected to high energy, electron beam radiation having suitable beam current to provide a radiation dose of from 5 to 10 megarads within a reasonable period of time.

Example: One hundred parts by weight of a noncured, nonfilled NBR rubber were banded onto one roll of a 2-roll mill; and then about 20 parts of carbon black were gradually added to the banded composition as the rolls turned. The milling action uniformly dispersed the filler throughout the rubber. An additional 20 parts of the carbon black, throughout which 8 parts of TMPT and 4 parts of glycidyl methacrylate (GMA) had been uniformly dispersed, were then added to and dispersed throughout the banded composition.

The noncured composition was then removed from the two-roll mill and tensile slabs (6.0 x 6.0 x 0.08 inch) were compression molded at a temperature of about 150°F. It is believed that no curing, i.e., crosslinking, occurred during this operation. These slabs were then placed on a cart and passed under the radiation beam produced by a 1.5 mev Dynamitron Electron Beam Accelerator having a maximum current capacity of 15 milliamps.

The total radiation dose was controlled by adjusting the cart speed, the current and the number of passes. In this specific example, the slabs were irradiated at a current of 4.2 milliamps and a cart speed of 40 centimeters per second, which is equivalent to about 0.25 megarad per pass. A total radiation dose of 10 megarads was required to cure this particular composition to the same extent as that of the conventional sulfur cured control compounds. The extent of cure was followed by measuring the equilibrium swelling of the material in toluene at 25°C. No attempt was made to relate this swell ratio to the actual crosslink density as this relationship is difficult to establish due to the nonhomogenous stress field generated by the restricted swelling at the rubber-carbon black interface.

The NBR elastomeric composition produced had an elongation at break of 440%. This is more than twice the elongation of a radiation cured elastomer which does not contain the noncrosslinking monomer. In accordance with the procedures outlined in this example, three additional series of experiments were run employing varying amounts of both TMPT and the GMA, and the elongation of these samples were compiled in Table 1 which reports: (1) the amount and type of both the crosslinking and noncrosslinking monomers, (2) the elongation of those compositions, and (3) the dosage required to cure them.

TABLE 1

Crosslinking Monomer	Parts/100 Parts of NBR	Extender Monomer	Parts/100 Parts of NBR	Dose, Mrad	Percent Elongation at Break
TMPT	8	None		10	200
	8	GMA	2	10	275
	8	GMA	4	10	440
	8	GMA	6	10	290
TMPT	12	None		8	125
	12	GMA	4	8	175
	12	GMA	6	8	220
	12	GMA	8	8	190
TMPT	16	None		5	125
	16	GMA	4	5	135
	16	GMA	6	5	150
	16	GMA	8	5	165
	16	GMA	10	5	135

In addition, using the methods described above, a series of compositions containing various noncrosslinking monomer candidates were prepared and evaluated and compared to a standard which contained no noncrosslinking monomer. The control composition contained 100 parts of Hycar 1032, which is a commercially available NBR rubber containing 34% by weight of acrylonitrile, 40 parts of carbon black and 8 parts of TMPT. The various compositions are listed in Table 2.

TABLE 2

	Noncrosslinking Monomer Candidate	Concentration, parts by weight per 100 parts NBR*	Percent Elongation
A	None	-	300
B	2-Methoxyethyl methacrylate	4.1	400
C	2-Methoxyethyl acrylate	3.7	350
D	2-Hydroxypropyl methacrylate	4.1	500
E	2-Hydroxypropyl acrylate	3.7	450
F	2-Ethoxyethyl acrylate	4.1	425
G	2-Ethylhexoxy-ethyl acrylate	6.4	470

*The concentrations were selected to provide an ethylenic unsaturation concentration equal to that provided by 4 parts of GMA.

Phosphonitrilic Fluoroelastomers

Phosphonitrillic fluoroelastomers and compositions based on the same, possessing outstanding solvent resistance, low temperature flexibility and good physical strength over a broad range of service conditions are described by *G.S. Kyker, T.A. Antkowiak and A.F. Halasa; U.S. Patent 3,970,533; July 20, 1976; assigned to The Firestone Tire & Rubber Company.*

The phosphonitrilic fluoroelastomer is represented by the general formula:

$$\left(\overset{\displaystyle Q}{\underset{\displaystyle Q'}{\overset{|}{\underset{|}{P}}}}=N-\overset{\displaystyle Q''}{\underset{\displaystyle Q}{\overset{|}{\underset{|}{P}}}}=N \right)_n$$

wherein each of Q and Q' represents a fluoroalkoxy group represented by the formula $-OCH_2C_nF_{2n}Z$ in which n is an integer from 1 to 10 and Z represents H or F; and Q and Q' are not required to be the same and there may be more than two such groups; and Q" represents an o-allylphenoxy group; the groups Q, Q' and Q" being randomly distributed along the chain. In the preferred embodiment, 0.1 to 5% by weight of o-allylphenol groups are randomly distributed among the groups attached to the chain.

The elastomers were prepared as follows: a suitable alcohol or fluoroalcohol or a mixture of alcohols and/or fluoroalcohols, together with o-allylphenol were slowly added to sodium metal in dried tetrahydrofuran under a nitrogen atmosphere in a 3 neck, 12 liter, round bottom flask fitted with a dropping funnel, a dry ice condenser with a H_2 vent line, and a motor driven glass stirrer. The flask was immersed in a thermostatically controlled liquid bath. After the reaction of the alcohols with sodium had proceeded to completion, a solution of linear poly(dichlorophosphazene) in benzene was added rapidly to the mixture of alkoxides and/or fluoroalkoxide(s) and o-allylphenoxide. The chlorine atoms of the poly(dichlorophosphazene) combined readily with the alkali metal of the alkoxides and as a result, the substituted polyphosphazene derivative was produced. The derivatized product was recovered by washing the reaction mixture free of salt until the test for chloride was negative. Terpolymers or copolymers containing the indicated unsaturation can be readily compounded and cured, i.e., vulcanized in much the same way as known rubbers.

They are also more amenable to curing than their saturated counterparts and it is possible to use peroxide curing agents in compositions containing carbon black fillers and to use sulfur accelerated cures as well as radiation cures. The properties of a compounded stock are set forth in the table which follows, compounding having been achieved by simply mixing the several constituents with polyphosphazene gums in a conventional mixer. Terpolymer (1) is a linear polyphosphazene with three randomly distributed substituents, namely: CF_3CH_2O—, $HF_2C(CF_2)_3CH_2O$— and o-allylphenoxy.

Radiation (High Energy Electron) Induced Vulvanization of Terpolymer

Stock A*							
Radiation dose (Mrad)	0	1	2	3	4	5	6
50% Modulus (psi)	75	175	240	290	500	630	655
100% Modulus (psi)	100	500	710	980			
Tensile strength (psi)	180	1,700	1,680	1,530	1,190	1,270	1,030
Elongation at break	405	175	150	125	85	75	65
Tension set (%)	40	14	11	9	4	2	3
Shore A hardness	45	65	65	70	74	75	75
Stock B**							
Radiation dose (Mrad)	0	1	2	3	4	5	6
50% Modulus (psi)	120	630	880	1,180			
100% Modulus (psi)	150						
Tensile strength (psi)	150	1,370	1,290	1,340	1,270	1,170	1,100
Elongation at break	110	95	65	55	40	35	30
Tension set (%)	1	1	1	1	1	1	1
Shore A hardness	45	60	65	70	73	75	75

*Stock A–Terpolymer (1), 100; Silanox 101, 30; Stan Mag ELC (MgO), 6 phr.
**Stock B–Terpolymer (1), 100; FEF Black, 30; Stan Mag ELC (MgO), 6 phr.

ADDITIVES AND CATALYSTS

Phosphite Ester Antioxidants for Styrene-Butadiene Elastomers

R.J. Eldred; U.S. Patent 3,888,752; June 10, 1975; assigned to General Motors Corporation has found that phosphite esters taken from the following group are effective antioxidants in radiation cured, styrene-butadiene elastomers when used in concentrations of preferably about 2 to 15% by weight of the composition prepared for curing.

The group consists of a pentaerythritol diphosphite having the general formula:

$$R_1{-}O{-}P\left\langle\begin{matrix}O{-}CH_2\\ \\ O{-}CH_2\end{matrix}\right\rangle C\left\langle\begin{matrix}CH_2{-}O\\ \\ CH_2{-}O\end{matrix}\right\rangle P{-}O{-}R_2$$

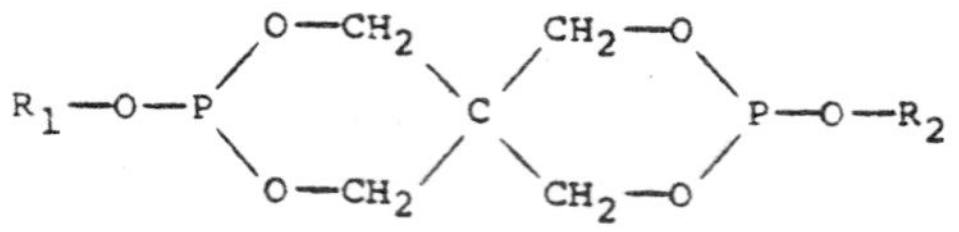

where R_1 and R_2 are alkyl groups containing from 12 to 20 carbon atoms, a trithiophosphite having the general formula:

$$R_1{-}S{-}P\left\langle\begin{matrix}S{-}R_2\\ \\ S{-}R_3\end{matrix}\right.$$

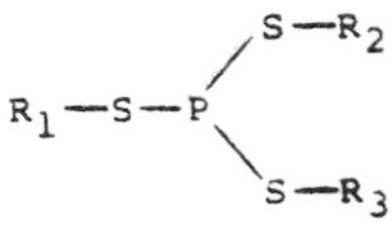

where R_1, R_2 and R_3 are alkyl groups containing from 10 to 15 carbon atoms and a hexathiodiphosphite having the general formula:

$$\left.\begin{matrix}R_1{-}S\\ \\ R_2{-}S\end{matrix}\right\rangle P{-}S{-}(CH_2)_x{-}S{-}P\left\langle\begin{matrix}S{-}R_3\\ \\ S{-}R_4\end{matrix}\right.$$

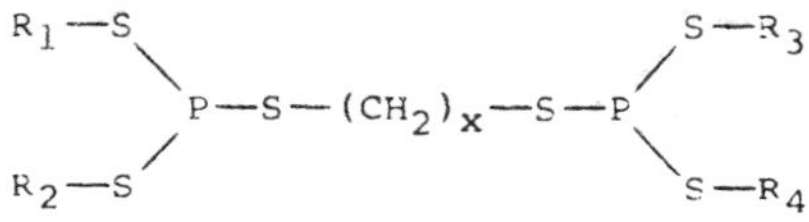

where R_1, R_2, R_3 and R_4 are alkyl groups containing from 10 to 15 carbon atoms and x is from 4 to 8. The styrene to butadiene weight ratio may vary from 15:85 to 50:50.

Example 1: This formulation consisted of 100 parts by weight styrene-butadiene copolymer with a styrene-butadiene weight ratio of 23.5 to 76.5, 40 parts by weight of carbon black N-550, 12 parts by weight trimethylolpropane trimethacrylate, and 2.4 parts by weight of commercially available trilauryl trithiophosphite (TLTTP).

The above ingredients were blended by a standard roll-milling operation; the rolls were 12 inches wide and 6 inches in diameter and were rotating at about 20 rpm. There were no steps taken to either heat or cool the formulation during the milling operation. After blending, the formulation was pressed into sheet form at a thickness of about 100 mils and subjected to a radiation dose of 14 megarads.

The equipment used was a 1.5 mev electron accelerator; the current was 4.2 milliamps. This elastomer had an original tensile strength of 1,900 psi, and after 70 hours aging at 257°F the tensile strength was 1,500 psi. This was deemed to be an acceptable loss and, therefore, the antioxidant was effective. The tensile strength after aging of these samples was about three times greater than that of either the radiation cured elastomers with conventional antioxidants or sulfur cured elastomers with the phosphite ester antioxidants.

Example 2: The formulation, blending, radiation and testing procedures were the same as described in Example 1 with the exception that 2.0 parts by weight of commercially available distearyl pentaerythritol diphosphite was used as the antioxidant. The tensile strength of the resulting elastomer before agins was 1,500 psi, and after aging for 70 hours at 257°F the tensile strength was 1,300 psi. This was also an acceptable performance.

Aluminum Chloride Catalyst for Beta-Acryloyloxypropanoic Acid

According to *M.D. Burguette; U.S. Patent 3,888,912; June 10, 1975; assigned to Minnesota Mining and Manufacturing Company* acrylic acid can be converted into relatively pure β-acryloyloxypropanoic acid (AOP). AOP has been found useful in radiation polymerization techniques. Acrylic acid is converted to β-acryloyloxypropanoic acid when heated in the presence of catalytic amounts of anhydrous aluminum chloride.

Example 1: Anhydrous aluminum chloride (13.0 g) was incrementally added to a vessel containing 300 g of acrylic acid. The acrylic acid was stirred continuously to insure that a solution of the acid and aluminum chloride was formed. The solution was heated to 156° to 158°C for over an hour in a round bottomed flask equipped with thermometer and reflux condenser. The solution was gently stirred during the heating. The reaction was stopped and the solution allowed to cool to about 100°C.

After suction filtering to separate any precipitate that may have formed, the unreacted acid and other volatile matter was distilled off under reduced pressure on a steam bath under a mild current of nitrogen. The residual oily liquid was fairly pure AOP, which can be easily purified further by fractional distillation, collecting that portion which distills at 101° to 103°C at 0.17 to 0.19 mm Hg. The AOP obtained is readily polymerizable in electron beam polymerization systems, and is soluble or miscible in most common solvents including water. Under high temperatures (above 175°C) at atmospheric pressure AOP decomposes into polyacrylic acid, and at pot temperatures of about 110°C at 0.17 to 0.19 mm Hg starts breaking down into acrylic acid.

The AOP may be polymerized by itself or with other monomeric or polymeric materials. Such comonomers include unsaturated polyesters such as the condensation product of maleic acid and propanediol, as well as any ethylenically unsaturated polymerizable monomer.

Example 2: The general configuration of the electron beam source involves a vacuum chamber, with a heated filament source of electrons within. The electrons are accelerated from the source toward the Lenard window with energy of from 100 kev to 10 mev so that the electrons pass through the window and retain sufficient energy to penetrate the thickness of the material to be cured. The material to be cured may be placed on a conveyor or other transporting apparatus in the path of the electrons. The electron flow or current and the time of electron impingement are related such that for an electron flow of 0.1 to 100 millamperes, a treatment duration of a fraction of a second may be required.

A 20 to 35 μ liquid film of AOP was deposited on a glass slide and exposed to an electron beam source set at 105 kev and 2.5 milliamps through nitrogen. A dry, clear, solid polymeric film was produced within one second of exposure. The film was insoluble in most solvents.

Example 3: One part of an olefinically unsaturated polyester, the commercially available Paraplex 70 was dissolved in 1.5 parts of AOP. This solution was deposited as a film (20 to 35 μ) on a glass slide and exposed as in Example 2. On exposure to the electron beam source a clear film formed which was insoluble in most solvents.

Fluorocarbon Polymer Crosslinking Agent

E.J. Aronoff, K.S. Dhami and T.-C. Shieh; U.S. Patent 3,894,118; July 8, 1975; assigned to International Telephone & Telegraph Corporation have found a class of crosslinking compounds for high temperature processing fluorocarbon polymers which are stable through all of the melt processing operations and readily form homogenous irradiation crosslinked systems. Processing temperature as high as 315°C (600°F) can be used without significant and detrimental thermal prereaction or volatilization during a melt processing procedure such as extrusion prior to irradiation activated curing.

The crosslinking coreactant compounds for use in the process are in general dimethacrylic acid esters corresponding to the structural formula

$$CH_2{=}\underset{\displaystyle CH_3}{\underset{|}{C}}-\overset{\displaystyle O}{\overset{\|}{C}}-O-\underset{\displaystyle R_2}{\underset{|}{\overset{\displaystyle R_1}{\overset{|}{R}}}}-O-\overset{\displaystyle O}{\overset{\|}{C}}-\underset{\displaystyle CH_3}{\underset{|}{C}}{=}CH_2$$

wherein R is a radical having from 5 to 14 carbon atoms and is selected from the group consisting of alkyl, cycloalkyl and aralkyl; R_1 and R_2 are independently selected from the group consisting of hydrogen, alkyl, cycloalkyl and aryl radicals and mixtures thereof; and the total number of carbon atoms in R, R_1 and R_2 is at least 10.

Example: Ninety parts of an essentially equimolar copolymer of ethylene and chlorotrifluoroethylene was blended and reblended in a mixer with 10 parts of decamethylene glycol dimethacrylate to give a 200 gram sample of a powder composition. A second 200 gram sample was fluxed in a mixing head for several minutes at 243°C and then compression molded at 255° to 260°C at about 10,000 psi ram pressure. Irradiations were done with a 2 mev electron beam accelerator. Physical properties and aging tests were carried out by conventional procedures. Test results are shown in Tables 1 and 2. Sample 1 was in powder form; Sample 2 a fluxed blend; Sample 3, as a control, contained no crosslinking additive.

TABLE 1

Dose (Mrad)	Mechanical Test*	Sample 1	Sample 2	Sample 3
4	Tensile strength	6,324	6,282	6,550
	Elongation	150	183	225
	Hot modulus	30	37	**

(continued)

TABLE 1 (continued)

Dose (Mrad)	Mechanical Test*	Sample 1	Sample 2	Sample 3
10	Tensile strength	6,365	6,705	6,416
	Elongation	100	100	216
	Hot modulus	11	18	**
25	Tensile strength	6,836	6,541	5,609
	Elongation	66	100	258
	Hot modulus	11	26	400

*Tensile strength in psi; elongation in percent at room temperature, 20" min; hot modulus in percent at 250°C, 50 psi load.
**Specimen broke, no elongation.

TABLE 2

Sample	Dose (Mrad)	Mechanical Test	. . Aging Period in Hours at 200°C . .			
			0	72	120	168
1	4	Tensile strength (psi)	6,324	6,052	5,576	6,131
		Elongation (%)	150	100	150	100
	10	Tensile strength (psi)	6,965	5,250	6,250	5,937
		Elongation (%)	100	100	112	75
	25	Tensile strength (psi)	6,836	6,025	5,266	5,128
		Elongation (%)	66	87	50	50
2	4	Tensile strength (psi)	6,282	5,640	5,855	5,320
		Elongation (%)	183	162	200	150
	10	Tensile strength (psi)	6,705	5,670	6,118	5,312
		Elongation (%)	100	87	112	100
	25	Tensile strength (psi)	6,541	5,125	5,562	5,127
		Elongation (%)	100	87	100	100

Table 1 shows that the additives significantly increase the degree of crosslinking at equivalent radiation dose levels and markedly improve deformation resistance as measured by hot modulus. It also shows decreased radiation degradation in the presence of the crosslinking additive with increasing radiation dose. Table 2 demonstrates that the compounds of the process retain a high degree of their original mechanical properties.

Additives to Accelerate Polyethylene Crosslinking

Tsutoma Kagiya, M. Hagiwara, Tsukasa Kagiya and M. Sohara; U.S. Patent 3,894,928; July 15, 1975; assigned to Kishimoto Sangyo Co., Ltd., Japan provide additives which are effective to accelerate radiation crosslinking of ethylenic polymers.

The crosslinking is effected by subjecting the ethylenic polymer to ionizing radiation in the presence of

(a) an acrylic monomer capable of being crosslinked with ethylenic polymer; and
(b) an acetylenic compound of the formula

$$(CH{\equiv}C)_nR$$

wherein R is an n-valent organic radical rendering said acetylenic compound capable of being blended with said ethylenic polymer at an elevated temperature.

Example 1: A commercial grade high density polyethylene (100 parts), Hizex 5000H, was mixed with hexamethylene diacrylate (2.0 parts by weight) and dipropargyl maleate (1.0 part) on the hot roller for 10 minutes at 150°C, and molded into a sheet 1.0 mm thick. The sheet was then exposed at room temperature to γ-rays from cobalt 60 in nitrogen or in air atmosphere. On the other hand, as a control sheets mixed with hexamethylene diacrylate (3.0 parts), and with no additives was irradiated under the same conditions.

		Gel Fraction (weight percent)			
		Air Atmosphere		Nitrogen Atmosphere	
Additives	**Parts**	**2.5 Mrad**	**5.0 Mrad**	**2.5 Mrad**	**5.0 Mrad**
Hexamethylene diacrylate	2				
and dipropargyl maleate	1	43.0	62.3	59.5	73.0
Hexamethylene diacrylate	3	31.0	37.0	29.0	49.5
No additives		0.6	19.3	0.7	26.0

Example 2: Mechanical properties of the crosslinked polyethylene obtained by a procedure described in Example 1 were as follows:

	Air Atmosphere			
Dose	**Tensile Strength (kg/cm²)**		**Elongation (percent)**	
(Mrad)	**HGA-DPM***	**No Additive**	**HGA-DPM***	**No Additive**
0	261	248	875	580
2.5	268	258	725	385
5.0	286	274	642	210

*Additives: HGA is hexamethylene diacrylate and DPM is dipropargyl maleate.

Propargyl-Containing Crosslinking Agent to Lower Radiation Dosage

N. Murayama, T. Katto and T. Ichii; U.S. Patent 3,923,621; December 2, 1975; assigned to Kureha Kagaku Kogyo K.K., Japan provide an improved vinylidene fluoride resin composition capable of crosslinking by application of a lesser amount of high energy ionizing radiation. The improvement comprises use of a crosslinking agent in the form of cyanurate or isocyanurate, having at least one propargyl radical functioning as the trifunctional radical.

The dosing ratio of the crosslinking agent to the polymer or copolymer may vary with kind, degree of polymerization and structure thereof, and with the amount of high energy ionizing radiation to apply. Remarkable results may be obtained with 0.05 to 20 phr, preferably 5 to 10 phr. The compounded composition may be shaped into any desired form, such as, for instance, sheet, film, tube, yarn or the like.

Beta rays and gamma rays may preferably be used as the high energy irradiation rays for the crosslinking reaction. As the irradiation dose necessary for the effective improvement of heat resisting and/or elastic properties of the resin products, 1 to 10 Mrad, preferably 2 to 8 Mrad, may be recommended.

Example: Ninety weight parts of powdered polyvinylidene fluoride were heated up to 80°C in a Henschel blender working at 1,000 rpm and liquefied crosslinking agent, diallyl propargyl cyanurate, 1 weight part, 40°C, was poured into the fluoride and the mixture was fully blended together for 3 minutes. This compound was extruded at 210°C from a head of an extruder to form a continuous bar of 2 mm diameter which was then cut into chips of 2 to 3 mm. These chips were extruded at 210°C from a head of a further extruder to form a continuous tube, i.d. 2.9 and o.d. 3.6 mm.

This tube was subjected to electron ray irradiation at a dose of 4 Mrad under ambient conditions. The remaining gel percent of this crosslinked tube was measured as 38% and the 100% modulus was measured at 200°C as 1.4 kg/cm^2. The remaining gel percent was measured upon extraction of the products in dimethyl amide, 100°C, for 24 hours. The crosslinked polyvinylidene fluoride tube was pressurized to have an inner pressure of 0.2 kg/cm^2 gauge, and set within a Teflon tube of i.d. 6.5 mm. The double tube assembly was then heated to 200°C at a speed of 50 cm/min from one end to the opposite end. With this heat-up operation, the inner tube was expanded to the inside wall surface of the outer tube gradually.

Then, the heated-up and expanded inner tube was cooled down to solidify while the inside gas pressure was maintained, and the thus solidified tube was taken out from the outer tube. The thus expanded and solidified tube had an i.d. of 6.0 mm and an o.d. of 6.4 mm. This tube was then immersed in a silicone oil bath of 220°C for 3 minutes and taken out therefrom and cast into a cold water bath. The tube was taken out and measured to have an i.d. of 3.0 mm and an o.d. of 3.7 mm, thus recovering substantially the original diametral dimensions. The axial contraction amounted to 6%. From these experiments, it was shown that a superior elasticity-memory material in the form of a thermally contractable tube was provided.

Additives to Prevent Bubble Formation and Odor

In the process of *A.B. Robertson and R.J. Schaffhauser; U.S. Patent 3,947,525; March 30, 1976; assigned to Allied Chemical Corporation* melt-processable, radiation crosslinkable ethylene/chlorotrifluoroethylene copolymer compositions are provided which contain about 0.1 to 5% by weight of the copolymer of a radiation crosslinking promoter, about 0.01 to 5% by weight of an antioxidant and about 0.1 to 30% by weight of an acid scavenger.

The radiation crosslinking promoter is selected from the group consisting of triallyl isocyanurate, triallyl cyanurate, triallyl phosphite, diallyl fumarate, diallyl isophthalate and diallyl terephthalate. The antioxidant is selected from the group consisting of (1) mixtures of a phosphite or an organic polyhydric phenol and a salt of a carboxylic acid and a metal of Group II of the Periodic Table, (2) alkylated phenols and bisphenols having 1 to 18 carbon atoms in the alkyl chain, (3) alkylidene bis, tris and polyphenols having 1 to 18 carbon atoms in the alkylidene chain, and (4) mixtures of (1), (2) or (3) with an ester or alkali metal salt of thiodipropionic acid. The acid scavenger comprises an oxide of metal of Group II of the Periodic Table. The compositions can be successfully melt processed in a variety of operations, without setting up in the extruder or mold, including wire and cable extrusion, sheet and film extrusion. The extrudates are bubble free and discoloration free. The compositions can be crosslinked at relatively low radiation levels without emitting an odor or forming bubbles in the extrudate.

The stabilizing compositions are particularly advantageous for use in about equimolar ethylene/chlorotrifluoroethylene copolymers containing between about 45 and about 55 mol percent of ethylene units and having melting points above about 220°C, preferably between about 220° and 265°C. The amount of radiation to which the copolymer is subjected to improve its mechanical and other properties is generally in the range of about 2 to 50 Mrad or higher, preferably 5 to 15 Mrad. Copolymer compositions of this process are prepared as follows.

Example: Dry powdered ethylene/chlorotrifluoroethylene copolymer of about 40 mesh particle size containing about 50% of ethylene units, having a melting point of 245°C and a melt index of about 1.0, is mixed with various additives in a ball mill for 1 hour.

Sample 1 contains 0.225% of phosphite of 4,4-n-butylidene-bis(6-tert-butyl-m-cresol) plus 0.075% of zinc 2-ethylhexoate plus 0.15% distearyl thiodipropionate plus 0.5% of CaO plus 0.25% triallyl isocyanurate. Sample 2 contains 0.225% of phosphite of 4,4-n-butylidene-bis(6-tert-butyl-m-cresol) plus 0.15% distearyl thiodipropionate plus 0.25% CaO plus 0.25% triallyl isocyanurate. Sample 3 contains 0.225% phosphite of 4,4-n-butylidene-bis(6-tert-butyl-m-cresol) plus 0.15% distearyl thiodipropionate plus 1.0% CaO plus 1.0% triallyl isocyanurate.

The compositions are compression molded into 0.060 inch sheets. The sheets are subjected to cobalt 60 radiation and electron beam radiation and various mechanical properties are measured. The results are summarized in the table below. Thermal stress cracking was determined by the mandrel wrap test, MIL-P-390C, Part H.

Sample	Temperature (°C)	Dosage Radiation (Mrad)		Yield Stress (psi)	Break Stress (psi)	Yield Elongation (%)	Break Elongation (%)	Tensile Modulus (psi) 10^5	Work to Break (psi) 10^3	Thermal Stress Cracking
1	23	electron beam	0	4,670	8,100	4.5	225	2.0	10,971	
			2	4,680	8,050	4.6	220	2.0	10,833	
			5	4,580	7,710	4.1	210	2.0	10,305	
			10	4,540	7,520	4.9	215	1.9	9,395	
			25	5,440	8,110	4.3	175	2.3	7,104	
	200		0	265	217	12	24	2.8	33	
			2	293	235	33	145	2.3	307	
			5	335	435	30	430	3.5	1,173	
			10	295	575	35	518	2.7	1,347	
			25	370	770	36	355	3.4	1,648	
	23	cobalt 60	0	4,250	7,740		234	1.77		
			2	4,500	7,670		217	1.71		
			5	4,650	7,530		201	1.83		
			15	4,520	6,860		171	1.73		
			30	4,470	4,590		104	2.8		

(continued)

Sample	Temperature (°C)	Dosage Radiation (Mrad)		Yield Stress (psi)	Break Stress (psi)	Yield Elongation (%)	Break Elongation (%)	Tensile Modulus (psi) 10^5	Work to Break (psi) 10^3	Thermal Stress Cracking
	200	cobalt 60	0	265	217		24	3.45		yes
			2	310	250		276	3.5		no
			5	316	626		488	3.68		no
			15	323	708		410	3.75		no
			30	310	580		390	3.61		
2	200	cobalt 60	0	177	244		17		14	yes
			5	179	307		25		75	yes
			10	266	271		230		559	no
		electron beam	0	177	244		17		14	
			5	-	135		9		39	
			10	225	115		36		75	
			25	250	390		565		1,457	
3	200	cobalt 60	0	265	217		33			yes
			2	310	250		721			no
			5	316	626		2,010			no
			15	323	708		1,567			no
		electron beam	0	265	217		33			
			2	293	235		307			
			5	335	435		1,173			
			10	295	575		1,347			
			25	370	770		1,648			

Inorganic Additive to Promote Curing

A process for curing an unsaturated polyester resin is provided by *K. Araki, T. Sasaki and K. Goto; U.S. Patent 3,971,711; July 27, 1976; assigned to Nitto Boseki Co. and Japan Atomic Energy Research Institute, Japan.* The process comprises the steps of:

(1) incorporating in said unsaturated polyester resin from about 50 to about 300 parts by weight, per 100 parts by weight of the resin, of one or more compounds of a finely powdered material of an inactive inorganic compound selected from the group consisting of calcium carbonate, clay, quartz powder, gypsum, cement, mica powder, glass powder, diatomaceous earth, talc, asbestos and red mud of bauxite; and

(2) irradiating the resin containing the compounds with an ionizing radiation applied thereto at a dose rate of 10^{-2} to 2×10^7 rads/sec for a total dose of 0.1 to 10 Mrad, and at a starting temperature of from about -20°C to about 60°C.

It is believed that the inactive inorganic additive momentarily becomes an active species which promotes the curing reaction due to elastic or nonelastic scattering of the energy of radiation.

Example: A mixture of 0.5 mol of phthalic anhydride, 0.5 mol of maleic anhydride and 1.1 mols of propylene glycol was placed in a reactor. While introducing nitrogen gas therein, the mixture was heated to 190°C to have one ingredient reacted with another, until the acid value of the mixture reached 50. To the product thus obtained was added 0.01 phr of hydroquinone at a temperature not higher than 70°C and, then, said product is poured in styrene kept at 25°C to be dissolved therein in order to obtain a product containing 70% by weight of solids.

To the unsaturated polyester resin prepared by the above procedure, 100 parts by weight of fine particles of calcium carbonate and 1.0 or 2.0 parts by weight of magnesium oxide powder per 100 parts by weight of resin were added. Then, for the purpose of curing, each composition was irradiated with an accelerated electron beam at the dose rate of 1.57×10^6 rads/sec at 2.0 mev. Remarkable promotion in curing said resin is confirmed by the data listed in the table below.

	A	B	C	D	E
Unsaturated polyester resin, g	100	100	100	100	100
$CaCO_3$, g	0	100	100	100	0
MgO, g	0	0	1	2	2
Radiation dose required for curing, Mrad	6.0	5.0	3.5	3.0	4.0
Time required for curing, sec	3.82	3.18	2.23	1.91	2.55
Acetone-extraction residue, %	98.0	97.9	98.5	98.8	98.5
Barcol hardness	40	46	46	45	40

Silicone Frothing Agents

J.M. Corbett; U.S. Patent 3,940,349; February 24, 1976; assigned to The Dow Chemical Company found that by the introduction of pendant phthalate, isophthalate or terephthalate half ester groups thermosettable vinyl ester can be successfully frothed using certain silicone frothing agents. The vinyl ester resins used herein are combined with a copolymerizable vinyl monomer or mixtures thereof. The resin comprises about 30 to 70 weight percent and the monomers about 70 to 30 percent. The vinyl ester resin is prepared by reacting about equivalent amounts of an unsaturated monocarboxylic acid and a polyepoxide and esterifying at least about half of the hydroxyl groups formed from the acid epoxide reaction to the half ester groups.

The frothing agent is a siloxane copolymer consisting of the following three units, namely, (a) SiO_2 units, (b) $(CH_3)_3SiO_{1/2}$ units, and (c) $(R)_2NR_1Si(R_2)_nO_{3-n/2}$ units. In the formula n has a value from 0 to 2. R_2 is an alkyl group having 1 to 6 carbon atoms. Preferably R_2 is methyl. When n is zero there are no R_2 substituents. R_1 is an alkylene radical of 1 to about 6 carbon atoms. R may be hydrogen, an alkyl radical having 1 to about 6 carbons or an aminoalkyl radical having 2 to about 6 carbon atoms. Preferably one of the R groups is hydrogen. In the siloxane copolymer the ratio of (b) units to the sum of the (a) and (c) units should be about 0.4:1 to 1.2:1. The silicone frothing agent is used in amounts of at least about 1 pph.

A convenient method of making the frothing agent consists of combining in an appropriate reactor 200 grams of a xylene solution (50% solids) of siloxane copolymer consisting essentially of SiO_2 units and $(CH_3)_3SiO_{1/2}$ units and varying amounts of an amine. The mixture will react at room temperature merely on standing for 16 to 24 hours. More usually the mixture is heated for 30 to 60 minutes at temperatures from about 130° to 190°C. The frothing agents used in the following examples are believed to have been prepared by reacting $H_2NCH_2CH_2NH(CH_2)_3Si(CH_3)(OCH_3)_2$ with the siloxane copolymer according to the above procedure in the following amounts: Frothing Agent 1, 20 g/100 g of polymer and Frothing Agent 2, 12.5 g/100 g of polymer.

Example: A vinyl ester resin prepared by first reacting a 2-hydroxypropyl acrylate half ester of maleic acid with D.E.R. 331 (1:1 equiv ratio) followed by reaction of 0.6 equivalent of phthalic anhydride per equivalent of hydroxyl was diluted to a 50% concentration in styrene.

A foam was prepared by adding 7.2 g of Frothing Agent 2 (50% solution) and 1.8 g of benzoyl peroxide to 120 g of the above resin and placed in the mixing bowl of Hobart N-50 mixer while maintaining an inert (N_2) atmosphere. N,N-dimethyl-p-toluidine (0.175 g) was added as a catalyst accelerator and the mixture whipped for 10 minutes at speed setting Number 6. The froth was poured into an open faced mold and allowed to cure. The resulting foam had a density of 19.9 lb/ft^3 and a screw holding strength of 254 lb/in of thread for a Number 6 wood screw.

Other means of curing may be employed, for example, the resin of this Example was frothed with 2.5 pph of Frothing Agent 1 and cured by exposure to a high energy electron beam from a Van de Graaff accelerator without any chemical catalysts.

PROCESSES

Irradiating Strand Material

J.R. Austin, M.J. Brown and L.D. Loan; U.S. Patent 3,925,671; December 9, 1975; assigned to Bell Telephone Laboratories and Western Electric Company, Inc. provide methods of and apparatus for the radiation of insulated conductors to uniformly dose the insulation and crosslink the insulation to produce a covering which possesses unusually desirable mechanical and electrical properties while optimizing the use of the source of radiation.

Conductors covered with insulation which is to be irradiated are passed between two groups of coaxial sheaves mounted rotatably individually. Successive sections of the conductors are advanced past the window of one accelerator head, around the associated sheave or sheaves, and then past the window of another accelerator head. The accelerators face in substantially opposite directions and are staggered along the paths of the conductors to avoid any substantial overlap of the electron beams associated therewith.

The windows extend vertically to encompass all the generally horizontal passes of the conductors as between the two groups of sheaves. Preferably, conductors are strung up between the sheaves in a modified figure eight pattern. The pattern is a figure eight modified to intermittently include a pass between the sheaves which is parallel to a line joining the axes of the two groups of sheaves. This reverses the direction of travel of the conductors and optimizes the uniformity of exposure of the cross sectional area of the insulation of the conductors to irradiation. The use of a figure eight path for the conductors causes the successive sections of the conductor to turn about the longitudinal axes thereof as they are advanced around the sheaves. In this way the insulation is more uniformly irradiated. In a preferred embodiment, twisted conductor pairs may be irradiated. The twist accentuates the longitudinal turning of the conductor pair. The irradiation of twisted pairs achieves obvious manufacturing economies while avoiding the necessity of having to twist irradiation crosslinked conductors.

Referring now to Figure 6.1, there is shown an apparatus, designated generally by the numeral **10**, which embodies the principles of this process, for irradiating the insulation covering a conductive element of successive sections of an insulated conductor **13**. The apparatus **10** generally includes facilities for moving successive sections of the conductor **13** in a path adjacent irradiation facilities and for impinging high energy electron beams on successive sections of the conductor.

FIGURE 6.1: IRRADIATING STRAND MATERIAL

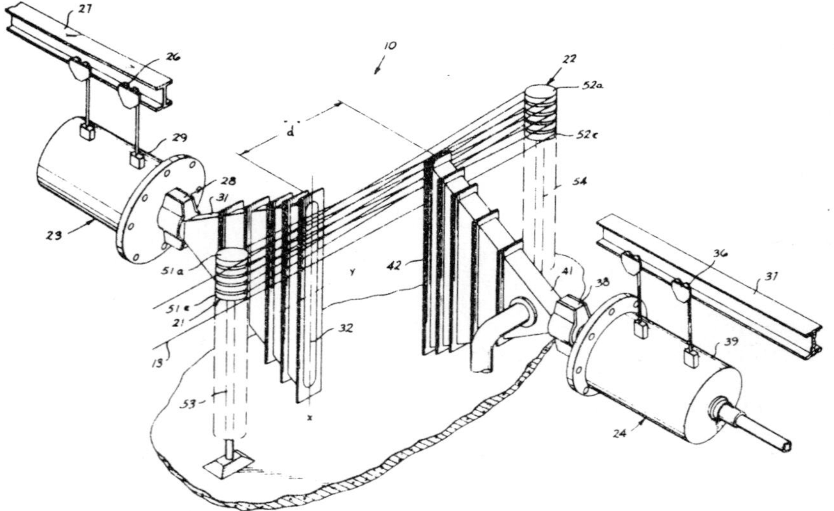

Source: U.S. Patent 3,925,671

Each of a plurality of conductors **13–13** (only one being shown in Figure 6.1 for purposes of simplicity) is strung up in superimposed modified figure eight paths between sheave banks, designated generally by the numerals **21** and **22**. Dual accelerators, designated generally by the numerals **23** and **24**, which are positioned between the sheave banks **21** and **22**, are used to irradiation crosslink the insulation of the successive sections of the strand material. The accelerators **23** and **24** face in opposite directions with the accelerators being staggered from each other so that the beams thereof do not overlap.

The conductor **13** is advanced in from the left, as viewed in Figure 6.1, into engagement with the sheave bank **21**, then around the sheave bank **22** in the plurality of groups of superimposed figure eight paths. As the conductor **13** is moved from one group of figure eight paths to another, the direction of travel in the figure eight path is reversed. The use of a figure eight path engenders a progressive screwing effect to cause the sections of the conductor **13** to turn about the longitudinal axes thereof to obtain a more uniform dosing of the insulation. In a preferred embodiment, twisted pairs of the conductors **13–13** are irradiation crosslinked. The progressive screwing effect is even more pronounced when using twisted pairs.

High-Melting Polyethylene by Irradiating and Orienting

Polyethylene products having many advantageous properties such as very high melting temperature, excellent mechanical properties and excellent transparency are obtained using the process of *R. Kitamaru, H.D. Chu and W. Tsuji; U.S. Patent 3,886,056; May 27, 1975.*

The process comprises irradiating a polyethylene with an ionizing radiation to produce crosslinked polyethylene having a gel content of at least one weight percent, extending the crosslinked polyethylene in at least one direction at a temperature of at least the anisotropic melting point of the crosslinked polyethylene and cooling the crosslinked polyethylene. The starting polyethylene has a viscosity average molecular weight of at least 1×10^5 and a melting point of 137° to 138°C when crystallized isothermally from the melt at a temperature of 130°C.

Example 1: A molecular weight fraction with a viscosity average molecular weight of 3.4×10^6 was obtained from a high density polyethylene by a liquid-liquid phase separation technique disclosed by H. Okamoto et al in *J Polymer Sci,* 55, 597 (1961). The high density polyethylene had a viscosity average molecular weight of 2.5×10^6 and a melting point of 137.2°C when crystallized isothermally at 130°C. An about 1.5 mm thick film of the above fraction was first irradiated to 0.25 megarad with x-ray from a 3 mev Van de Graaff accelerator at 150°C in high vacuum.

A gel sample was obtained by removing the soluble fraction of this irradiated sample with boiling xylene. The gel content of the irradiated sample was 95%. The gel sample thus obtained was next compressed between two polished metal plates at 180°C until the sample film was reduced to ⅛ and $^1/_{10}$ the original thickness. Each film was then cooled to room temperature for one hour in the compressed state. The samples thus compressed at ratios of ⅛ and $^1/_{10}$ respectively exhibited very high melting temperature and excellent heat stability and transparency. The thermodynamic and crystallographic data of the sample of a compression ratio of 1:10 are as follows.

Macroscopic density at 25°C	0.942
Melting temperature by DSC, °C	155.0
Degree of crystallinity	
determined from density	0.65
determined by DSC	0.56
Unit-cell density	1.002

As listed above, although the macroscopic density and accordingly the degree of crystallinity are not so high as crystalline polyethylene, the melting temperature is among the highest of all the values reported for polyethylene. Furthermore, the unit cell density determined by x-ray diffraction technique is a very high value for polyethylene. This indicates the existence of a very pure crystalline phase in the sample. Furthermore, x-ray studies revealed that this sample has a very special orientation of the crystalline phase so that the crystal planes (110) or (200) in this sample are oriented almost parallel to the film plane. The mechanical properties and the heat stabilities of the sample of a compression ratio of 1:8 are as follows.

	Process Sample	Reference Sample
Dynamic modulus (dynes/cm^2)	2.27×10^{10}	1.84×10^{10}
Young's modulus (kg/cm^2)	1.87×10^4	1.35×10^4
Strength at break (kg/cm^2)	1.02×10^3	0.30×10^3
Elongation at break (%)	84	1,100
Shrinking temperature (°C)	125	75
Shrinking (%)		
at 120°C	0	17.5
at 130°C	12.5	melt
at 150°C	60	melt

The reference sample above was prepared by compressing the molecular weight gel fraction obtained in the same manner as above directly at 190°C with a pressure of 50 kg/cm^2 to produce a polyethylene film of about 0.19 mm in thickness. In comparison with the reference sample, the present sample has very high moduli, and high strength with low elongation at the break. Particularly, its heat stability is excellent. The shrinking starts at 125°C and it shrinks only 12.5% even at 130°C. Thus this product can be used practically in a temperature range up to 125°C.

Example 2: Fibers 1.5 mm in diameter were made from unfractionated Marlex 50 (high density polyethylene) through a melt spinning procedure at 200°C with a screw extruder type spinning equipment without draft. The fibers thus produced were next irradiated with electron beam from Van de Graaff accelerator in air to dosages of 10 and 20 megarads. The gel contents of the irradiated samples thus obtained were 13 and 44% for samples irradiated to 10 and 20 megarads, respectively. The irradiated fibers were next drawn to 14.3 or 10.0 times the original length at 180°C by a continuous drawing equipment for fibers. The properties of the fibers thus obtained are listed in the following table.

Irradiation Dosage (Mrad)	. . . . . Drawing . . . . . Temperature (°C)	Ratio (times)	. . . . . Tensile. . . . . . Strength (g/d)	Elongation (%)	Young's Modulus (kg/mm^2)	Shrinkage in Boiling Water (%)
0	100*	8.5*	3.2*	30*	314.4*	8.3*
10	180	14.3	2.8	15.8	430.7	2.1
	100*	9.5*	3.4*	15.8*	333.3*	8.5*
20	180	10.0	3.2	21.1	322.8	2.3
	100*	7.8*	2.8*	23.4*	257.8*	8.9*

*Reference value data.

Curing Intermediate Articles

C.R. Stine, W.J. Herbert and B.E. Klipec; U.S. Patents 3,911,202; October 7, 1975 and 3,990,479; November 9, 1976 both assigned to Samuel Moore & Company describe methods for processing curable thermoplastic elastomers and flexible, shaped articles made therefrom which have improved high strength, radiation and temperature resistance and electrical characteristics useful as fluid transmission tubing and/or as electrical insulation.

Such articles are made from uncured, thermoplastic elastomer materials, in the form of selectively hard, radiation-sensitive, flowable solids, which are thermoplastically molded to provide an essentially uncured, dimensionally stable, intermediate shaped article. Such intermediate articles are radiation cured to provide finished crosslinked articles having such characteristics.

The preferred thermoplastic elastomers are halosulfonated polyethylene, poly(ethylene-propylene-ethylidene-norbornene) polymer, poly(ethylene-propylene-hexadiene) polymer, poly(ethylene-propylene-methylene norbornene) polymer, and poly(ethylene-propylene-dicyclopentadiene) polymer.

The initial components of the desired uncured polymer composition are blended to uniformly distribute the components into a uniform composition. This is preferably done by first mixing the respective components of the composition in an intensive internal batch mixer having sigma blades such as a Banbury mixer, in order to flux the components at a high temperature and then continuing the mixing by masticating the mixture on an open mill, such as a two roll mill, which also converts the lump form discharged by the batch mixer into a strip form.

After open milling, the composition can be optionally sieved to mechanically remove physical impurities and undesired lumps. The milled composition is stripped from the mill in sheet or strip form and passed through a water bath to cool it. After cooling, it is subjected to an air wipe to remove or strip retained water. Finally, the composition is physically subdivided into dice or pellets by feeding the strip into either a dicing machine or a pelletizing machine. In its subdivided form, the composition is a flowable solid having an extended storage life at room temperature. However, the pellets or dice can be stored at room temperature until required, as desired.

The uncured, thermoplastic polymer compositions are thermoplastically processed so as to shape them into an intermediate article which has the configuration of the desired finished article. It has been found that the compositions are particularly suitable for shaping by thermoplastic extrusion, although other techniques such as thermoplastic injection molding might be employed. The composition, in its subdivided form as a flowable solid, is fed into a thermoplastic extruder which fuses the subdivided composition into a plastic fluid under compressive and shear stress and externally applied heat.

The uncured, but curable, polymeric composition of the process is thermoplastically extruded at an elevated temperature with a long residence time in the extruder. In other words, high temperatures and long barrel and screen packs are used in the extrusion. Thermoplastic extrusion of the process utilizes extruders having a barrel length to diameter ratio (L/D ratio) of 20:1 or greater, with about 32:1 being the present maximum size. This longer relative barrel length to barrel diameter provides a longer residence time for the uncured, but curable composition within the extruder and provides greater homogeneity and better mixing action.

Thermoplastic extrusion is conducted at relatively high temperatures with sufficient heat being used to reach and maintain them. The olefinic terpolymer composition of the process has a preferred stock melt temperature in the extruder from 300° to about 350°F and can, in fact, be increased to about 380°F without scorching (undesired curing). The halosulfonated polyolefin composition has a preferred stock melt temperature in the extruder from 250° to 300°F.

In the thermoplastic extrusion, production speed can be increased by increasing melt temperature and thereby decreasing melt viscosity. External heat increases the melt temperature above that obtainable from internal work heat alone. The ratio of external heat to internal heat is from approximately 0.2:1 to approximately 0.3:1 and possibly as high as 0.5:1. If desired, scrap from the extruder can be reextruded and used without significant change in melt rheology or physical properties of the composition being extruded. In view of the high melt temperatures and long residence time in the extruder, it is preferable to exclude or minimize chemical curing agents or initiators from this composition.

In forming the foregoing extruded articles, a die is affixed to the terminal end of the extruder to shape the composition as it exits from the extruder into an atmospheric pressure environment. This die may be either a pressure die using elevated extrusion pressures or a drawdown die. A drawdown die providing an annular solid configuration with a generally uniform thickness is preferred. The drawdown die provides an extruded polymeric article having a relatively thin-walled construction of uniform thickness, thereby minimizing nonuniform characteristics in the finished article.

In addition, where the intermediate article is electrical primary insulation or cable jacketing, the drawn down uniform thickness follows the configuration of the conductor being insulated, thereby facilitating stripping of the jacketing, if desired.

Due to the shearing action of the extruder barrel on the composition as the screw advances it toward the extrusion die and the long residence time produced by the length of the extruder barrel, it is possible to mix additional nonreactive materials, preferably in a flowable solid form such as powder or pellet, into the uncured composition as it is fed into the extruder. Suitable materials for addition at the extrusion stage are colorants, radiation sensitizers, lubricants and thermal stabilizers. By this arrangement, the composition can be formulated as a natural color, master composition and then colored, or otherwise modified, as desired, during the extrusion step.

The intermediate article thus provided by the foregoing thermoplastic forming technique is: (a) elastomeric, (b) still thermoplastic (or heat fusible), (c) still radiation sensitive, and (d) still curable (but essentially uncured). As such, this intermediate article has an extended shelf life so that it can be stored for extended periods at room temperature before it is finally cured.

In order to produce the cured article, the uncured, intermediate article is radiation cured, such as by irradiation with high energy, ionizing radiation. This irradiation cures, or crosslinks, the article's polymeric composition and irreversibly fixes the configuration of the finished article. While alpha, beta, gamma or neutron irradiation might be used, the preferred means of irradiation is high energy electrons which may be conveniently controlled. The source of the high energy electrons may be an electron accelerator, such as a Van de Graaff accelerator. The energy of the electrons should be from 100,000 electron volts to 3 mev.

Preferably, the total dosage is from 5 Mrad to 12 Mrad and optimally from 10 Mrad to 12 Mrad. The curing is done at room temperature and atmospheric pressure, eliminating the need for temperature and pressure control. The curing by high energy electron irradiation is virtually instantaneous so that the intermediate article may be passed continuously under the electron beam at a relatively high speed, thereby providing a high speed curing operation, for example, of the order of 500 feet per minute or more for individual conductors. In fact, thermoplastic forming and radiation curing may be done as a single high speed operation.

Example: Using the aforesaid blending and subdividing techniques, a poly(ethylene-propylene-ethylidene-norbornene) composition suitable for a tubular article was prepared using the following materials and proportions, expressed in parts by weight.

	Parts by Weight
Epsyn 4506 poly(ethylene-diene-ethylidene-norbornene) base polymer	75
Epsyn 5509 poly(ethylene-diene-ethylene-norbornene) base polymer	25
Burgess KE clay	125
St. Joe 20 zinc oxide	5
A-172 vinyl silane	1
DYNH-3 polyethylene polymer	20
Agerite Resin D poly(trimethyl-dehydroquinoline)	0.75
SR-230	3
Sunpar 2280 paraffinic oil	10

The uncured compositions had the characteristics listed below.

Heat distortion, 300°F IPCEA-S-61-402 (conditioned for 5 minutes; under load for 15 minutes)	98 %
Tensile modulus (tensile strength at 50% elongation) (ASTM-D-412)	271 psi
Ultimate tensile strength (ASTM-D-412)	725 psi
Ultimate tensile elongation	610 %
Hardness (Shore)	A56

The aforegoing composition was thermoplastically extruded using the aforesaid extrusion technique to provide a tubular article with an annular configuration of indeterminate length and having a nominal internal diameter of 0.375 inch and a nominal wall thickness of 62 mils. This intermediate shaped article was cured with high energy electrons having a kinetic energy of 1 mev at a dosage of 20 Mrads. The final dimensions of the article were the same as those of the intermediate shaped article. The cured composition had the characteristics listed below.

Heat distortion, 300°F, IPCEA-S-61-402 (conditioned for 5 minutes; under load for 15 minutes)	1.8 %
Tensile modulus (tensile strength at 50% elongation, ASTM D-412)	770 psi
Ultimate tensile strength (ASTM D-412)	1,654 psi
Ultimate tensile elongation (ASTM D-412)	125 %
Hardness (Shore)	A61

Batch/Continous Process for Tetrafluoroethylene-Propylene Polymers

Thermoplastic elastomeric copolymers of tetrafluoroethylene and propylene and terpolymers of tetrafluoroethylene, propylene and a cure site monomer are disclosed by *R.F. Foerster; U.S. Patent 3,933,773; January 20, 1976; assigned to Thiokol Corporation.*

The polymers have a uniform composition, that is, a substantially uniform molar ratio of monomeric units and a relatively high molar ratio of tetrafluoroethylene units to propylene units within the range 1.0:0.11 to 1.0:0.54. The combination of high molar ratio of tetrafluoroethylene/propylene units, uniform composition and good elastomeric properties is achieved by using a hybrid batch-continous process wherein a reactor is initially charged with a mixture of tetrafluoroethylene, propylene and, optionally, a cure site monomer having a TFE/propylene molar ratio substantially higher than that of the polymer to be produced, that is, in the range 1.0:0.01 to 1.0:0.87.

The polymerization reaction is then initiated and a mixture of the same monomers is fed to the reactor at such a rate and in such proportions as to maintain the molar ratio of the unreacted monomers substantially the same as that in the initial charge to the reactor.

Example: A stirred stainless steel reactor of 86 liter capacity was charged to 75% of its volume with deionized water (64.5 liters) containing 0.5 pph of technical grade ammonium perfluorooctanoate (319 g), and 0.5 pph of sodium hydroxide (319 g). The reactor was sealed, flushed with nitrogen, evacuated, and then pressurized with a gas mixture containing 99.0 mol % of tetrafluoroethylene and 1.0 mol % of propylene, to approximately 300 psig at 145°F. Polymerization was initated by injecting 31.4 g of ammonium persulfate and 31.1 g of sodium sulfite as separate, approximately 6% aqueous solutions. As the polymerization proceeded, the pressure in the reactor was maintained essentially constant by continuously feeding a gas mixture containing 76 mol % TFE and 24 mol % propylene until the polymer solids content of the latex was about 22%. The feed composition of 76/24 by mol % was chosen so as to replenish each of the monomers in the reactor at approximately the same rate and in the same mol ratio as they were being consumed, since separate experiments had established that the 99/1 monomer mixture originally charged would produce a copolymer with an initial composition of about 76/24. The product was isolated by coagulation of the latex, filtering, washing and drying. Analysis indicated a copolymer composition of 76.5 mol % TFE and 23.5 mol % propylene. The thermoplastic, rubbery material appeared to be highly homogeneous.

This material was blended with another batch of material made in a similar run and the blend was molded to produce virtually colorless, transparent, flexible sheets by subjecting it to a pressure of 700 psi at 325°F for 15 minutes. Some specimens from these sheets were tested in the original form, that is, not crosslinked. The remainder of the sheets were exposed to gamma radiation from a cobalt 60 source for 7.5 hours, at an average dose rate of about 2.2 Mrad per hour, at ambient temperature, in a nitrogen atmosphere. Properties before and after irradiation are listed in the table below.

The irradiated polymer showed significantly improved modulus retention, dimensional stability, elastic recovery, and compression set resistance at elevated temperatures (up to about 400°F). It also offered greatly improved resistance to stress cracking in hot oil, for example, ASTM Oil No. 3 at 250°F.

	Original	Irradiated
Tensile strength, psi	2,400	2,470
Elongation, %	395	275
Hardness, Shore A	95	97
100% modulus, psi	890	1,045
200% modulus, psi	1,125	1,450
300% modulus, psi	1,675	-
Permanent set after 100% extension, %	43	48
Compression set after 22 hr at 250°F	87	27
Solvent swell (volume increase after 1 wk at room temperature) %		
Butyl acetate	8	9
Toluene	4	2
Tetrachloroethylene	4	1
Properties after heat aging in air for 1 wk at 450°F		
Tensile strength, psi	2,790	2,280
Elongation, %	425	285
100% modulus, psi	760	995
200% modulus, psi	925	1,350
300% modulus, psi	1,405	-
Permanent set after break, %	400	200

Silicone Rubber Coated with Collagen

S. Okamura and T. Hino; U.S. Patent 3,955,012; May 4, 1976; assigned to Zaidan Hojin, Seisan Kaihatsu Kagaku Kenkyusho, Japan describe a method of manufacturing medical articles composed of silicone rubber coated with collagen and to be used in a living body. The method comprises:

(1) subjecting a surface of shaped articles composed of silicone rubber in an oxygen-containing atmosphere to a spark discharge wherein the product of the spark length in centimeters and the discharge time per square centimeter is in the range of 9 to 12 cm-sec/cm^2;

(2) coating the thus treated surface with an acidic aqueous solution of collagen;

(3) drying the collagen coated shaped article at a temperature lower than the denaturation temperature of collagen to form a collagen layer; and

(4) irradiating the shaped article coated with collagen layer with gamma ray or electron beam to a dosage of 1 to 5×10^6 roentgens or the equivalent dosage of ultraviolet light under an atmosphere having a humidity such that the water content of the coated collagen becomes greater than 20% by weight, thus fixing the collagen layer.

Example: After a surface of a silicone rubber sheet (Phycon) was washed with acetone and then with hot water, the sheet was dried in air. A spark discharge having a spark length of 4 cm was applied to the surface in air by means of a spark discharge generator (supply voltage AC 100 v; input current 0.1 to 0.8 amp; frequency 5 to 100 kc; spark length 1 to 5 cm). During the discharge, the end of spark was always moved so as to apply the discharge uniformly as far as possible to the entire surface, and the discharge time was 3 sec/cm^2. Then, the surface of the sheet was easily wettable.

A 0.5% acidic aqueous solution of collagen (N/400 HCl) was applied on the surface of the above treated sheet, and the sheet was dried in air at 30°C. After drying, the sheet was dipped in a 1% aqueous solution of NH_4OH for about 1 hour to neutralize the collagen layer, dipped in cold water 3 times to remove the salt, and again dried in air at 30°C.

The sheet was irradiated with γ-rays of 1.0×10^6 roentgens at 20°C under 100% humidity and gaseous nitrogen atmosphere. The surface of the resulting silicone rubber sheet was tightly coated with a collagen layer in a thickness of about 6 μ. The collagen layer was not peeled off in a peel-off test by means of an adhesive tape. Even after the sheet was dipped in water for 10 days, collagen was not dissolved out in water.

Surface Treatment of Tires to Reduce Flash

In the normal building operation of a pneumatic tire the various components are assembled on a tire building and shaping drum. This operation may include building the tire completely in a cylindrical form or partially building the tire in a cylindrical form and then expanding the tire to a generally torus form followed by the application of a belt or breaker structure and tread rubber. In accordance with the process of *G.H. Fisk and J.A. Loulan, Jr.; U.S. Patent 3,959,053; May 25, 1976; assigned to The Goodyear Tire & Rubber Company,* the surface of those portions of the tire at which there is a desire to eliminate or reduce the vent or flash is treated with a critical dosage of electron radiation. This dosage is selected to provide a sufficient penetration of the rubber and a degree of cure on the outer surface of the rubber to prevent flow of the rubber into the cracks or junctures between the sections of the mold or into the vent holes of the mold.

It is important, however, that the dosage of radiation treatment is not too large such that the rubber will not flow completely into the matrix of the mold and result in void portions in the tread or sidewall pattern. Thus, the amount of dosage of electron radiation is critical to the complete formation of the desired tread and sidewall pattern without the production of unwanted flash or vents.

It has been found that a dosage of electron radiation between 1 and 10 Mrad is suitable for most applications. The preferred dosage is between 3 and 6 megarads. The radiation source should be between 500,000 kv and 3,000,000 kv in order to obtain the proper penetration. If these parameters are followed there will be effective carbon-carbon crosslinking of polymer chains to a depth of between about 0.02 inch and about 0.45 inch.

The tire, before it is placed in the mold and heat cured does not have any significant amount of carbon-sulfur-carbon crosslinked polymer chains and thus will not be susceptible to overcure due to the prior radiation treatment. By being free of any significant amount of carbon-sulfur-carbon crosslinked polymer chains it is meant that the rubber has not been subjected to heat cure and the only such carbon-sulfur-carbon crosslinked polymer chains that exist are those caused by passage of time in ambient conditions in the tire plant.

COMPANY INDEX

The company names listed below are given exactly as they appear in the patents, despite name changes, mergers and acquisitions which have, at times, resulted in the revision of a company name.

INVENTOR INDEX

U.S. PATENT NUMBER INDEX

NOTICE

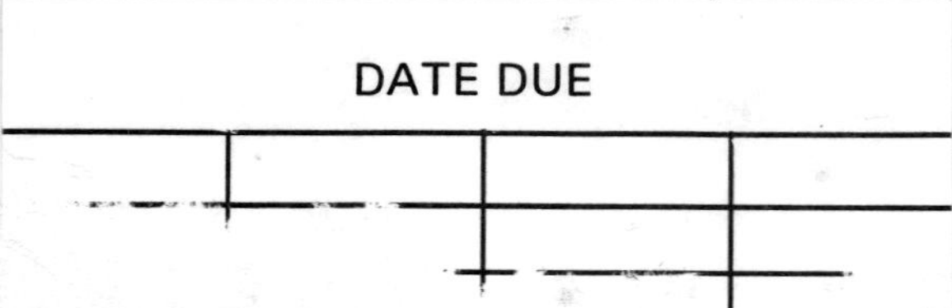